Introduction to

Numerical and Analytical Methods with MATLAB® for Engineers and Scientists

Introduction to
Numerical and Analytical Methods with MATLAB® for Engineers and Scientists

William Bober

CRC Press
Taylor & Francis Group
Boca Raton London New York

CRC Press is an imprint of the
Taylor & Francis Group, an **informa** business

First Indian Reprint, 2015

CRC Press
Taylor & Francis Group
6000 Broken Sound Parkway NW, Suite 300
Boca Raton, FL 33487-2742

© 2013 by Taylor & Francis Group, LLC
CRC Press is an imprint of Taylor & Francis Group, an Informa business

Printed in India by Manipal Technologies Ltd, Manipal

International Standard Book Number-13: 978-1-4665-7602-5

Library of Congress Cataloging-in-Publication Data

Bober, William, 1930-
    Introduction to numerical and analytical methods with MATLAB for engineers and scientists / author, William Bober.
        pages cm
    Includes bibliographical references and index.
    ISBN 978-1-4665-7602-5 (hardback)
    1. Engineering mathematics. 2. Numerical analysis. 3. MATLAB. I. Title.

TA335.B63 2013
518.0285'53--dc23   2013018873

Visit the Taylor & Francis Web site at
http://www.taylorandfrancis.com

and the CRC Press Web site at
http://www.crcpress.com

For sale in India, Pakistan, Nepal, Bhutan, Bangladesh and Sri Lanka only.

Contents

v

Contents

Preface

I have been teaching two courses in computer applications for engineers at Florida Atlantic University for many years. The first course is usually taken in the student's sophomore year while the second course is usually taken in the student's junior or senior year. Both computer classes are run as lecture-laboratory courses, and the MATLAB® software program is used in both courses. I have collaborated with several colleagues in writing two textbooks. The first was a textbook was titled *Numerical and Analytical Methods with MATLAB®*, with coauthors Drs. Chi-Tay Tsai and Oren Masory. This textbook was primarily oriented toward mechanical, civil, and aeronautical engineers. The second textbook was titled *Numerical and Analytical Methods with MATLAB® for Electrical Engineers*, with coauthor Dr. Andrew Stevens. This latest book is a composite of both textbooks and would be appropriate for all engineering students. The primary objectives of the textbook are to

1. Teach engineering students how to write computer programs (or scripts) on the MATLAB platform that solve engineering-type problems
2. Demonstrate various mathematical concepts that can be used to solve engineering-type problems, such as matrices, roots of equations, integration, ordinary differential equations, curve fitting, algebraic linear equations, and others
3. Demonstrate the use of numerical and analytical method for solving the mathematical problems associated with engineering-type problems, such as Simpson's rule on integration, Gauss elimination method for solving a system of linear algebraic equations, the Runge–Kutta method for solving a system of ordinary differential equations, the iteration method for solving pipe flow problems, the Hardy Cross method for solving flow rates in a pipe network, the method of separation of variables to solve partial differential equations, the use of Laplace transforms to solve both ordinary and partial differential equations, and others

After receiving some feedback on the first textbook, I realized that for a first course in computer programming under the MATLAB platform, concepts in computer programming needed to be substantially expanded. This was undertaken in the second textbook as well as in this new textbook. The material covered in Chapter 2 of the second textbook is now covered in Chapters 2 and 3. Furthermore, I also decided not to cover topics that require additional MATLAB tool boxes, and as a result, I have eliminated the chapters on finite elements and controls. However, I did keep the chapter on optimization, which also requires an additional tool box.

The advantage of using the MATLAB software program over other software programs is that it contains built-in functions that numerically solve systems of linear equations, systems of ordinary differential equations, roots of transcendental equations, definite integrals, statistical problems, optimization problems, and many other types of problems encountered in engineering. A student version of the MATLAB program is available at a reasonable cost. However, to students, these built-in functions are essentially black boxes. By combining a textbook on MATLAB with basic numerical and analytical analysis (although I am sure that MATLAB uses more sophisticated numerical techniques than the ones that are described in this text), the mystery of what these black boxes might contain is somewhat alleviated. The text contains many sample MATLAB programs that should provide guidance to the student on completing the exercises and projects that are listed in each chapter. Projects are at the end of the chapters and are usually more difficult than the exercises. Many of the projects are nontrivial.

In recent times, I have used several exercise problems as in-class exams in which students submit their MATLAB programs and results to me on blackboard. Projects are given as take-home exams to be submitted to me within one or two weeks, depending on the difficulty of the project.

The advantage of running these courses (especially the first course) as a lecture-laboratory course is that the instructor is in the computer laboratory to help the student debug his or her program. This includes the example programs as well as the exercises and the projects. Although this textbook is suitable for a first course in computer application for engineers, say at the sophomore level, there is enough material in the textbook that makes it suitable for a higher level course or as a reference book in higher level courses. All three textbooks contain many sample programs to teach the student programming techniques.

For a first course in computer applications, Chapters 1 through 6 would be appropriate. These chapters include review sections, which may be used by the course instructor to ask the class questions on the material that was recently covered. The topics covered in Chapters 1 through 6 are as follows:

Chapter 1—Numerical Modeling for Engineers
Chapter 2—MATLAB® Fundamentals
Chapter 3—Taylor Series, Self-Written Functions, and MATLAB®'s `interp1` Function
Chapter 4—Matrices

Chapter 5—Roots of Algebraic and Transcendental Equations
Chapter 6—Numerical Integration

Chapters 7 through 14 can be used for a second MATLAB course at the junior or senior level or as a reference in upper division undergraduate courses. The topics covered in these chapters are as follows:

Chapter 7—Numerical Integration of Ordinary Differential Equations
Chapter 8—Boundary Value Problems of Ordinary Differential Equations
Chapter 9—Curve Fitting
Chapter 10—Simulink®
Chapter 11—Optimization
Chapter 12—Iteration Method
Chapter 13—Partial Differential Equations
Chapter 14—Laplace Transforms

The governing equations for most projects are derived either in the main body or in the project description itself or in the appendices. All example scripts are available for download on the CRC Press website at http://www.crcpress.com/product/isbn/9781466576025 (see Section 1.11).

MATLAB® is a registered trademark of The MathWorks, Inc. For product information, please contact:

The MathWorks, Inc.
3 Apple Hill Drive
Natick, MA 01760-2098 USA
Tel: 508-647-7000
Fax: 508-647-7001
E-mail: info@mathworks.com
Web: www.mathworks.com

Acknowledgments

I would like to thank Jonathan Plant of CRC Press for his confidence and encouragement in writing this textbook. I wish to thank Dr. Andrew Stevens for allowing me to extract many electrical engineering concepts and projects from our joint textbook titled *Numerical and Analytical Methods with MATLAB® for Electrical Engineers* and for proof reading the manuscript. I also wish to thank Jennifer Stair and Laurie Schlags for guiding me through the textbook submission process. I also wish to thank the following people for their graphic contributions: Husayn El Sharif, Marques Johnson, and Fernando Lascano. Finally, I wish to express my deep gratitude to my wife for tolerating the many hours I spent on preparation of this manuscript—time which otherwise would have been devoted to my family.

Author

William Bober, PhD, received his BS in civil engineering from the City College of New York (CCNY), his MS in engineering science from Pratt Institute, and his PhD in engineering science and aerospace engineering from Purdue University. At Purdue University, he was on a Ford Foundation Fellowship and was assigned to teach one engineering course each semester. After receiving his PhD, he worked as an associate engineering physicist in the Applied Mechanics Department at Cornell Aeronautical Laboratory in Buffalo, New York. After leaving Cornell Labs, he was employed as an associate professor in the Department of Mechanical Engineering at the Rochester Institute of Technology (RIT) for 12 years. Since leaving RIT, he has been an associate professor at Florida Atlantic University (FAU) in the Department of Mechanical Engineering. More recently, he transferred to the Department of Civil Engineering at FAU. While at RIT, he was the principal author of a textbook, *Fluid Mechanics*, published by John Wiley & Sons. He has also written several papers for *The International Journal of Mechanical Engineering Education* (*IJMEE*) and more recently coauthored several textbooks on *Analytical and Numerical Methods with MATLAB®*.

Chapter 1

Computer Programming with MATLAB® for Engineers

1.1 Introduction

All disciplines of science and engineering use numerical methods for the analysis of complex problems. The field of engineering, in particular, lends itself to computational solutions due to the highly mathematical nature of the field and the fact that analytical methods alone are unable to solve many complex mathematical problems.

In this book we will describe various methods and techniques for numerically solving a variety of common engineering applications, including problems in statics, dynamics, strength of materials, thermodynamics, fluid mechanics, electrical circuits, and electromagnetic field theory. Classical engineering curricula teach a variety of methods for solving these problems using concepts from linear algebra, root extraction of a transcendental equation, integration, differential equations, etc.

1.2 Computer Usage in Engineering

Some of the ways that the computer is used in engineering are

    a. Solving mathematical models of physical phenomena
    b. Storing and reducing experimental data
    c. Controlling machine operations
    d. Communicating with other engineers and technicians on a particular project

This textbook is mostly concerned with part (a).

The engineer's interest lies in

a. Designing of new products or improving existing ones
b. Improving manufacturing efficiency
c. Minimizing cost, power consumption, and nonreturnable engineering cost (NRE)
d. Maximizing yield and return on investment (ROI)
e. Minimizing time to market
f. Research on developing new products

These can be accomplished by

a. Full-scale experiments. May be prohibitively expensive.
b. Small-scale model experiments. Still very expensive, and extrapolation is frequently questionable.
c. A mathematical model that is the least expensive and faster. It can provide more detailed answers and different cases under different conditions can be run quickly. If there is confidence in a mathematical model, it will be used in preference to experiment.

1.3 Mathematical Model

Physical phenomena are described by a set of governing equations.

Numerical methods are frequently used to solve the set of governing equations, since closed-form solutions for many types of problems are not available. Even when closed-form solutions are available, the solution may be sufficiently complicated that the computer is needed to calculate the desired answer. Numerical methods invariably involve the computer. The computer performs arithmetic operations upon discrete numbers in a defined sequence of steps. The sequence of steps is defined in the program. A useful solution is obtained if

a. The mathematical model accurately represents the physical phenomena; that is, the model has the correct governing equations.
b. The numerical method is accurate.
   Note: If the governing equations aren't correct, the solution will be worthless regardless of the accuracy.
c. The numerical method is programmed correctly.

This book is mainly concerned with items (b) and (c).

1.4 Computer Programming

The advantage of using the computer is that it can carry out many calculations in a fraction of a second; at the time of this writing, computer speeds are measured in teraflops (trillions of floating-point operations per second). However, to leverage this power, we need to write a set of instructions, that is, a program (or script). For the problems of interest in this book, the digital computer is only capable of performing arithmetic, logical, and graphical operations. Therefore, arithmetic procedures must be developed for solving differential equations, evaluating integrals, determining roots of an equation, solving a system of linear equations, etc. The arithmetic procedure usually involves a set of algebraic equations. A computer solution for such problems involves developing a computer program that defines a step-by-step procedure for obtaining an answer to the problem of interest. The method of solution is called an *algorithm*. Depending on the particular problem, we might write our own algorithm or, as we shall see, we can also use the algorithms built into a package like MATLAB® in order to perform well-known algorithms for solving many types of mathematical problems.

1.5 Why MATLAB®?

MATLAB was originally written by Dr. Cleve Moler at the University of New Mexico in the 1970s and was commercialized by MathWorks in the 1980s. It is a general-purpose numerical package that allows complex equations to be solved efficiently and subsequently generate tabular or graphical output. While there are many numerical packages available to engineers, many are very highly focused toward a particular application, for example, ANSYS for modeling structural problems using the finite element method. Also, originally, MATLAB was a command-line program that ran on MS-DOS and UNIX hosts. As computers have evolved, so has MATLAB, and modern editions of the program run in windowed environments. As of the time of this writing, MATLAB R2013a runs natively on Microsoft Windows, Apple Mac OS, and Linux. In this text, we will assume that you are running MATLAB on your local machine in a Microsoft Windows environment. It should be straightforward for non-Windows users to translate the usage descriptions to their preferred environment. In any case, these differences are largely limited to the cosmetics and presentation of the program, and not the MATLAB commands themselves. All versions of MATLAB (on any platform) use the same command set, and the Command Window on all platforms should behave identically.

MATLAB is offered with accompanying "toolboxes" at additional cost to the user. A wide variety of toolboxes are available in fields such as partial differential equations (PDEs), optimization, and control systems. However, in this text, we will largely focus on fundamental numerical concepts and will limit ourselves (except for optimization) to basic MATLAB functionality without requiring the purchase of any additional toolboxes.

1.6 Programming Methodologies

There are many methodologies for computer programming, but the tasks at hand boil down to:

a. Studying the problem to be programmed including the geometry of the problem.
b. Listing the algebraic equations to be used in the program based on the known physical phenomena.
c. Selecting the most efficient numerical method for obtaining a solution.
d. Creating an outline or a flow chart for the program flow. You might even consider carrying out a sample calculation by hand to prove the program flow.
e. Writing the program using the list of algebraic equations and the outline or flow chart.
f. Debugging the program by running it and fixing any syntax errors (programming language errors).
g. Testing the program by running it using parameters with a known (or intuitive) solution.
h. Refining and further debugging the algorithm and program flow.

Experienced programmers often omit some of these steps (or do them in their head), but the overall process resembles any engineering project: design, create a prototype, test, and iterate the process until a satisfactory product is achieved.

1.7 MATLAB® Programming Language

MATLAB may be considered a programming or scripting language unto itself, but like every programming language, it has the following core components:

a. Data types, that is, formats for storing numbers and text in the program (e.g., integers, floating-point numbers, strings, vectors, matrices)
b. Operators and built-in functions (e.g., commands for addition, subtraction, multiplication, division, trigonometric functions, log function)
c. Control flow directives for making decisions and performing repeated operations (e.g., loops, alternate paths, functions)
d. Input/output ("I/O") commands for receiving input from a user or a file and for generating output to a file or to the screen (e.g., read and print statements)

MATLAB borrows many constructs from other languages. For example, the `for`, `while`, `switch`, and `fprintf` commands are from the C programming language (or its descendents C++, Java, and Perl). However, there are some fundamental differences as well. For example, MATLAB stores *functions* (known in other languages as "subroutines") in separate files. The first entry in a *vector* (known in most other languages as an *array*) is indexed by the number one and not zero. However, the biggest difference is that all MATLAB variables are vectors or matrices, thus providing the ability to manipulate large amounts of data with a terse syntax and allowing for the solution of complicated problems in just a few lines of code. In addition, because MATLAB is normally run interactively, it is also very rich in presentation functions to display sophisticated plots and graphs.

1.8 Building Blocks in Writing a Program

Most engineering programs will include some or all of the following building blocks in program development:

- a. Arithmetic statements
- b. I/O statements
- c. Loop statements (`for` and `while` loops)
- d. Alternative path statements also known as conditional operators (`if`, `elseif`)
- e. Functions (built-in and self-written)

Building blocks (a)–(d) are introduced in Chapter 2. Building block (e) is introduced in Chapter 3. Example programs containing these program building blocks are given throughout the book.

1.9 Conventions in This Book

We will use the following typographical conventions in this text:

- a. All I/O to and from MATLAB will be printed in `typewriter` font.
- b. In cases where you are typing directly into the computer, the typed text will be displayed in **bold**.

We illustrate this in the following example where we use MATLAB to find the value $x = \sin(\pi/4)$:

```
>> x = sin(pi/4)
x =
    0.7071
```

In this case, `>>` represents MATLAB's prompt, `x = sin(pi/4)` represents text typed into the MATLAB command window, and `x = 0.7071` represents MATLAB's response.

1.10 Example Programs

The example programs in this book may be downloaded from the publisher's website at http://www.crcpress.com/product/isbn/9781466576025. Students may then run the example programs on their own computer and see the results. It also may be beneficial for students to type in a few of the sample programs (along with some inevitable syntax and typographical errors), thereby giving them the opportunity to see how MATLAB responds to program errors and subsequently learn what they need to do to fix the problem.

Review 1.1

1. List several ways engineers use the computer.
2. List several areas of interest for engineers.
3. List several methods that can be used in the design of new products.
4. Which method mentioned in item 3 is the least expensive?
5. For engineers, what is the principle advantage of using MATLAB over some of the other computer programming platforms?
6. List several items that are recommended in developing a computer program.
7. List several items that can be considered building blocks available in developing a computer program in MATLAB.

Chapter 2

MATLAB® Fundamentals

2.1 Introduction

This chapter and the chapters that follow cover many concepts in computer programming and how to apply them to solving engineering-type problems. We do this on the MATLAB® platform. Recall that at the end of Chapter 1, it was mentioned that a number of building blocks in any programming language are (a) the arithmetic statement or assignment; (b) input/output (I/O) statements, including graphs and tables; (c) loops; (d) conditional operators which lead to alternate paths in the program, and (e) functions, both built-in and self-written. Items (a)–(d) are covered in this chapter as well as some of the elementary built-in functions of item (e). Examples of self-written functions and MATLAB's functions that solve particular types of mathematical problems are covered throughout this book.

MATLAB is a software program for numeric computation, data analysis, and graphics. One advantage that MATLAB has for engineers over programming languages such as C or C++ is that the MATLAB program includes functions that numerically solve large systems of linear algebraic equations, systems of ordinary differential equations, roots of transcendental equations, definite integrals, statistical problems, optimization problems, control systems problems, and many other types of problems encountered in engineering. MATLAB also offers toolboxes (which must be purchased separately) that are designed to solve problems in specialized areas.

In this chapter, the following items are covered:

- The MATLAB desktop environment
- Constructing a program (also called a script) in MATLAB

7

- MATLAB fundamentals and basic commands, including `clear`, `clc`, colon operator, arithmetic operators, trigonometric functions, logarithmic and exponential functions, and other useful functions such as `max`, `min`, and `length`
- I/O in MATLAB, including the `fprintf` and `input` statements
- MATLAB program flow, including `for` loops, `while` loops, `if` and `elseif` statements, and `switch` group statement
- MATLAB graphics, including the `plot` and `subplot` commands
- Debugging a program

Many example scripts are included throughout this chapter to illustrate these various topics.

2.2 MATLAB® Desktop

Mathworks, the company that develops MATLAB, normally updates their version of MATLAB twice a year. In 2012, Mathworks introduced MATLAB version R2012b whose desktop interface was very different from their MATLAB version R2012a. Thus, getting started in MATLAB version R2012b is very different than getting started in earlier versions of MATLAB. Sections 2.2 and 2.3 in this chapter, which discusses the MATLAB desktop windows and how to construct a script in MATLAB is based on MATLAB version R2012b. At the time of this writing, Mathworks came out with MATLAB version R2013a, which has the same desktop as MATLAB version R2012b. Since many schools may still be using earlier versions

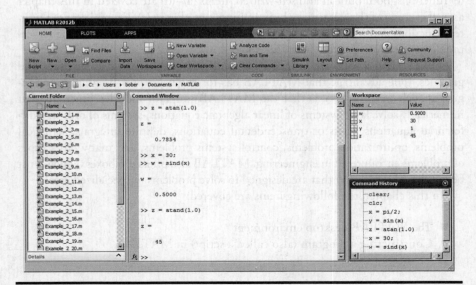

Figure 2.1   MATLAB® desktop.

of MATLAB, we thought it would be appropriate to discuss getting started with MATLAB version 2012a and earlier in Appendix E.

Under Microsoft Windows, MATLAB may be started via the Start menu or by clicking on the MATLAB icon on the desktop. Upon startup, a new window will open containing the MATLAB "desktop" (not to be confused with the Windows desktop), and one or more MATLAB windows will open within the MATLAB desktop (see Figure 2.1 for the default configuration).

The main windows are: the Command Window, Command History, Current Folder, and Workspace. You can customize the MATLAB windows that appear upon startup by opening clicking on *Layout* in the Toolstrip and checking (or unchecking) the windows that you wish to appear on the MATLAB desktop. Figure 2.1 shows the Command Window (in the center), the Current Folder Window (on the left), the Workspace Window (on the top right), the Command History window (on the bottom right), and a long narrow box containing the path to the Current Folder (just below the Toolstrip and just above the Command Window). MATLAB designates this long narrow box as the Current Folder Toolbar. These windows and the Current Folder Toolbar are summarized as follows:

- *Command Window*: In the Command window, you can enter commands and data, make calculations, and print results. You can write a script in the Command window and execute the script. However, writing a script directly into the Command window is discouraged because it will not be saved, and if an error is made, the entire script must be retyped. By using the up arrow (↑) key on your keyboard, the previous command can be retrieved (and edited) for reexecution.
- *Command History Window:* This window lists a history of the commands that you have executed in the Command Window. You can click on a command in this window and it will be reexecuted.
- *Current Folder Toolbar:* This toolbar gives the path to the Current Folder. **To run a MATLAB script, the script needs to be in the folder listed in this toolbar.**
- *Current Folder Window* (on the left): This window lists all the files in the Current Folder whose path is listed in the Current Folder Toolbar. By double clicking on a file in this window, the file will open within MATLAB.
- *Script Window*: To open this window, click on the *New Script* icon in the Toolstrip in MATLAB's desktop. This will open the Script window (see Figure 2.2a). The Script window may be used to create, edit, and execute MATLAB scripts (programs). Scripts are then saved as *M-Files*. These files have the extension *.m*, such as *heat.m*. To execute the script, you can click the *Save and Run* icon (the green arrow) in the Script window (see Figure 2.2b) or return to the Command window and type in the name of the program (without the *.m* extension).

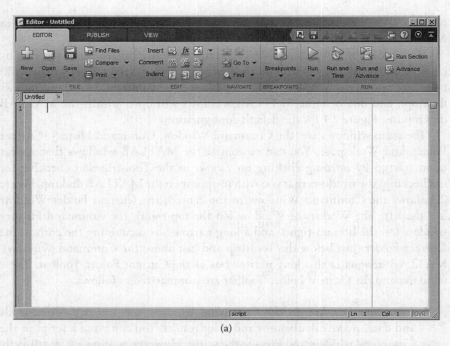

(a)

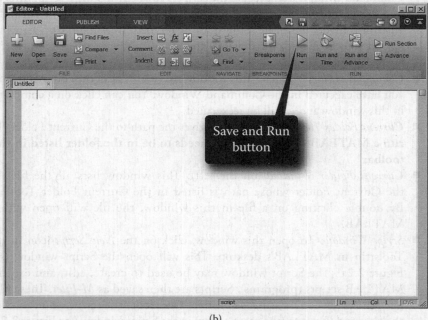

(b)

Figure 2.2 (a) Script window. (b) Save and Run button.

2.3 Constructing a Program in MATLAB®

The following list summarizes the steps for writing a MATLAB script:

1. If available, start the MATLAB program by double-clicking on the MATLAB icon on the Window's desktop. If not available, go to the Window's Start menu, click on *All Programs*, find the *MATLAB* program among the list of available programs, and double-click on it. This will open up the MATLAB desktop.
2. Click on the *New Script* icon in the Toolstrip in MATLAB's desktop. This brings up a new Script window (see Figure 2.2a).
3. Type your program into the Script window.
4. When you are finished typing in the program, save the script by clicking on the *Save* icon in the Toolstrip. A dialog box will appear in which you are to type in the name of the script that you which to save (see Figure 2.3). It is best to use a folder that contains **only** your own MATLAB scripts.
5. You may then run the script in the script window by clicking on the *Run* green arrow in the Toolstrip (see Figure 2.2b) or alternatively, from the command window by typing the script name (without the *.m* extension) after

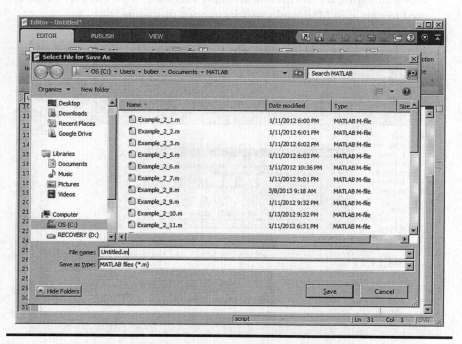

Figure 2.3 **The dialog box in which you are to type in the name of the script to be saved.**

the MATLAB prompt (>>). For example, if the program has been saved as *heat.m,* then type heat after the MATLAB prompt (>>), as shown below:

```
>> heat
```

When you open the MATLAB program, MATLAB assigns a default folder to the Current Folder Toolbar. Most likely, your saved script will not be in that folder. When you try to run your script by clicking on the Run icon in the script window, a dialog box will appear giving you the option of changing the MATLAB Current Folder or adding your folder to MATLAB's path (see Figure 2.4). Selecting the Change Folder option will change the folder listed in the Current Folder Toolbar, allowing you to run your program.

If you need additional help on getting started or with any MATLAB topic, you can click on The *Help* icon ⓸ in the Toolstrip in MATLAB's desktop. In the window that opens (see Figure 2.5) you have several choices:

1. You can type in a subject of interest the Search Documentation box.
2. In the center section there are a list of topics related to the number of MATLAB licenses that goes with the computer. New users should click on the MATLAB option, which opens the window shown in Figure 2.6, which contains additional topic choices.
3. You can click on the PDF Documentation option, which opens the window shown in Figure 2.7, which also contains additional topic choices. You may need to sign in to Mathworks website to open this figure.

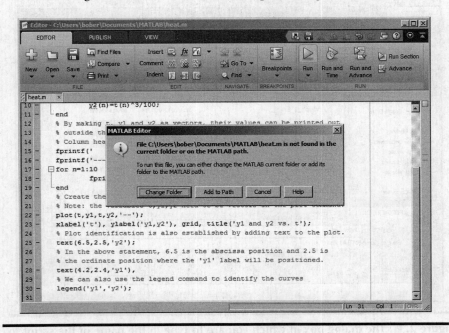

Figure 2.4 Dialog box for changing folder or path.

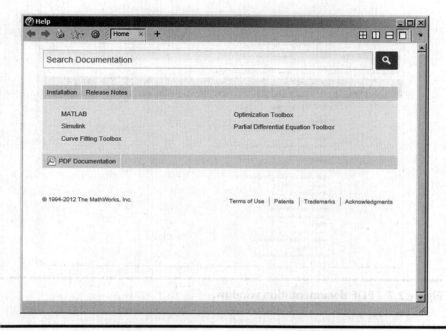

Figure 2.5   Product help window.

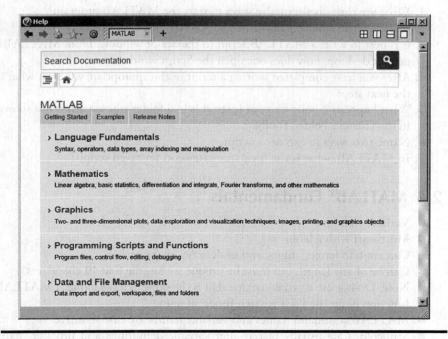

Figure 2.6   MATLAB help window.

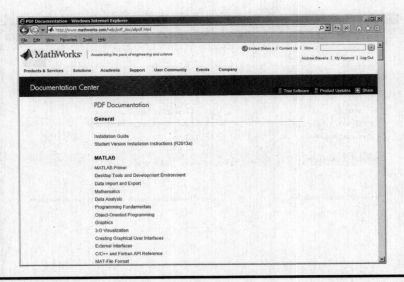

Figure 2.7   PDF documentation window.

Review 2.1

1. What are the two alternative ways to start the MATLAB program?
2. What are the windows in the MATLAB's default desktop?
3. It is best to write a MATLAB script in the Script window. From MATLAB's default desktop, how does one open the Script window?
4. After you have completed writing a script in the appropriate window, what is the next step?
5. What happens if your script of interest is in a folder whose path is not shown in the Current Folder Toolbar?
6. Name two ways to execute a script.
7. In MATLAB, what is the file name extension for saved scripts?

2.4 MATLAB® Fundamentals

- Variable names:
  Must start with a letter
  Can contain letters, digits, and underscore character
  Can be of any length but must be unique within the first 19 characters
  Note: Do not use a variable name that is the same as a file name, a MATLAB function name, or a self-written function name.
- MATLAB command names and variable names are case sensitive.
- Semicolons are usually placed after variable definitions and program statements when you do not want the command echoed to the screen. In the

absence of a semicolon, the defined variable appears on the screen, for example, if you entered the following assignment in the Command Window:

```
>> A = [3 4 7 6]
```

In the Command window, you would see

```
A =
    3 4 7 6
>>
```

Alternatively, if you add the semicolon after the assignment, then your command is entered, but there is nothing printed to the screen, and the prompt immediately appears for you to enter your next command:

```
>> A = [3 4 7 6];
>>
```

■ The percent sign (%) is used for a comment line.
■ A separate Graphics window opens to display plots and graphs.
■ MATLAB's clear and interrupt commands.
  `clear`: removes all variables and data from the Workspace.
  `clc`: clears the Command window.
  `clf`: clears the Graphics window.
  ctrl-C: aborts a program that may be running in an infinite loop.
■ Commands are case sensitive. Use lowercase letters for commands.
■ The `quit` command or `exit` command terminates MATLAB.
■ The `save` command saves variables or data in the Workspace of the Current Folder. The data file name will have the *.mat* extension.
■ User-defined functions (also called *self-written* functions) are also saved as M-files.
■ Scripts and functions are saved as ASCII text files. Thus, they may be written either in the built-in Script window or in Notepad or in any word processor (saved as a text file). Be aware that the single quotation mark in Microsoft Word is not the same as the one in MATLAB and will need to be changed in the MATLAB program.
■ The basic data structure in MATLAB is a matrix.
■ A matrix is surrounded by brackets and may have an arbitrary number of rows and columns; for example, the matrix $\mathbf{A} = \begin{bmatrix} 1 & 3 \\ 6 & 5 \end{bmatrix}$ may be entered into MATLAB as

```
>> A = [1 3 <enter>
        6 5]; <enter>
```

or

```
>> A = [1 3; 6 5]; <enter>
```

where the semicolon within the brackets indicates the start of a new row within the matrix.

■ A matrix of 1 row and 1 column is a scalar, for example,

```
>> A = [3.5];
```

Alternatively, MATLAB also accepts **A** = **3.5** (without brackets) as a scalar.

■ A matrix consisting of 1 row and several columns or 1 column and several rows is considered a vector, for example,

```
>> A = [2 3 6 5]  (row vector)
```

```
>> A = [2

3

6

5]  (column vector)
```

A column vector may also be entered by placing a semicolon after each element in the row, for example,

```
>> A = [2; 3; 6; 5]
```

A matrix can be defined by including a second matrix as one of the elements, for example,

```
>> B = [1.5 3.1];
```

```
>> C = [4.0 B];  (thus C = [4.0 1.5 3.1])
```

A specific element of matrix C can be selected by writing

```
>> a = C(2);  (thus a = 1.5)
```

If you wish to select the last element in a vector, you can write

```
>> a = c(end);  (thus a = 3.1)
```

■ The colon operator (:) may be used to
  1. Create a new matrix from an existing matrix, for example,

$$\text{if } A = \begin{bmatrix} 5 & 7 & 10 \\ 2 & 5 & 2 \\ 1 & 3 & 1 \end{bmatrix},$$

typing $x = A(:,1)$ gives $x = \begin{bmatrix} 5 \\ 2 \\ 1 \end{bmatrix}$.

The colon in the expression A(:,1) implies all the rows in matrix **A**, and the 1 implies column 1:

$$\text{Typing } x = A(:,2:3) \quad \text{gives } \mathbf{x} = \begin{bmatrix} 7 & 10 \\ 5 & 2 \\ 3 & 1 \end{bmatrix}$$

The first colon in the expression A(:,2:3) implies all the rows in **A**, and the 2:3 implies columns 2 and 3.

We can also write

$$y = A(1,:) \quad \text{which gives } \mathbf{y} = \begin{bmatrix} 5 & 7 & 10 \end{bmatrix}$$

The 1 implies the first row and the colon implies all the columns.

2. Colon can also be used to generate a series of numbers. The syntax is n = starting value : step size : final value. If the step size is omitted, the default step size is one. For example,

$$n = 1:8 \quad \text{which gives } \mathbf{n} = \begin{bmatrix} 1 & 2 & 3 & 4 & 5 & 6 & 7 & 8 \end{bmatrix}$$

To increment in steps of 2, use

$$n = 1:2:7 \quad \text{which gives } \mathbf{n} = \begin{bmatrix} 1 & 3 & 5 & 7 \end{bmatrix}$$

These types of expressions are often used in a for loop, which is discussed later.

■ Arithmetic operators

+	Addition
−	Subtraction
*	Multiplication
/	Division
^	Exponentiation

For arithmetic statements containing several of these arithmetic operators, MATLAB has a specific order in carrying out the operations. First, all expressions within parentheses will be carried out first in the following order: exponentiation, then multiplication and division, and then addition and subtraction. Expressions outside parentheses will be carried out in the same order. Knowing this order may help you in deciding where parentheses are required when you write arithmetic statements. For example, for the expression $y = c/2m$, you might be tempted to write the expression in the MATLAB Command window (after defining c and m) as

```
c = 5.2;  m = 24.6;
y = c/2*m
```

This would give the wrong answer for y. MATLAB would divide c by 2 and multiply the result by m. The correct ways to write the expression are

```
y = c/(2*m)  or  c/2/m
```

In the first expression, MATLAB will first carry out the expression within the parentheses, so that the 2*m becomes one number and then c is divided by this one number. In the second expression, there are no parentheses, so MATLAB, proceeding from left to right, will calculate c/2 first and then divide the result by m. Try typing the above expressions in the Command window and observe the two different answers you get for y.

To display a variable value, just type the variable name without the semi-colon, and the variable will appear on the screen.

Try typing these statements into the Command window:

```
clc;
x = 5;
y = 10;
z = x + y
w = x - y
z = y/x
z = x*y
z = x^2
```

■ Special values

pi	π
i or j	$\sqrt{-1}$
inf	∞
ans	The last computed unassigned result to an expression typed in the Command window

Try typing these statements in the Command window:

```
x = pi;
z = x/0   (gives the infinity symbol)
```

■ Trigonometric functions

sin	sine
sinh	hyperbolic sine
asin	inverse sine
asinh	inverse hyperbolic sine
cos	cosine
cosh	hyperbolic cosine
acos	inverse cosine
acosh	inverse hyperbolic cosine
tan	tangent
tanh	hyperbolic tangent
atan	inverse tangent
atan2	four-quadrant inverse tangent
atanh	inverse hyperbolic tangent

The arguments of these trigonometric functions are in radians. However, the arguments can be made in degrees if a "d" is placed after the function name, such as sind(x).

Try typing these statements into the Command window:

```
clc;
x = pi/2;   % (x in radians)
y = sin(x)
z = atan(1.0)
x = 30;     % (x in degrees)
w = sind(x)
z = atand(1.0)
```

Note: x(radians) = x(degrees) × $\pi/180$

■ Exponential, logarithm, square root, and error functions

exp	exponential
log	natural logarithm
log10	common (base 10) logarithm
sqrt	square root
erf	error function

Try typing these statements into the Command window:

```
clc;
x = 2.5;
y = exp(x)
z = log(y)
w = sqrt(x)
```

Suppose we had a problem involving the following arithmetic statement that we needed to evaluate:

$$y = \cos\left(\sqrt{\frac{k}{m} - \left(\frac{c}{2m}\right)^2}\, t\right)$$

To make it easier to write the MATLAB statement corresponding to the above arithmetic statement, we could break up the argument of the cos term as follows (type the following in the Command window):

```
k = 200;  c = 5;  m = 25;  t = 5;
arg = sqrt(k/m - (c/(2*m))^2);
y = cos(arg*t)
```

■ Complex numbers
Complex numbers may be written in two forms: Cartesian, for example, z = x + yj, or polar, for example, z = r * exp(j*theta). Note that we will use j for $\sqrt{-1}$ throughout this chapter. However, MATLAB allows the use of i for $\sqrt{-1}$ as well.

Note: i and j are also legal MATLAB variable names that are often used within loops. To avoid confusion, programs that involve complex numbers should not use i or j as variable names.

■ Other special values

abs	Absolute value (magnitude)
angle	Phase angle (in radians)
conj	Complex conjugate
imag	Complex imaginary part
real	Complex real part

Try typing these statements into the Command window:

```
clc;
z1 = 1 + j;
z2 = 2 * exp(j * pi/6)
y = abs(z1)
w = real(z2)
v = imag(z1)
phi = angle(z1)
```

■ Other useful functions

size (X)	Gives the size (number of rows and the number of columns) of matrix X.
length(X)	For vectors, length(X) gives the number of elements in X.
linspace (X1,X2,N)	Generates N points between X1 and X2
sum(X)	For vectors, sum(X) gives the sum of the elements in X. For matrices, sum(X) gives a row vector containing the sum of the elements in each column of the matrix.
max(X)	For vectors, max(X) gives the maximum element in X. For matrices, max(X) gives a row vector containing the maximum in each column of the matrix. If X is a column vector, it gives the maximum absolute value of X.
min(X)	Same as max(X) except it gives the minimum element in X.
sort(X)	For vectors, sort(X) sorts the elements of X in ascending order. For matrices, sort(X) sorts each column in the matrix in ascending order.
factorial(n)	$n! = 1 \times 2 \times 3 \times \cdots \times n$
mod(x,y)	Modulo operator gives the remainder resulting from the division of x by y. For example, mod(13,5) = 3, that is, $13 \div 5$, gives 2 plus remainder of 3 (the 2 is discarded). In another example, mod(n,2) gives zero if n is an even integer and one if n is an odd integer.

If you wish to obtain the index of the element in vector A that is a maximum (or minimum), use [Amax, i] = max(A), where Amax is the maximum element value in vector A and i is the index number of that element. Try typing these statements into the Command window:

```
clc;
A = [2 15 6 18 4 12 16];
length(A)
Amax = max(A)
[Amax,i] = max(A)
z = sum(A)
A = [2 15 6 18; 15 10 8 4; 10 6 12 3];
x = max(A)
y = sum(A)
size(A)
mod(21,2)
mod(20,2)
```

A list of the complete set of elementary math functions can be obtained by typing **help elfun** in the Command window.

Sometime, it is necessary to preallocate a matrix of a given size. This can be done by defining a matrix of all zeros or ones; examples are as follows:

$$A = zeros(3) = \begin{bmatrix} 0 & 0 & 0 \\ 0 & 0 & 0 \\ 0 & 0 & 0 \end{bmatrix}$$

$$B = zeros(3,2) = \begin{bmatrix} 0 & 0 \\ 0 & 0 \\ 0 & 0 \end{bmatrix}$$

$$C = ones(3) = \begin{bmatrix} 1 & 1 & 1 \\ 1 & 1 & 1 \\ 1 & 1 & 1 \end{bmatrix}$$

$$D = ones(2,3) = \begin{bmatrix} 1 & 1 & 1 \\ 1 & 1 & 1 \end{bmatrix}$$

The function to generate the identity matrix (main diagonal of ones, all other elements are zero) is eye, for example,

$$I = eye(3) = \begin{bmatrix} 1 & 0 & 0 \\ 0 & 1 & 0 \\ 0 & 0 & 1 \end{bmatrix}$$

Review 2.2

1. Are command statements and variable names case sensitive?
2. What is the purpose of placing a semicolon at the end of a command statement or a variable assignment?
3. How does one establish a comment line in a script?
4. What is the command that will clear the Command window?
5. What is the basic data structure in MATLAB?
6. Name two uses of the colon operator.
7. List the arithmetic operators in MATLAB.
8. What is MATLAB's command for
   a. π
   b. e
   c. ln
   d. sine function in radians
   e. sine function in degrees
   f. $\sin^{-1}$ function
   g. The number of elements in a vector
   h. The size of a matrix (the number of rows and columns)
   i. The sum of the elements in a vector
   j. The maximum of the elements in a vector
   k. Preallocating storage for a 3 × 3 matrix

2.5 MATLAB® Input/Output

2.5.1 Output

To display a vector X, just type X without the semicolon, and vector X will be displayed on the screen. For example, first, define X:

```
>> X = [0 1 2 3 4 5];
>>
```

Now enter vector X without the semicolon.

```
>> X
```

The following will be displayed on the screen:

```
X =
     0     1     2     3     4     5
>>
```

The disp command prints only the items that are enclosed within the parentheses, which can be a variable or alphanumeric information. Alphanumeric information must be enclosed by singe quotation marks. For example (assuming that vector **X** earlier has already been entered in the Command window), type in

```
>> disp(X); disp('m/s');
```

The following will be displayed on the screen:

```
0    1    2    3    4    5
m/s
>>
```

The fprintf command prints formatted text to the screen or to a file, for example,

```
>> V = 2.2;
>> fprintf('The velocity is %f m/s \n', V);
```

The following will appear on the screen:

```
The velocity is 2.200000 m/s
```

The \n (newline) tells MATLAB to move the cursor to the next line. Similarly, a \t (tab) tells MATLAB to move the cursor several spaces along the same line. The %f refers to a formatted floating-point number that is assigned to variable V, and the default is six decimal places (though in earlier versions of MATLAB, the default for %f format was four decimal places).

The command fprintf uses format strings based on the C programming language [1]. Thus, you can specify the number of spaces for the printed variable as well as the number of decimal places. For example, to specify eight spaces for the printed variable to two decimal places, use %8.2f. You can also just specify the number of decimal places, for example, three decimal places, then let MATLAB decide the number of spaces for the variable by using %.3f. However, to create neat-looking tables, it is best to specify the number of spaces in the format statement that allows for several spaces between variables in adjacent columns, such as %10.3f.

Other formats

%i or %d	Used for integers.
%e	Scientific notation (e.g., 6.02e23), default is six decimal places.
%g	Automatically uses the briefest of %f or %e format.
%s	Used for a string of characters.
%c	Used for a single character.

Unlike C, the format string in MATLAB's fprintf statement must be enclosed by single quotation marks (and **not** double quotes).

It is often useful to print the results of a MATLAB program to a file, possibly for inclusion in a report. In addition, program output that is printed to a file can be subsequently edited within the file, such as aligning or editing column headings in a table. Before you can print to a file, you need to open a file for printing with the command, fopen. The syntax for fopen is

```
fo = fopen('filename','w')
```

In this statement, fo is a pointer to the file named filename, and the w indicates writing to the file. To print to filename, use

```
fprintf(fo,'format',var1,var2,..);
```

where the format string contains the text format for var1, var2, etc.

Example 2.1

Printing to a file is illustrated in this example:

```
% Example_2_1.m
% This program is an example for printing to a file.
clear; clc;
V = 12;   % velocity
F = 50.2; % force
fo = fopen('output.txt','w');
fprintf(fo,'V=%4i m/s, F=%5.2f N \n',V,F);
fclose(fo);
```

The file "output.txt" contains:

```
V = 12 m/s, F = 50.20 N
```

The extension on the output file should be *.txt* (otherwise, when you try to open the file, MATLAB will start the import wizard). The resulting output file can be opened from either the Script Window or the Command Window. To access the output file, in the script window click on the *Open* icon in the Toolstrip which brings up the screen shown in Figure 2.8. Note, if you click on the word *Open* instead of the icon, you will get a listing of all files that were recently used. In the box labeled *File name*, type in **.txt*. This will bring up all the files with the extension *.txt* in the Current Folder. To open the file of interest, double click on the name of the output file (in this example, the file name is *output.txt*). In the earlier versions of MATLAB, you would

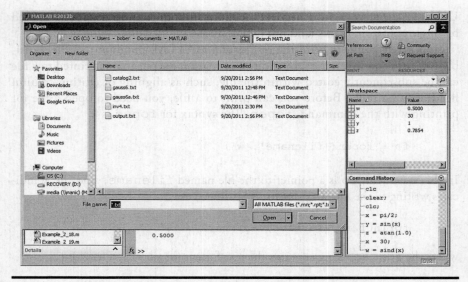

Figure 2.8 Window for accessing on output file with .txt extension.

not be able to open the output file without having run the `fclose(fo)` statement in the program. But it is still a good practice to include the `fclose` statement after all the output statements in the program or at the end of the program itself.

2.5.2 *Input*

If you wish to have your program pause to accept input from the keyboard, use the `input` statement; for example, to enter a 2 by 3 matrix, use

```
Z = input('Enter values for matrix Z enclosed by brackets \n');
```

You will see the following on the screen:

```
Enter values for matrix Z enclosed by brackets
```

As an example, type in

```
[5.1 6.3 2.5; 3.1 4.2 1.3]
```

Thus, $Z = \begin{bmatrix} 5.1 & 6.3 & 2.5 \\ 3.1 & 4.2 & 1.3 \end{bmatrix}$.

Note that the argument to `input()` is a character string enclosed by single quotation marks. The character string will be printed to the screen as shown above.

If the response to the input statement is a character or a string, you also need to enclose the character or the string with single quotation marks. However, you can avoid this requirement by entering a second argument of 's' to the input statement as shown in the following statement:

```
response = input('Print Z to a file? (y/n):\n', 's');
```

In this case, the user can respond with either a y or n (without single quotation marks).

An example using this concept will be given later in this chapter.

The fscanf command may be used to read from an existing file as shown in the following:

```
A = zeros(n,m);
fi = fopen('filename.txt','r');
[A] = fscanf(fi,'%f',[n,m]);
```

where n and m correspond to the number of columns and rows in the data file (n and m must be assigned numerical values). The 'r' in the fopen statement tells MATLAB that this file is for reading in data that are contained in the file. The $n \times m$ matrix is filled in column order. Thus, rows become columns and columns become rows. To use the data in their original order, transpose the read-in matrix. To transpose a matrix A in MATLAB, simply type in A'. This changes columns to rows and rows to columns.

An existing data file can also be entered into a program by the load command. The load command, unlike the fscanf command, leaves rows as rows and columns as columns.

Example

```
load filename.txt
x = filename(:,1);
y = filename(:,2);
```

The input file must have the same number of rows in each column (see Example 2.11).

Review 2.3

1. If you enter a variable assignment without placing a semicolon after the assignment, what happens?
2. Name two commands that will result in printing to the screen.
3. What is the format that will move the cursor to the next line?
4. What is the format that will print a variable to 10 spaces and to three decimal points?
5. What are the commands necessary to print to a file?
6. Name the three methods that can be used in a script to import data into the Workspace for the execution of the script.

2.6 Loops

Loops provide the means to *repeat* a series of statements with just a few lines of code. For example, suppose we had an equation for the x position of a vehicle as a function of time, such as

$$x(t) = x_o - vt$$

where
x_o is the initial position of the vehicle
v is the vehicle's speed

We wish to create a table or a plot of the x position of the vehicle at various times. In our program, we will create a loop, assign or calculate a time, calculate the x position based on the assigned or calculated time, print the time and the position, return to the top of the loop, calculate a different time, calculate the x position based on the new time, and print the new time and the new position. Continue the process until you have reached the last time that you wish to determine the x position. Since x_o and v, are constants, you would not want those variables to be inside the loop.

- The for loop
  The syntax for the for loop is

  for index variable = starting value : step size : final value

  The step size may be omitted and then MATLAB will take the step size to be 1.
  As an example, we will take the index variable as m and the starting value as 1. We will omit the step size and take the final value as 20. Then our for loop will be

  ```
  for m = 1:20
          statement;
          ⋮
          statement;
  end
  ```

MATLAB sets the index m to 1, carries out the statements between the for and end statements, then returns to the top of the loop, changes m to 2, and repeats the process. After the process has been carried out 20 times, the program exits the loop without further executing any of the statements within the loop. All statements that are not to be repeated should not be within the for loop. For example, table headings that are not to be repeated should be outside the for loop. Also notice that statements within the for loop are indented. MATLAB does this to make it easier to read and debug a script containing for loops. You can have MATLAB do final indenting by highlighting your entire script and then entering Ctl-I.

Example 2.2

In this example, we illustrate the use of a for loop. The indices in the for loop are nonintegers:

```
% Example_2_2.m
% This program determines the x position of a vehicle as a
% function of time, t.
% the governing equation is x = xo+v*cos(theta)*t
% t varies from 0 to 200 seconds.
% v = 10 m/s and theta = 30 degrees
clear; clc;
xo = 0.0; theta = pi/6; v = 10.0;
fprintf('           t(s)              x(m)              \n');
fprintf('------------------------------------------\n');
for t = 0:0.5:10
    x = xo+v*cos(theta)*t;
    fprintf('%10.1f  %10.1f \n',t,x);
end
```

Program results

t(s)	x(m)
0.0	0.0
0.5	4.3
1.0	8.7
1.5	13.0
2.0	17.3
⋮	⋮
8.0	69.3
8.5	73.6
9.0	77.9
9.5	82.3
10.0	86.6

`>>`

Note: If the index in a for loop is used to select an element of a matrix, then the index *must* be an integer. However, if you are not using the for indices to select an element of a matrix, then the for indices need not be an integer as was shown in the previous example.

Example 2.3

In this example, the indices, *j*, in the for loop are integers and are used to select an element of a vector, *t*:

```
% Example_2_3.m
% This program is an example of the use of a for loop in which
% the indices of the for loop select an element of a vector.
% The program creates a table of y1 and y2 vs t.
% 0 <= t <= 10
clear; clc;
% Table headings:
fprintf(' t            y1          y2        \n');
fprintf('---------------------------------\n');
t = 0:0.5:10;
for j = 1:length(t)
        y1 = t(j)^2/10;
        y2 = t(j)^3/100;
        fprintf(' %5.1f      %10.3f      %10.3f \n',t(j),y1,y2);
end
```

Program results

t	y1	y2
0.0	0.000	0.000
0.5	0.025	0.001
1.0	0.100	0.010
1.5	0.225	0.034
2.0	0.400	0.080
⋮	⋮	⋮
8.0	6.400	5.120
8.5	7.225	6.141
9.0	8.100	7.290
9.5	9.025	8.574
10.0	10.000	10.000
>>		

Example 2.4

In this example, we will calculate the position and velocity of a ball thrown vertically upward in a gravitational field (neglecting drag) as a function of time *t*. The approximate path of the ball is shown in Figure 2.9.

Recall that velocity is the rate of change of distance with respect to time and acceleration is the rate of change of velocity with respect to time. For motion in the *y* direction only, with the *y*-axis pointing upward, the governing equations are

$$V = V_o - g t \tag{2.1}$$

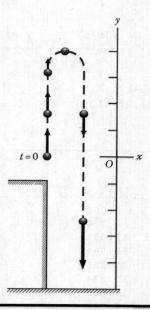

Figure 2.9   Ball in a gravitational field.

$$y = V_o t - \frac{gt^2}{2} \tag{2.2}$$

where

V is the velocity

V_o is the velocity at $t = 0$

y is the position of the ball at time t

g is the acceleration of gravity

t is the time

Equations 2.1 and 2.2 are based on the initial conditions $V(0) = V_o$ and $y(0) = 0$.

The following MATLAB program calculates V and y vs. t, for $0 \leq t \leq 5$ s in steps of 0.5 s. We have taken $V_o = 20$ m/s and $g = 9.81$ m/s².

The program follows:

```
% Example_2_4.m
% This program calculates the velocity and position of a free
% falling body vs. time.
% The velocity = Vo-gt
% the position y = Vo*t-0.5*g*t^2
% Vo = 20 m/s, g = 9.81 m/s^2
% The output goes to a file named output.txt.
clear; clc;
Vo = 20.0; g = 9.81;
fo = fopen('output.txt','w');
```

```
% Table headings
fprintf(fo,'   t(s)        v(m/s)        y(m)        \n');
fprintf(fo,'-------------------------------------------\n');
for t = 0:0.5:5
        v = Vo-g*t;
        y = Vo*t-0.5*g*t^2;
        fprintf(fo,'%6.2f %10.3f %10.3f \n',t,v,y);
end
fclose(fo)
```

Program results

t(s)	v(m/s)	y(m)
0.00	20.000	0.000
0.50	15.095	8.774
1.00	10.190	15.095
1.50	5.285	18.964
2.00	0.380	20.380
2.50	-4.525	19.344
3.00	-9.430	15.855
3.50	-14.335	9.914
4.00	-19.240	1.520
4.50	-24.145	-9.326
5.00	-29.050	-22.625

You might think that the statement $t = 0:0.5:5$ as part of the for loop statement will produce a vector of t values. However, that is not the case. As the program progresses back to the start of the for loop, old values of t are overwritten by the new value of t. Try running the following example:

```
% for_loop_assignment.m
% This program tests an assignment in a for loop as compared
% to the same assignment outside the for loop.
clear; clc;
Vo = 10; g = 9.81;
for t = 0:0.5:5
    v = Vo-g*t;
    y = Vo*t-0.5*g*t^2;
    fprintf('%6.2f  %10.3f  %10.3f \n',t,v,y);
end
fprintf('This value for t is from the for loop \n');
fprintf('t=%6.2f \n\n',t);
fprintf('We see that the assignment of t=0:0.5:5 \n');
fprintf('within the for loop does not produce \n');
fprintf('a vector. In fact, only the last value \n');
fprintf('of t is printed. \n\n');
```

Now type $t = 0:0.5:5$ in the Command window without the semicolon. See that t is now a vector.

Example 2.5

Although atmospheric conditions vary from day to day, it is convenient for design purposes to have a model for atmospheric properties with altitude. The U.S. Standard Atmosphere, modified in 1976, is such a model. For altitudes less than or equal to 11,000 m, the governing equations for the air temperature, pressure, and density are as follows:

$$p = p_o \left(1 - \frac{\lambda z}{T_o} \right)^{g/\lambda R} \tag{2.3}$$

$$T = T_o - \lambda z \tag{2.4}$$

$$\rho = \frac{p}{RT} \tag{2.5}$$

where
- z = altitude (m)
- T = air temperature (K)
- p = air pressure (Pa)
- ρ = air density (Kg/m³)
- T_o = 288.15 K (temperature at $z = 0$)
- p_o = 1.01325 × 10⁵ Pa (pressure at $z = 0$)
- R = 287 J/(kg-K) (gas constant for air)
- g = 9.81 m/s² (gravitational constant)
- λ = 0.0065 K/m (the lapse rate)

In the following example, we calculate T, p, and ρ every 1000 m from $z = 0$ (sea level) to $z = 11,000$ m and print the results to a file in a table format.

The program follows:

```
% Example_2_5.m
% This program determines atmospheric properties of
% temperature, T, pressure, p, and density, rho, every
% 1000 m of altitude and prints these values to a file
% in table format. The governing equations are
% T=To-lamda*z, p=po*(1-lamda*z/To)^ex,
% where ex=g/(lamda*R), and rho=p/(R*T).
clear; clc;
To = 288.15;
po = 1.01325e5;
R = 287.0;
g = 9.81;
lamda = 0.0065;
z = 0:1000:11000;
ex = g/(lamda*R);
```

```
fo = fopen('output.txt','w');
fprintf(fo,'Atmospheric Properties \n');
% Table headings
fprintf(fo,'    z           T           p           rho \n');
fprintf(fo,'   (m)         (K)         (Pa)         (kg/m^3) \n');
fprintf(fo,'------------------------------------------------------\n');
for i = 1:length(z)
    T = To-lamda*z(i);
    p = po*(1-lamda*z(i)/To)^ex;
    rho = p/(R*T);
    fprintf(fo,'%6.0f %8.2f %10.4e %10.4f \n', z(i),T,p,rho);
end
```

Program results

```
Atmospheric Properties
    Z           T           p           rho
   (m)         (K)         (Pa)         (kg/m^3)
------------------------------------------------
      0       288.15     1.0133e+005    1.2252
   1000       281.65     8.9869e+004    1.1118
   2000       275.15     7.9485e+004    1.0065
   3000       268.65     7.0095e+004    0.9091
   4000       262.15     6.1624e+004    0.8191
   5000       255.65     5.4002e+004    0.7360
   6000       249.15     4.7162e+004    0.6596
   7000       242.65     4.1041e+004    0.5893
   8000       236.15     3.5580e+004    0.5250
   9000       229.65     3.0723e+004    0.4661
  10000       223.15     2.6418e+004    0.4125
  11000       216.65     2.2614e+004    0.3637
```

■ The while loop

An alternative to the for loop is the while loop. In the while loop, MATLAB will carry out the statements between the while and end statements as long as the condition in the while statement is satisfied. If an index in the program is required, the use of the while loop statement (unlike the for loop statement) requires that the program generate its own index, as shown in the following example:

```
n = 0;
while n < 10
        n = n+1;
        y = n^2;
    end
```

Note that the above statement "n = n+1" may not make sense algebraically but does make sense in the MATLAB language. The " = " operator in

MATLAB (as in many computer languages) is the *assignment* operator that tells MATLAB to fetch the contents in the memory cell containing the variable n, put its value into the arithmetic unit of the CPU, increment the variable n by 1, and put the new value back into the memory cell designated for the variable n. Thus, the old value of n has been replaced by the new value for n.

Review 2.4

1. What is the objective in using a for loop?
2. What is the syntax of a for loop?
3. Should table headings that are not to be repeated be inside or outside a for loop?
4. If the index of a for loop is used to select an element of a vector or a matrix, what variable type should the for loop index be?
5. What other statement type can be used to create a loop?
6. What is the major difference between a for loop and a while loop?

Exercises

Mechanical Engineering

E2.1 The motion of a piston in an internal combustion engine is shown in Figure 2.10. The piston's position, s, as seen from the crankshaft center is determined to be

$$s(t) = r \cos(2\pi\omega t) + \sqrt{b^2 - r^2 \sin^2(2\pi\omega t)} \qquad (2.6)$$

where
   b is the length of the piston rod
   r is the radius of the crankshaft
   ω is rotational speed of the crankshaft in revolutions per second

Develop a MATLAB program that determines s vs. t for $0 \le t \le 0.01$ s. Use 20 subdivisions on the t domain. Take $r = 9$ cm, $\omega = 100$ revolutions per second, and $b = 14$ cm. Create a table of s vs. t and print the results to both the screen and to a file.

E2.2 The position, y, of a mass in a mass–spring–dashpot system (for a derivation of Equation 2.7, see Project P2.5) is given by

$$y = \exp\left(-\frac{c}{2m}t\right)\left\{ A\cos\left(\sqrt{\frac{k}{m} - \left(\frac{c}{2m}\right)^2}\ t\right) + B\sin\left(\sqrt{\frac{k}{m} - \left(\frac{c}{2m}\right)^2}\ t\right) \right\} \qquad (2.7)$$

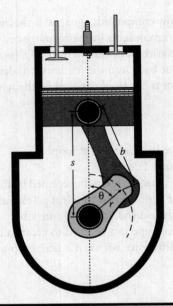

Figure 2.10   Piston motion in an internal combustion engine.

Take
  m = 25.0 kg
  c = damping factor = 5.0 N s/m
  k = spring constant = 200.0 N/m
  A = 5.0 m
  B = 0.25 m

a.   Determine $y(t)$ for $0 \leq t \leq 10$ s in steps of 0.1 s.
b.   Create a table of y vs. t every 1 s and print the results both to the screen and to a file.

There are several exercises that follow that involve Newton's second law of motion, which briefly stated is, the sum of the forces acting on a body is equal to the body's mass times its acceleration. Mathematically, it is:

$$\sum_i \vec{F}_i = m\vec{a} = m\frac{d\vec{V}}{dt} \qquad (2.8)$$

where
  $\vec{F}_i$ is the ith force acting on the body
  m is the body's mass
  $\vec{a}$ is its acceleration
  $\vec{V}$ is its velocity

Note that the equation is a vector equation. In most cases we will decompose the vector equation into its (x, y) components.

E2.3 A basketball player shoots the ball when he or she is 6 m from the center of the hoop as shown in Figure 2.11. The ball is released at a velocity, V_o, and makes an angle of 40° with the horizontal. We will neglect the drag on the basketball in this analysis.

Using Newton's second law and taking the x component of Equation 2.8 gives

$$\sum F_{x,i} = 0 = m\frac{dV_x}{dt} \rightarrow V_x = \text{constant} = V_o \cos(\theta)$$

$$\frac{dx}{dt} = V_o \cos(\theta) \rightarrow x = V_o \cos(\theta)t + c_1, \quad x(0)=0 \rightarrow c_1 = 0$$

Thus,

$$x = V_o \cos(\theta)t \qquad (2.9)$$

The y component of Equation 2.8 is

$$\sum F_{y,i} = -W = \frac{W}{g}\frac{dV_y}{dt} \rightarrow V_y = -gt + c_2 \rightarrow V_y(0) = V_o \sin(\theta) = c_2$$

where W is the weight of the basketball and g is the gravitational constant.

Thus,

$$V_y = \frac{dy}{dt} = -gt + V_o \sin(\theta) \rightarrow y = -g\frac{t^2}{2} + V_o \sin(\theta)t + c_3$$

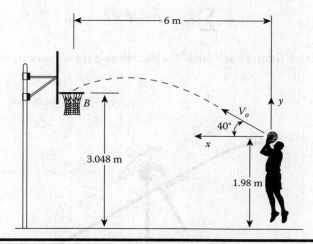

Figure 2.11 Player shooting basketball. (From Beer, F.P. and Johnson, E.R., *Vector Mechanics for Engineers: Statics and Dynamics,* **McGraw-Hill, New York, 2007, Figure P11.110. With permission.)**

$$y(0)=0 \quad \rightarrow \quad c_3 = 0$$

$$y = V_o \sin(\theta)t - \frac{g}{2}t^2 \tag{2.10}$$

Take the (x, y) position of the center of the hoop to be $(x_f, y_f) = (6.0, 3.048 - 1.98)$ and $\theta = 40°$:

a. Determine the time, t_f, that it takes for the ball to reach the center of the hoop. Time, t, equals zero when the ball leaves the player's hands.
b. Determine the velocity, V_o, that will result in the ball reaching the center of the hoop.
c. Create a table consisting of t, x, y for $0 \leq t \leq t_f$ in steps of 0.01 s.

E2.4 A boy on a snowboard, initially at rest, slides down a smooth hill that makes an angle θ with the horizontal (see Figure 2.12). The boy weighs 650 N, and the friction coefficient, μ, between the snowboard and the snow is 0.05.

Neglect the drag on the boy and his snowboard. The x and y components of Equation 2.8 are

$$\sum F_{x,i} = m\,a_x = m\frac{dV_x}{dt} \quad \text{and} \quad \sum F_{y,i} = N - W\cos\theta = 0 \tag{2.11}$$

where N is normal force on the snowboard. The sum of the forces in the x direction is given by

$$\sum F_{x,i} = W\sin\theta - f \tag{2.12}$$

The friction force, $f = \mu N$ and $W = mg$, where g is the gravitational constant.

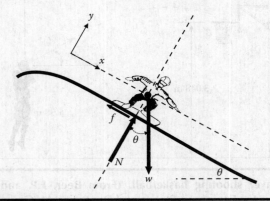

Figure 2.12   Boy on snowboard sliding down a hill.

Thus,

$$\sum F_{x,i} = W \sin\theta - \mu W \cos\theta = \frac{W}{g}\frac{dV_x}{dt} \qquad (2.13)$$

where V_x is the velocity of the boy-snowboard unit in the x direction. Determine an expression for $V_x(t)$, taking $V_x(0) = 0$.

The x position of the boy-snowboard unit is determined by

$$\frac{dx}{dt} = V_x \qquad (2.14)$$

Determine an expression for $x(t)$, taking $x(0) = 0$.

Develop a MATLAB program that calculates position, x, and velocity, V_x, as a function of time, t, for $0 \le t \le 10$ s in steps of 0.5 s. Print the results both to the screen and to a file. Take $\theta = 10°$ and $g = 9.81$ m/s². Also run the program for 1 min to see if the boy reaches a terminal velocity. If not, why not?

E2.5 A small sphere moving though a fluid at a slow velocity will have a drag force acting on it which is described by Stokes' law. The sphere could be a dust particle or a raindrop moving in air or a ball bearing moving in oil. The drag force described by Stokes' law is

$$D = 6\pi R\mu V \qquad (2.15)$$

where
   D is the drag
   R is the radius of the sphere
   μ is the viscosity of the fluid
   V is the velocity of sphere

Let us consider a ball bearing dropped in oil (see Figure 2.13) with an initial velocity of zero. We want to determine the velocity and position as a function of time. The forces acting on the ball bearing are the gravitational force, W; buoyancy force, B; and the drag force, D. The buoyancy force is equal the weight of the fluid displaced.

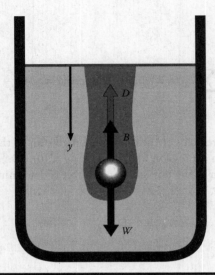

Figure 2.13   Ball bearing in a viscous liquid column.

Applying Newton's second law to the sphere gives

$$W - B - D = \frac{W}{g}\frac{dV}{dt} \tag{2.16}$$

where

    W is the weight of sphere = $\rho \mathbb{V} g$
    B is the buoyancy = weight of fluid displaced
    ρ is the mass density
    g is the gravitational constant = 9.81 m/s²
    $\mathbb{V}$ is the volume of sphere = $(4/3)\pi R^3$

Terminal velocity (no change in the velocity of the moving sphere) occurs when

$$W - B - D = 0$$

$$W - B - 6\pi R \mu V_T = 0 \quad \rightarrow \quad W - B = 6\pi R \mu V_T \tag{2.17}$$

$$W = \rho_{steel}\, g \times \frac{4}{3}\pi R^3, \quad B = \rho_{oil}\, g \times \frac{4}{3}\pi R^3, \tag{2.18}$$

$$W - B = (\rho_{steel} - \rho_{oil}) \times \frac{4}{3}\pi R^3 \tag{2.19}$$

Substituting Equations 2.15 and 2.17 into Equation 2.16 gives

$$6\pi R\mu(V_T - V) = \frac{W}{g}\frac{dV}{dt} \tag{2.20}$$

Separating variables and integrating gives

$$\int_0^V \frac{dV'}{(V' - V_T)} = -\int_0^t \frac{6\pi R\mu g}{W}dt \tag{2.21}$$

Integrating Equation 2.21 gives

$$\left[\ln(V' - V_T)\right]_0^V = \ln(V - V_T) - \ln(-V_T)$$

$$= \ln\left(\frac{V - V_T}{-V_T}\right) = \ln\left(\frac{V_T - V}{V_T}\right) = -\frac{6\pi R\mu g}{W}t \tag{2.22}$$

Then,

$$1 - \frac{V}{V_T} = e^{-\frac{6\pi R\mu g}{W}t} \tag{2.23}$$

or

$$V = V_T\left(1 - e^{-\frac{6\pi R\mu g}{W}t}\right) \tag{2.24}$$

Take μ = 3.85 (N-s)/m^2, R = 0.01 m, ρ_{steel} = 7910 kg/m^3, ρ_{oil} = 899 kg/m^3, and g = 9.81 m/s^2.

Create a MATLAB program that will

a. Determine the buoyancy force, B.
b. Determine the weight of the ball bearing, W.
c. Determine the terminal velocity, V_T.
d. Use a while loop to determine V as a function of time, for $0 \leq t \leq 0.05$ s in steps of 0.005 s.
e. Create and print to a file a table containing t and V. Also print values for W, B, and V_T to the same file.

Electrical Engineering

E2.6 The voltage in a parallel resistance, inductor, and capacitor (RLC) circuit (for a derivation of Equation 2.25, see Project P2.7) is given by

$$v = \exp\left(-\frac{1}{2RC}t\right)\left\{A\cos\left(\sqrt{\frac{1}{LC}-\left(\frac{1}{2RC}\right)^2}\,t\right) + B\sin\left(\sqrt{\frac{1}{LC}-\left(\frac{1}{2RC}\right)^2}\,t\right)\right\}$$

(2.25)

Take
$R = 100\ \Omega$
$L = 10^{-3}$ H
$C = 10^{-6}$ F
$A = 6.0$ V
$B = -8.9$ V

a. Determine $v(t)$ for $0 \le t \le 5.0 \times 10^{-4}$ s. Use 100 subdivisions on the time domain.
b. Print out a table of v vs. t every fourth subdivision.

E2.7 In North America, residential electrical service is usually delivered as two 340 $V_{peak\text{-}to\text{-}peak}$ sinusoids at 60 Hz with no DC component, each 180° out of phase with each other (see Figure 2.14), referred to as *one-phase* service. For high-power appliances (e.g., air conditioners and stoves), the current is drawn from both "legs" of the service.

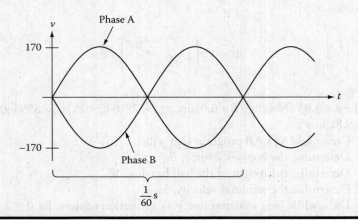

Figure 2.14   One-phase electrical service.

1. For phase A, use MATLAB's `sin()` function to create a table of v vs. t, for $0 \leq t \leq 100$ ms in steps of 4 ms.
2. Compute and print to the screen the root-mean-square (RMS) voltage for this waveform by performing the following operations in MATLAB:
   a. Squaring it
   b. Computing the mean value of the squared waveform by averaging it over the 100 ms time interval
   c. Taking the square root of the average

2.7 MATLAB® Graphics

■ Plot commands

MATLAB provides many different types of plots. Clicking on the PLOTS tab in MATLAB's desktop graphically lists the various types of plots that are available (see Figure 2.1). The commands for creating linear plots, semilog plots, and log-log plots are as follows:

`plot(x,y)`	Linear plot of y vs. x
`semilogx(x,y)`	Semilog plot (log scale for x-axis, linear scale for y-axis)
`semilogy(x,y)`	Semilog plot (linear scale for x-axis, log scale for y-axis)
`loglog(x,y)`	Log-log plot (log scale for both x- and y-axes)

The variable arguments in the plot commands need to be vectors. In addition, the vectors need to be of the same length. If the arguments in the plot command are scalars, the plot commands will produce just a single point.

■ Simple linear plot

Suppose we have created vectors y and t, where $y = f(t)$, and we wish to create a linear plot with the command `plot(t, y)`. We can label the t-axis and y-axis and add a title and a grid with the following commands:

```
xlabel('t');
ylabel('y');
title('y vs. t');
grid;
```

Example 2.6

A table and a plot of a cubic function are created in the following example:

```
% Example_2_6.m
% This program creates a simple table and a simple plot.
% First a table of y = x^3+3.2x^2-3.4x-20 is created.
% Then y vs. x is plotted.
% To plot y vs. x both variables need to be vectors of
% same length.
clear; clc;
x = -5:0.5:5;
% Column headings
fprintf('        x        y        \n');
fprintf('------------------------------\n');
for n = 1:length(x)
        y(n) = x(n)^3+3.2*x(n)^2-3.4*x(n)-20.2;
        fprintf('%8.1f   %10.1f    \n',x(n),y(n));
end
% Create the plot of y vs. x.
plot(x,y), xlabel('x'), ylabel('y'), grid,
title('y vs. x');
% Plot identification is also established by adding text to
% the plot.
```

Program results

x	y
-5.0	-48.2
-4.5	-31.2
-4.0	-19.4
-3.5	-12.0
-3.0	-8.2
⋮	⋮
3.0	25.4
3.5	50.0
4.0	81.4
4.5	120.4
5.0	167.8

\>>
See Figure 2.15.

If your program involves creating more than one plot, you need to include the statement figure after each plot statement, except for the last; otherwise, only the last figure will appear. The following example program produces two separate plots available for viewing:

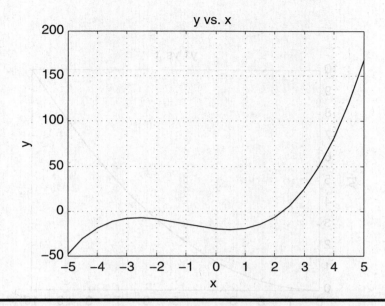

Figure 2.15 Plot of *y* vs. *x* for a cubic function.

Example 2.7

```
% Example_2_7.m
% This program creates two separate plots.
% First y1 = t^2/10 is plotted with 0 <= t <= 10,
% then y2 = t^3/100 is plotted over the same t range.
% To plot y1 and y2 vs. and t, y1, y2 and t need to be made
% vectors of the same length.
clear; clc;
t = 0:0.5:10;
for n = 1:length(t)
        y1(n) = t(n)^2/10;
        y2(n) = t(n)^3/100;
end
plot(t,y1), xlabel('t'), ylabel('y1'), grid,
   title('y1 vs. t');
figure;
plot(t,y1), xlabel('t'), ylabel('y2'), grid,
   title('y2 vs. t');
```

Program results

See Figures 2.16a and 2.16b.

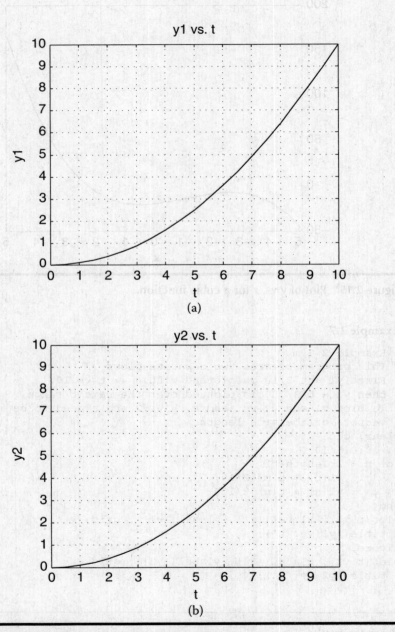

Figure 2.16   (a) Plot of *y*1 vs. *t*. (b) Plot of *y*2 vs. *t*.

Multiple plots

Suppose in matrix **A**, shown below, we wished to plot column 2 vs. column 1, column 3 vs. column 1, and column 4 vs. column 1:

$$
\mathbf{A} = \begin{bmatrix} t_1 & y_1 & z_1 & w_1 \\ t_2 & y_2 & z_2 & w_2 \\ \vdots & \vdots & \vdots & \vdots \\ t_n & y_n & z_n & w_n \end{bmatrix}
$$

We could let $\mathbf{T} = \mathbf{A}(:,1)$, $\mathbf{Y} = \mathbf{A}(:,2)$, $\mathbf{Z} = \mathbf{A}(:,3)$, and $\mathbf{W} = \mathbf{A}(:,4)$, giving

$$
\mathbf{T} = \begin{bmatrix} t_1 \\ t_2 \\ \vdots \\ t_n \end{bmatrix}, \quad \mathbf{Y} = \begin{bmatrix} y_1 \\ y_2 \\ \vdots \\ y_n \end{bmatrix}, \quad \mathbf{Z} = \begin{bmatrix} z_1 \\ z_2 \\ \vdots \\ z_n \end{bmatrix}, \quad \mathbf{W} = \begin{bmatrix} w_1 \\ w_2 \\ \vdots \\ w_n \end{bmatrix}
$$

Then to plot y vs. t, z vs. t, and w vs. t all on the same graph, we would write

```
plot(T,Y,T,Z,T,W);
```

Of course, we could have avoided the additional steps by writing

```
plot(A(:,1),A(:,2),A(:,1),A(:,3),A(:,1),A(:,4))
```

To identify which curve goes with which variable, you can add text to the plot with the command

```
text(x,y,'text statement');
```

where (x, y) are the coordinates on the graph where the text statement will start.

Multiple curves on the same graph can be distinguished by color coding the curves.

Available color types:

black	`'k'`
blue	`'b'`
green	`'g'`
red	`'r'`
cyan	`'c'`
yellow	`'y'`

Multiple curves on the same graph can also be distinguished by using different types of lines.

Available line types:

solid	(default)
dashed	`'--'`
dashed-dot	`'-.'`
dotted	`':'`

Alternatively, you can create a marker plot of discrete points (without a line) by using one of these marker styles:

point	'.'
plus	'+'
star	'*'
circle	'o'
x-mark	'x'
diamond	'd'

The `legend` command may also be used in place of the text command to identify the curves. The format for the `legend` command is:

```
legend('text1', 'text2')
```

The legend box may be moved by clicking on the box and dragging it to the desired position.

Review 2.5

1. What is the command that will produce a linear graph?
2. What are the commands that will label the *x*- and *y*-axis and provide a title to the graph?
3. When there is more than one function plotted on a graph, what are the ways to identify which curve corresponds with which function?

Example 2.8

The following example illustrates a multiple plot program:

```
% Example_2_8.m
% This program creates a simple table and a multiple plot.
% First a table of y1 = t^2/10 and y2 = t^3/100 is created.
% To plot y1, y2 vs. and t, they need to be made vectors.
% y1 and y2 vs. t are plotted on the same graph.
clear; clc;
t = 0:10;
for n = 1:length(t)
        y1(n) = t(n)^2/10;
        y2(n) = t(n)^3/100;
end
% By making t, y1 and y2 as vectors, their values can be
% printed out outside the for loop that created them.
% Column headings
fprintf('     t           y1           y2 \n');
fprintf('-----------------------------\n');
for n = 1:length(t)
        fprintf('%8.1f %10.2f %10.2f \n',t(n),y1(n),y2(n));
end
% Create the plot, y1 as a solid line, y2 as a dashed line.
plot(t,y1,t,y2,'--');
xlabel('t'), ylabel('y1,y2'), grid, title('y1 and y2 vs. t');
```

```
% Plot identification is also established by adding text to
% the plot.
text(6.5,2.5,'y2');
% In the above statement, 6.5 is the abscissa position and
% 2.5 is the ordinate position where the 'y2' label will be
% positioned.
text(4.2,2.4,'y1'),
% We can also use the legend command to identify the curves
% legend('y1','y2');
```

Program results

t	y1	y2
0.0	0.0000	0.0000
1.0	0.1000	0.0100
2.0	0.4000	0.0800
3.0	0.9000	0.2700
4.0	1.6000	0.6400
5.0	2.5000	1.2500
6.0	3.6000	2.1600
7.0	4.9000	3.4300
8.0	6.4000	5.1200
9.0	8.1000	7.2900
10.0	10.0000	10.0000

```
>>
```

The resulting plot is shown in Figure 2.17.

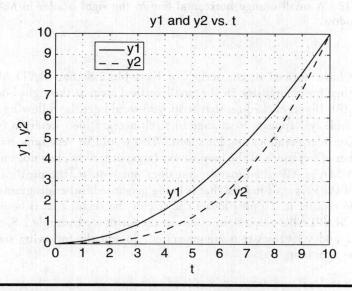

Figure 2.17   Plot of *y*1 and *y*2 vs. *t* on the same graph.

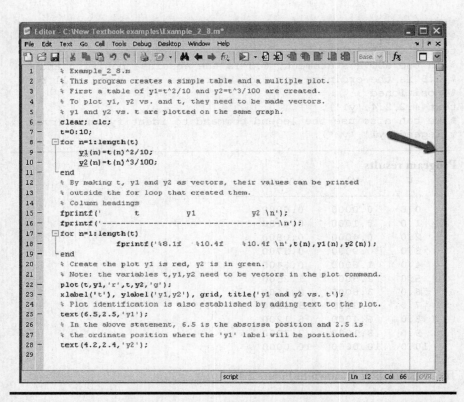

Figure 2.18 **A small orange horizontal line in the right border in MATLAB®'s script window.**

You may have noticed in the script for Example 2.8 that MATLAB had a small orange horizontal line in the small vertical strip at the right border (see Figure 2.18). If you clicked on that strip, you would get the following message "The variable 'y1' appears to change in size on every loop iteration (within a script). Consider preallocating for speed." This would be very important when the number of iterations of the loop is very large; otherwise, it is not important. Although MATLAB recommends, but does not require, the preallocation of the size of the vector or matrix that is being generated, other programs such as C/C++ do require it. To preallocate the size of the vector that is being generated, use MATLAB's `zeros` function. In the script for Example 2.8, there are eleven `y1` and `y2` values that are generated. So add the following statements before the `for` loop:

`y1=zeros(11,1)` and `y2=zeros(11,1)`

Example 2.9

The following example illustrates the plotting of trigonometric functions:

```
% Example_2_9.m
% This script calculates both sin(2x/3), sin(2x/3)^2 and
% cos(2x/3+pi) for -pi <= x <= pi. The x domain is
% subdivided into 50 subdivisions.
% The script plots the 3 functions and determines the
% absolute maximum values of the vectors fsin, fsinsq and
% fcos and prints those values to the screen.
clear; clc;
xmin = -pi; dx = 2*pi/50;
for i = 1:51
    x(i)  = xmin+(i-1)*dx;
    arg1  = 2*x(i)/3;
    arg2  = 2*x(i)/3+pi;
    fsin(i) = sin(arg1);
    fsinsq(i) = sin(arg1)^2;
    fcos(i) = cos(arg2);
end
fsin_max = max(abs(fsin)); fcos_max = max(abs(fcos));
fsinsq_max = max(fsinsq);
fprintf('\n fsin_max = %10.5f, fcos_max = %10.5f \n',...
        fsin_max,fcos_max);
fprintf('fsinsq_max = %10.5f \n',fsinsq_max);
plot(x,fsin,x,fcos,'--',x,fsinsq,'-.'), xlabel('x'),
ylabel('fsin,fcos,fsinsq'), grid
title('fsin, fcos, fsinsq vs. x'),
legend('fsin','fcos','fsinsq');
```

Program results

From the Command window,

```
fsin_max = 0.99978, fcos_max = 1.00000, fsinsq_max = 0.99956
>>
```

See Figure 2.19.

Exercises

E2.8 In the following exercises, carry out the prescribed modifications:
  a. Modify Example 2.5 program to include creating three separate plots of T vs. z, p vs. z, and ρ vs. z for $0 \le z \le 11{,}000$ m.
  b. Modify Exercise E2.1 to include a plot of s vs. t for $0 \le t \le 0.01$ s.
  c. Modify Exercise E2.2 to include a plot of y vs. t for $0 \le t \le 10$ s.

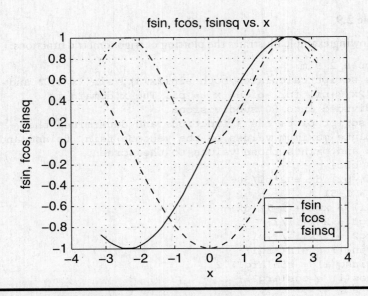

Figure 2.19   Plot of trigonometric functions (fix).

d. Modify Exercise E2.3 to include a plot of y vs. x at $0 \le t \le t_f$ in steps of 0.01 s; use small circles to define the basketball path.

e. Modify Exercise E2.4 to include a plot of x vs. t and V vs. t on two separate graphs.

f. Modify Exercise E2.5 to include a plot of V vs. t. Also, print to the screen values of W, B, and V_T.

g. Modify Exercise E2.6 to include a plot of v vs. t.

h. Modify Exercise E2.7 to include a plot of v vs. t.

E2.9 In this exercise, we consider two cars on a collision course. Each car's initial position and the angle its path makes with the x axis is specified below. The speed of car_1 is also specified.

Car_1: $x_1 = 500$ m, $y_1 = 100$ m, and $|\vec{V}_1| = 40$ m/s. Car_1 moves in a straight line that makes an angle of 60° with the x-axis.

Car_2: $x_2 = 2000$ m and $y_2 = 200$ m. Car_2 moves in a straight line and makes an angle of 45° with the $(-x)$-axis.

The collision coordinates are (x_c, y_c). See Figure 2.20.

We can determine the coordinates of the collision point by writing the equation for the tangent of each line, solving each equation for y_c, equating the two y_c expressions, then solving for x_c as shown below.

$$\frac{y_c - y_1}{x_c - x_1} = \tan(60°), \quad \frac{y_c - y_2}{x_2 - x_c} = \tan(45°) \qquad (2.26)$$

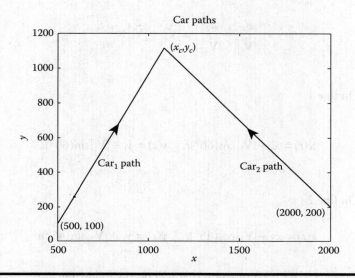

Figure 2.20   Plot of vehicle paths.

$$y_c = y_1 + (x_c - x_1) \times \tan(60°) \tag{2.27}$$

$$y_c = y_2 + (x_2 - x_c) \times \tan(45°) \tag{2.28}$$

$$y_1 + (x_c - x_1) \times \tan(60°) = y_2 + (x_2 - x_c) \times \tan(45°) \tag{2.29}$$

$$x_c = \frac{y_2 - y_1 + x_1 \tan(60°) + x_2 \tan(45°)}{\tan(60°) + \tan(45°)} \tag{2.30}$$

$$d_1 = \sqrt{(x_c - x_1)^2 + (y_c - y_1)^2} = \left|\vec{V}_1\right| t_c \tag{2.31}$$

$$d_2 = \sqrt{(x_2 - x_c)^2 + (y_c - y_2)^2} = \left|\vec{V}_2\right| t_c \tag{2.32}$$

where

$$t_c = \text{time of collision} = \frac{d_1}{\left|\vec{V}_1\right|} \tag{2.33}$$

$$\frac{d_1}{|\vec{V}_1|} = \frac{d_2}{|\vec{V}_2|} \rightarrow |\vec{V}_2| = |\vec{V}_1| \times \frac{d_2}{d_1} \tag{2.34}$$

On line 1

$$x(t) = x_1 + |\vec{V}_1|\cos(60°)t, \quad y(t) = y_1 + |\vec{V}_1|\sin(60°)t \tag{2.35}$$

On line 2

$$x(t) = x_2 - |\vec{V}_2|\cos(45°)t, \quad y(t) = y_2 + |\vec{V}_2|\sin(45°)t \tag{2.36}$$

Create a MATLAB program that will do the following:
a. Create a plot of the intersecting lines of lengths d_1 and d_2.
   Note: You only need to specify two points on the line to plot the line.
b. Determine $|\vec{V}_2|$ that will cause the collision to take place.
c. Take $0 \le t \le t_c$ in steps of $\frac{t_c}{5}$.
d. Plot the two lines and the two car positions at t_i, shown as small circles, all on the same graph. Your second plot should look like Figure 2.21.

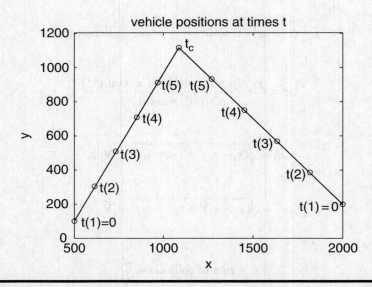

Figure 2.21 **Vehicle paths on a collision course.**

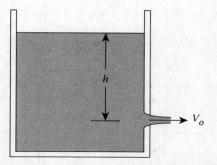

Figure 2.22   Fluid discharging through orifice.

E2.10 A formula describing the fluid level, $h(t)$, in a tank as the fluid discharges through a small orifice (see Figure 2.22) is

$$\sqrt{h} = \sqrt{h_o} - \frac{C_d A_o}{2 A_T} \sqrt{2g}\, t \qquad (2.37)$$

where

C_d is the discharge coefficient
h_o is the fluid level in the tank at time, $t = 0$
A_o is the circular area of the orifice
A_T is the circular cross-sectional area of the tank

Create a MATLAB program that will

a. Determine vectors h and t, for $0 \le t \le 80$ s in steps of 1 s.
b. Create a table containing 20 values of t and h (every fourth time step) and print the table to a file.
c. Create a plot of h vs. t and print it.

Use the following parameters: $h_o = 0.3$ m; the tank diameter, $D = 0.8$ m; and the orifice diameter, $d = 0.05$ m, $g = 9.81$ m/s², and $C_d = 0.7$.

E2.11 Figure 2.23 shows a differential amplifier using bipolar junction transistors (BJTs) that takes an input V_{in} and generates an output V_{out}. The relationship between the output and input voltages for this amplifier corresponds to the equation

$$V_{out} = \alpha I_{EE} R_L \tanh\left(-\frac{q}{2kT} V_{in}\right) \qquad (2.38)$$

where

q is the electron charge
k is Boltzmann's constant
T is absolute temperature
α is the collector–emitter current ratio (and is dependent on the process used to fabricate the transistors)
V_{in}, V_{out}, R_L, and I_{EE} are defined in Figure 2.23

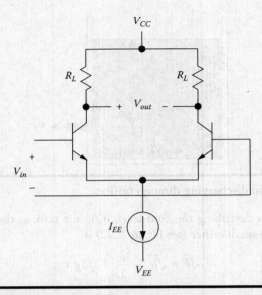

Figure 2.23 **Differential amplifier using BJTs.**

The gain, A, of the amplifier is defined as the slope of the curve of V_{out} vs. V_{in}, which we can approximate as

$$A \approx \frac{\Delta V_{out}}{\Delta V_{in}} \qquad (2.39)$$

However, we can also obtain the slope by taking the derivative of Equation 2.38, that is,

$$\frac{dV_{out}}{dV_{in}} = \frac{d}{dV_{in}}\left(\alpha I_{EE} R_L \tanh\left(-\frac{qV_{in}}{2kT} \right) \right) \qquad (2.40)$$

Note

$$\frac{d \tanh x}{dx} = \mathrm{sech}^2 x = \frac{1}{\cosh^2 x} \qquad (2.41)$$

Create a MATLAB program that
a. Determines V_{out} and A for $-200 < V_{in} < 200$ mV in steps of 2 mV. Assume $\alpha = 0.98$, $I_{EE} = 2$ mA, $R_L = 10$ kΩ, and $T = 300$ K.
b. Plots V_{out} vs. V_{in} and A vs. V_{in}, both by Equations 2.39 and 2.40.

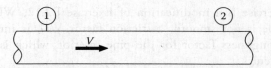

Figure 2.24 Fluid flowing in a pipe.

E2.12 When a fluid flows through a pipe, there is a pressure drop that is proportional to the pipe's length (see Figure 2.24). For a pipe having a circular cross section, the pressure drop, $p_1 - p_2$, is given by

$$p_1 - p_2 = \frac{\rho V^2}{2} \frac{L}{D} f \tag{2.42}$$

where
  ρ is the fluid density
  V is the average fluid velocity in the pipe
  D is the pipe diameter
  L is the pipe length between points 1 and 2
  f is the friction factor

The friction factor has been determined by experiment. For smooth pipes, a formula that approximates the experimental data is [2]

$$f = (1.82 \log_{10} \text{Re} - 1.64)^{-2} \tag{2.43}$$

where
  $$\text{Re} = \frac{\rho V D}{\mu} \text{ (Reynolds number)} \tag{2.44}$$

  μ is the absolute fluid viscosity

Develop a MATLAB program that will calculate f vs. Re. Take

```
Re = [5e3 7.5e3 1e4 2.5e4 5e4 7.5e4 1e5 2.5e5 5e5 7.5e5 ...

      1e6 2.5e6 5e6 7.5e6 1e7 2.5e7 5e7 7.5e7 1e8]
```

Plot log(Re) on the x-axis and f on the y-axis (semilog plot).

E2.13 This exercise is a modification of Exercise E2.12. When the interior of a pipe is not smooth, a friction factor, *e/D*, is introduced, where *e* is a roughness factor for the pipe interior, which is determined by experiment.

The Swamee–Jain formula has been used to approximate the experimental data. The Swamee–Jain formula [3] is

$$f = \frac{1.325}{\left[\log\left(\frac{1}{3.7} \times \frac{e}{D} + \frac{5.74}{Re^{0.9}}\right)\right]^2} \tag{2.45}$$

Develop a MATLAB program that will calculate *f* as a function of Re and *e/D*. Take

```
Re = [5e3 7.5e3 1e4 2.5e4 5e4 7.5e4 1e5 2.5e5 5e5 ...

      7.5e5 1e6 2.5e6 5e6 7.5e6 1e7 2.5e7 5e7 7.5e7 1e8]
```

$$\frac{e}{D} = [0.0 \ 0.0001 \ 0.001 \ 0.005];$$

Create a semilog plot of *f* vs. log(Re) for the different $\frac{e}{D}$ values all on the same graph. Make log(Re) on the *x*-axis and *f* on the *y*-axis.

2.8 Conditional Operators and Alternate Paths

■ The `if` statement
Syntax:

```
if logical expression
     statement;
        ⋮
     statement;
else
     statement;
        ⋮
     statement;
end
```

If the logical expression is true, then only the upper set of statements are executed. If the logical expression is false, then only the bottom set of statements are executed.

■ Logical expressions are of the form

```
a == b;    a <= b;
a < b;     a >= b;
a > b;     a ~= b;  (a not equal to b)
```

■ Compound logical expressions

```
a > b  &&  a ~= c    (a > b and a ≠ c)
a > b  ||  a < c     (a > b or a < c)
```

The following example illustrates the use of the if statement:

Example 2.10

```
% Example_2_10.m
% This program uses an input and an if statement to
% determine if the output is to go to the screen or to a
% file. The variables y1 and y2 are made vectors so that
% these variables can be printed outside the for loop that
% created them. As vectors, they can also be plotted.
clear; clc;
t = 0:0.5:5
for j = 1:length(t)
        y1(j) = t(j)^2/10;
        y2(j) = t(j)^3/100;
end
fprintf('Do you wish to print the output to \n');
fprintf('the screen or to a file? \n');
response = input('enter S for screen or F for file \n','s');
% Note, since we entered 's' in the input statement, do not
% enclose your answer in single quotation marks.
if response =='S'
    % Table headings:
    fprintf('   t           y1             y2              \n');
    fprintf('---------------------------------------------\n');
    for j = 1:length(t)
            fprintf(' %5.1f     %10.3f    %10.3f \n',...
            t(j),y1(j),y2(j));
    end
end
if response =='F'
    fo = fopen('output.txt','w');
    % Table headings:
    fprintf(fo,'   t             y1            y2           \n');
    fprintf(fo,'-------------------------------------------\n');
    for j = 1:length(t)
            fprintf(fo,' %5.1f    %10.3f    %10.3f \n',...
            t(j),y1(j),y2(j));
    end
end
```

Program results (either from the screen or from the file "output.txt")

t	y1	y2
0.0	0.000	0.000
0.5	0.025	0.001
1.0	0.100	0.010
1.5	0.225	0.034
2.0	0.400	0.080
2.5	0.625	0.156
3.0	0.900	0.270
3.5	1.225	0.429
4.0	1.600	0.640
4.5	2.025	0.911
5.0	2.500	1.250

■ The `if-elseif` ladder
  Syntax:

```
if logical expression 1
    statement(s);
elseif logical expression 2
    statements(s);
elseif logical expression 3
    statement(s);
else
    statement(s);
end
```

The `if-elseif` ladder works from top down. If the top logical expression is true, the statements related to that logical expression are executed, and the program will leave the ladder. If the top logical expression is not true, the program moves to the next logical expression. If that logical expression is true, the program will execute the group of statements associated with that logical expression and leave the ladder. If that logical expression is not true, the program moves to the next logical expression and continues the process. If none of the logical expressions are true, the program will execute the statements associated with the `else` statement. The `else` statement is not required. In that case, if none of the logical expressions are true, no statements within the ladder will be executed.

■ The `break` command may be used with an `if` statement to end a loop, for example,

```
for m = 1:20
        statement(s);
        if m > 10
            break;
        end
end
```

In the previous example, when m becomes greater than 10, the program leaves the for loop and moves on to the next statement outside the for loop.

Suppose in the atmospheric model described in Example 2.5, we did not have the governing equations that produced the atmospheric property table but just had the table itself and we wished to determine atmospheric properties at altitudes not in the table. This situation occurs in thermodynamics, where thermodynamic properties of various substances are tabulated. The simplest way to determine the atmospheric properties would be to interpolate between table values. If we assume that the atmospheric properties vary linearly with altitude between table values, then we can use linear interpolation. In this case, we should use table values that are closest to the altitude of interest. One way to accomplish this (although not the most efficient way) would be to use the if-elseif ladder. The following example illustrates this concept.

The general linear interpolation formula, based on similar triangles, in terms of y and x is as follows:

$$y = y_1 + \frac{(y_2 - y_1) \times (x - x_1)}{x_2 - x_1} \qquad (2.46)$$

where x_1 and x_2 are the values of x that enclose x and y_1 and y_2 are the values of y at x_1 and x_2, respectively. First, we need to create the table listed below and save it in a file named *atm_properties.txt*. This file will be loaded into the program of Example 2.11. The first column in the table is the altitude, the second column is the temperature, the third column is the pressure, and the fourth column is the density.

0	288.15	1.0133e+005	1.2252
1000	281.65	8.9869e+004	1.1118
2000	275.15	7.9485e+004	1.0065
3000	268.65	7.0095e+004	0.9091
4000	262.15	6.1624e+004	0.8191
5000	255.65	5.4002e+004	0.7360

Example 2.11

```
% Example_2_11.m
% This program loads data from a file named atm_propeties.txt
% The program asks the user to enter an elevation at which
% atmospheric properties are to be determined by linear
% interpolation.
% The altitude range is from 0 to 5000 m.
% Then the atmospheric properties are printed to the screen.
% The program uses the if-elseif ladder to select the
% closest interval to the entered altitude. The properties
% in this interval will be used in the interpolation
% formula.
```

```
% Temperature is in degrees Kelvin (K), pressure is in
% Pascal (Pa) and density is in kg/m^3.
clear; clc;
load atm_properties.txt;
% establishing variable names to loaded data.
zt = atm_properties(:,1);
Tt = atm_properties(:,2);
pt = atm_properties(:,3);
rhot = atm_properties(:,4);
fprintf('Enter the altitude at which atmospheric \n');
fprintf('properties are to be determined. \n');
z=input('Altitude range is from 0 to 5000 m \n');
if z >= zt(1)&& z < zt(2)
        z1 = zt(1); z2 = zt(2); T1 = Tt(1); T2 = Tt(2);
        p1 = pt(1); p2 = pt(2);
        rho1 = rhot(1); rho2 = rhot(2);
elseif z >= zt(2)&& z < zt(3)
        z1 = zt(2); z2 = zt(3); T1 = Tt(2); T2 = Tt(3);
        p1 = pt(2); p2 = pt(3);
        rho1 = rhot(2); rho2 = rhot(3);
elseif z >= zt(3)&& z < zt(4)
        z1 = zt(3); z2 = zt(4); T1 = Tt(3); T2 = Tt(4);
        p1 = pt(3); p2 = pt(4);
        rho1 = rhot(3); rho2 = rhot(4);
elseif z >= zt(4)&& z < zt(5)
        z1 = zt(4); z2 = zt(5); T1 = Tt(4); T2 = Tt(5);
        p1 = pt(4); p2 = pt(5);
        rho1 = rhot(4); rho2 = rhot(5);
elseif z >= zt(5)&& z < zt(6)
        z1 = zt(5); z2 = zt(6); T1 = Tt(5); T2 = Tt(6);
        p1 = pt(5); p2 = pt(6);
        rho1 = rhot(5); rho2 = rhot(6);
end
T = T1+(T2-T1)*(z-z1)/(z2-z1);
p = p1+(p2-p1)*(z-z1)/(z2-z1);
rho = rho1+(rho2-rho1)*(z-z1)/(z2-z1);
fprintf('T=%6.2f(K), p=%10.4e(Pa) rho=%6.4f(kg/m^3) \n', ...
        T,p,rho);
```

Program results

```
Enter the altitude at which atmospheric properties are to be
determined. Altitude range is from 0 to 5000 m
4380
T = 259.68(K), p = 5.8728e+04(Pa) rho = 0.7875(kg/m^3)
>>
```

It is always prudent to examine the results of a program to see if they make sense. In this case, do the obtained properties lie within the proper interval?

An alternative to loading the data in the file *atm_properties.txt* into the previous script is to enter the data directly into the program as vectors. To accomplish this, replace the following lines in Example 2.11:

```
load atm_properties.txt
% establishing variable names to loaded data.
zt = atm_properties(:,1);
Tt = atm_properties(:,2);
pt = atm_properties(:,3);
rhot = atm_properties(:,4);
```

with

```
zt = [0 1000 2000 3000 4000 5000];
Tt = [288.15 281.65 275.15 268.65 262.15 255.65];
pt = [10.133 8.9869 7.9485 7.0095 6.1624...
        5.4002]*1.0e+004;
rhot = [1.2252 1.1118 1.0065 0.9091 0.8191 0.7360];
```

One additional thought is as follows: it is possible to develop the program containing the if-elseif ladder using noncompound logical expressions. This is demonstrated in Example 2.11b:

```
% Example_2_11b.m
% This program enters the data shown in atm_properties.txt
% directly into the program as vectors.
% The program asks the user to enter an elevation at which
% atmospheric properties are to be determined by linear
% interpolation.
% The altitude range is from 0 to 5000m.
% Then the requested atmospheric properties are printed to
% the screen.
% The program uses the if-elseif ladder to select the
% closest interval to the entered altitude. The properties
% in this interval will be used in the interpolation
% formula.
% Temperature is in degrees Kelvin (K), pressure is in
% Pascal (Pa) and density is in kg/m^3.
clear; clc;
zt = [0 1000 2000 3000 4000 5000];
Tt = [288.15 281.65 275.15 268.65 262.15 255.65];
pt = [10.133 8.9869 7.9485 7.0095 6.1624 5.4002]*1.0e+004;
rhot = [1.2252 1.1118 1.0065 0.9091 0.8191 0.7360];
fprintf('Enter the altitude at which atmospheric \n');
fprintf('properties are to be determined. \n');
z=input('Altitude range is from 0 to 5000 m \n');
if z < zt(2)
    % if true, z lies between 0 & 1000 m.
    z1 = zt(1); z2 = zt(2); T1 = Tt(1); T2 = Tt(2); p1 = pt(1);
    p2 = pt(2); rho1 = rhot(1); rho2 = rhot(2);
elseif z < zt(3)
    % if this condition is tested and found true, then the first
    % condition was false so z lies between 1000 & 2000 m.
```

```
    z1 = zt(2); z2 = zt(3); T1 = Tt(2); T2 = Tt(3); p1 = pt(2);
    p2 = pt(3); rho1 = rhot(2); rho2 = rhot(3);
elseif z < zt(4)
    % if this condition is tested and found true, then the first
    % & second conditions were false so z lies between 2000 &
    % 3000 m.
    z1 = zt(3); z2 = zt(4); T1 = Tt(3); T2 = Tt(4); p1 = pt(3);
    p2 = pt(4); rho1 = rhot(3); rho2 = rhot(4);
elseif z < zt(5)
    % if this condition is tested and found true, then the
    % first, second & third conditions were false so z lies
    % between 3000 & 4000 m.
    z1 = zt(4); z2 = zt(5); T1 = Tt(4); T2 = Tt(5); p1 = pt(4);
    p2 = pt(5); rho1 = rhot(4); rho2 = rhot(5);
elseif z < zt(6)
    % if this condition is tested and found true, then the
    % first, second, third & fourth conditions were false so
    % z lies between 4000 & 5000 m.
    z1 = zt(5); z2 = zt(6); T1 = Tt(5); T2 = Tt(6); p1 = pt(5);
    p2 = pt(6); rho1 = rhot(5); rho2 = rhot(6);
end
T = T1+(T2-T1)*(z-z1)/(z2-z1);
p = p1+(p2-p1)*(z-z1)/(z2-z1);
rho = rho1+(rho2-rho1)*(z-z1)/(z2-z1);
fprintf('T=%6.2f(K), p=%10.4e(Pa) rho=%6.4f(kg/m^3)\n',...
        T,p,rho);
```

Example 2.12

An *uninterruptible power supply* (UPS) is used to prevent a computer from crashing during a temporary power outage. A UPS contains a battery that is normally charged by house current. During a power outage, the battery is used to generate substitute line power.

Although electricity delivered by a power company is usually sinusoidal (as shown in Figure 2.25a), some low-cost UPS models generate an approximated sinusoid by taking the UPS voltage to be zero for the first eighth of the period, then a high for the next one-fourth, then zero for the next fourth, and then a low for the next fourth as shown in Figure 2.25b. (Note: The 50Hz waveform shown in Figure 2.25a is used in some regions of Japan.) We wish to construct a MATLAB program that demonstrates that both waveforms have nearly the same RMS voltage.

The RMS is obtained by the following procedure:

a. Squaring the voltage
b. Computing the mean value of the squared waveform by averaging it over one cycle
c. Taking the square root of the average

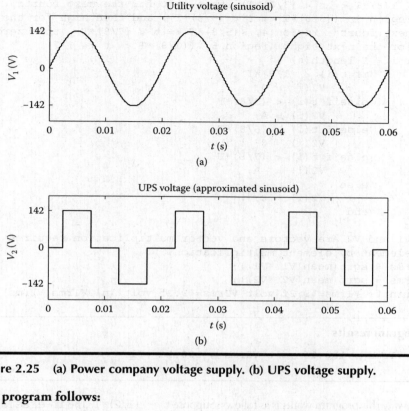

Figure 2.25 (a) Power company voltage supply. (b) UPS voltage supply.

The program follows:

```
% Example_2_12.m
% This program determines the root-mean-square of two different
% waveforms as shown in Figure 2.25a and b. One waveform is
% a sinusoidal Signal. The period and amplitude are read off
% Figure 25a:
% T is in second(s) and A is in volt(V).
clear; clc;
A = 142; T = 0.02;
% choose an arbitrary step size which is much less than
% the signal period:
step = 0.0001;
% Define timepoints over a single period of the waveform:
t = 0:step:T;
% Calculate V1
for i = 1:length(t)
        V1(i) = A * sin(2*pi * (1/T) * t(i));
end
% Define V2 piece-by-piece. V2 will be zero for the first
% eighth of the period, region 1, (0 <= t < (1/8)*T), then
% high for the next one fourth of a period, region 2,
```

```
% ((1/8)*T <= t < (3/8)*T), then zero for the next fourth,
% region 3, ((3/8)*T <= t < (5/8)*T), and then high for the
% next fourth, region 4, ((5/8)*T <= t < (7/8)*T), then zero
% for the next eight, region 5, ((7/8)*T <= t <= 1*T).
for i = 1:length(t)
         if t(i) < (1/8)*T
               V2(i) = 0;
         elseif t(i) < (3/8)*T
               V2(i) = A;
         elseif t(i) < (5/8)*T
               V2(i) = 0;
         elseif t(i) < (7/8)*T
               V2(i) = -A;
         else
               V2(i) = 0;
         end
end
% V1 and V2 are vectors and vector multiplication requires
% element by element multiplication.
V1rms = sqrt(mean(V1.*V1));
V2rms = sqrt(mean(V2.*V2));
fprintf('V1rms=%9.2f volt V2rms=%9.2f volt \n',V1rms,V2rms);
```

Program results

```
   V1rms = 100.16 volt V2rms = 100.16 volt
>>
```

The way this program works is as follows. Suppose t = (1/2)*T. The if-elseif ladder works from top of the ladder down. For the first logical expression, is t < (1/8)*T? No, so the first region is eliminated. Next logical expression, is t < (3/8)*T? No, so the second region is eliminated. Next logical expression, is t < (5/8)*T? Yes. Since we have already eliminated regions 1 and 2, t is in region 3.

As mentioned earlier, the programs in Examples 2.11, 2.11b, and 2.12 are not the most efficient way to solve the problem. We can use a for loop to determine the closest interval to the entered altitude in Examples 2.11 and 2.11b, thus reducing the number of lines in the program. This is demonstrated in the following example. This becomes important when the number of lines in the data is large.

Example 2.13

```
% Example_2_13.m
% This program enters the data shown in atm_properties.txt
% directly into the program as vectors.
% The program then asks the user to enter an elevation at
% which the atmospheric properties are to be determined by
% linear interpolation.
```

```
% The atmospheric properties are then printed to the screen.
% The program uses a for loop and a compound if statement to
% determine the closest interval to the entered altitude.
% The properties in this interval will be used in the
% interpolation formula.
% Temperature is in degrees Kelvin (K), pressure is in
% Pascal (Pa) and density is in (kg/m^3).
clear; clc;
zt = [0 1000 2000 3000 4000 5000];
Tt = [288.15 281.65 275.15 268.65 262.15 255.65];
pt = [10.133 8.9869 7.9485 7.0095 6.1624 5.4002]*1.0e+004;
rhot = [1.2252 1.1118 1.0065 0.9091 0.8191 0.7360];
fprintf('Enter the altitude at which atmospheric \n');
fprintf('properties are to be determined. \n');
z=input('Altitude range is from 0 to 5000 m \n');
for i = 1:length(zt)-1
    if z >= zt(i)&& z < zt(i+1)
        z1 = zt(i); z2 = zt(i+1); T1 = Tt(i); T2 = Tt(i+1);
        p1 = pt(i); p2 = pt(i+1); rho1 = rhot(i);
        rho2 = rhot(i+1);
        break;
    end
end
T = T1+(T2-T1)*(z-z1)/(z2-z1);
p = p1+(p2-p1)*(z-z1)/(z2-z1);
rho = rho1+(rho2-rho1)*(z-z1)/(z2-z1);
fprintf('T=%6.2f(K), p=%10.4e(Pa) rho=%6.4f(kg/m^3) \n',...
        T,p,rho);
```

Program results

```
Enter the altitude at which atmospheric properties are to be
determined. Altitude range is from 0 to 5000 m
1350
T = 279.38(K), p = 8.6235e+04(Pa) rho = 1.0749(kg/m^3)
>>
```

An alternative to the `load` command is the `dlmread` command. This command will read an ASCII delimited file. All data in the file must be numeric. In Example 2.11, we could replace the lines starting with `load atm_properties.txt;` and ending with `rhot = atm_properties(:,4);` with

```
Y = dlmread('atm_properties.txt');
zt = Y(:,1);
Tt = Y(:,2);
pt = Y(:,3);
rhot = Y(:,4);
```

Exercises

E2.14 Carbon dioxide properties of temperature, T, specific volume, v, and pressure, p, based on the Redlich–Kwong equation are tabulated in Table 2.1. We wish to determine by interpolation v and p at a temperature between table values. The general interpolation formula in terms of y and x was given in Equation 2.41, which is

$$y = y_1 + \frac{(y_2 - y_1) \times (x - x_1)}{x_2 - x_1}$$

where

x_1 and x_2 are the values of x that enclose x

y_1 and y_2 are the values of y at x_1 and x_2, respectively

Develop a MATLAB program that does the following:

1. Using Table 2.1, create vectors T, v, and p.
2. Ask the user to enter a temperature, T, from the keyboard that is not in the table.
3. Using the `if-elseif` ladder, determine the interval enclosing the entered T value.
4. Interpolate for v and p.
5. Print to the screen the interpolated values of v and p.

Table 2.1 Carbon Dioxide Properties

T (K)	v (m³/kmol)	p (bars)
350	0.28	7.650
400	0.32	8.574
450	0.36	9.159
500	0.40	9.547
550	0.44	9.813
600	0.48	10.001
650	0.52	10.136
700	0.56	10.236
750	0.60	10.310

E2.15 Repeat Exercise E2.14, but this time use a `for` loop and a single `if` statement to determine the interval that encloses the temperature entered from the key board.

■ The `switch` group

In some cases, the `switch` group may be used as an alternative to the `if-elseif` ladder.

Syntax

```
switch(var)
      case var1
             statement(s);
      case var2
             statement(s);
      case var3
             statement(s);
      otherwise
             statement(s);
end
```

where `var` takes on the possible values `var1`, `var2`, `var3`, etc.

If `var` equals `var1`, those statements associated with `var1` are executed, and the program leaves the switch group. If `var` does not equal `var1`, the program tests if `var` equals `var2`, and if yes, the program executes those statements associated with `var2` and leaves the switch group. If `var` does not equal any of `var1`, `var2`, etc., the program executes the statements associated with the `otherwise` statement. If `var1`, `var2`, etc., are strings, they need to be enclosed by single quotation marks. It should be noted that `var` cannot be a logical expression, such as `var1 >= 80`.

The following example illustrates the use of the switch group in a MATLAB program:

Example 2.14

```
% Example_2_14.m
% This program is a test of the switch statement.
clear; clc;
var = 'a';
x = 5;
switch(var)
      case 'b'
             z = x^2;
      case 'a'
             z = x^3;
      otherwise
             z = 0;
end
fprintf('z=%6.1f \n',z);
```

Program results

```
z= 125.0
>>
```

Review 2.6

1. What statement is frequently used to establish two alternate paths?
2. What series of statements is used to establish several alternate paths?
3. List the various types of logic statements that can be used with the `if-else` and `if-elseif-else` ladder.
4. Is the `else` statement required with the `if-else` and the `if-elseif-else` ladder?
5. What statement group is an alternative to the `if-elseif-else` ladder?

Exercise

E2.16 Many gases under certain conditions behave as an ideal gas. The governing equation for an ideal gas is

$$pv = RT \tag{2.48}$$

where

  R is the gas constant (R is different for different gases)
  p is the gas pressure
  v is the specific volume
  T is the absolute temperature

Create a MATLAB program that will plot v vs. T for the following four gases: air, hydrogen, oxygen, and carbon dioxide. Although air is a mixture of different gases, it is frequently treated as a pure gas. Use the switch group to determine the proper gas constant and to create the requested plots. Take $p = 2$ atm and $300 \le T \le 600$ K in steps of 50 K.

Note: 1 atm = 1.01325×10^5 N/m²

The gas constant, R, for each of the aforementioned listed gases is given in Table 2.2.

2.9 Working with Built-In Functions with Vector Arguments

MATLAB allows the built-in functions, such as $\sin()$, $\cos()$, and $\exp()$, as well as functions in general to have vectors as arguments. The result will also be a vector. This is demonstrated in the next example.

Table 2.2   Gas Constant for Several Gases

Gas	R (N-m/kg-K)
Air	287.0
Hydrogen	4124.0
Oxygen	259.8
Carbon dioxide	188.9

Example 2.15

```
% Example_2_15.m
% This program demonstrates that if the argument in a built
% in function, such as MATLAB's sine function, is a vector,
% the result will also be a vector.
clear; clc;
% Define vector x;
x = 0:30:360;
% Let y1 be the sine of a vector x where x is in
% degrees.
% Running sind with vector x as an argument will return a
% vector:
y1 = sind(x);
% Let y2(n) be the sine of the nth element of x. We will use
% a for loop to calculate each value y2(n) and then compare
% y1 and y2.
for n = 1:length(x)
    y2(n) = sind(x(n));
end
% Table headings
fprintf('   x         y1        y2      \n');
fprintf('----------------------------------------- \n');
for n = 1:length(x)
    fprintf('%5.1f %10.5f %10.5f \n', x(n),y1(n),y2(n))
end
```

Program results

```
    x         y1          y2
-----------------------------------
    0.0     0.00000     0.00000
   30.0     0.50000     0.50000
   60.0     0.86603     0.86603
   90.0     1.00000     1.00000
  120.0     0.86603     0.86603
```

```
150.0     0.50000     0.50000
180.0     0.00000     0.00000
210.0    -0.50000    -0.50000
240.0    -0.86603    -0.86603
270.0    -1.00000    -1.00000
300.0    -0.86603    -0.86603
330.0    -0.50000    -0.50000
360.0     0.00000     0.00000
>>
```

In the generated output, does $y1 = y2$?

We see that in some scripts, we could replace the use of a for loop by using a vector argument in many built-in functions, which produces a vector result, thus reducing the number of lines in the script. This concept was demonstrated in the above example. However, if the script requires a mathematical operation of two functions (such as a product of two vector functions), then the operation will require an element-by-element operation. Element-by-element operations are discussed later in this book.

2.10 More on MATLAB® Graphics

■ **The subplot command**

Suppose you want to plot each of several curves as a separate plot but all on the same page. The subplot command provides the means to do so. The command subplot(m,n,p) breaks the page into an m by n matrix of small plots, and p selects the matrix position of the plot. The following example demonstrates the use of the subplot command.

■ **The sprintf command**

Suppose you created several plots and you wish to print a variable number in either the xlabel, ylabel or title commands. The sprintf command writes formatted data to a string which can then be used to print a variable number in these command. This is done in Example 2.16.

Example 2.16

```
% This program is an example of the use of the subplot and
% sprintf commands, and the if-elseif ladder. Values of y1,
% y2, y3 and y4 are constructed as vectors. Separate plots
% of y1 vs. t, y2 vs. t, y3 vs. t and y4 vs. t are plotted on
% the same page.
% The sprintf command is used in the title to identify the n in
% n*pi*t(
clc; clear;
t = 0:0.01:pi;
```

```
for i = 1:length(t)
    y1(i) = sin(pi*t(i));
    y2(i) = sin(2*pi*t(i));
    y3(i) = sin(3*pi*t(i));
    y4(i) = sin(4*pi*t(i));
end
for n = 1:4
    subplot(2,2,n)
    if n == 1
        plot(t,y1),xlabel('t'), ylabel('y'), grid,
        title(sprintf('sin %d*pi*t vs. t',n));
    elseif n == 2
        plot(t,y2),xlabel('t'), ylabel('y'), grid,
        title(sprintf('sin %d*pi*t vs. t',n));
    elseif n == 3
        plot(t,y3),xlabel('t'), ylabel('y'), grid,
        title(sprintf('sin %d*pi*t vs. t',n));
    elseif n == 4
        plot(t,y4),xlabel('t'), ylabel('y'), grid,
        title(sprintf('sin %d*pi*t vs. t',n));
    end
end
```

Program results

See Figure 2.26

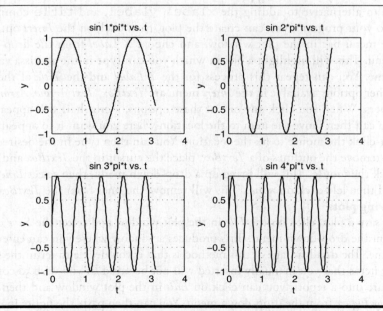

Figure 2.26   Subplots of y1, y2, y3, and y4 vs. t.

E2.17 Repeat Exercise 2.16, but this time use the if-elseif ladder to determine the proper gas constant. Create the four requested plots, all on the same page.

■ **Greek letters and mathematical symbols**
Greek letters and mathematical symbols can be used in `xlabel`, `ylabel`, `title`, and `text` by spelling out the Greek letter and preceding it with a "\" (backslash character). Thus, to display ω, use \omega, and to display β, use \ beta.

Example

```
ylabel(' \omega'), title(' \omega vs. \beta'),
text(10,5,' \omega');
```

For a complete list of Greek symbols and additional special characters, see Appendix A. You may also occasionally need to print a "'" character in your label or title. In this case, use a double-apostrophe to "escape" the single-quote character in your string. Thus, to generate the plot title "Signal 'A' vs. Signal 'B'," you would type

```
title('Signal ''A'' vs. Signal ''B''').
```

■ **Interactively annotating plots**
As an alternative to adding the `xlabel`, `ylabel`, and `title` commands into your program, you can create the plot, then click on the *Insert* option in the menu bar in the plot window, and choose *X Label* from the drop-down menu. This will highlight a box in which you can type in the abscissa variable name. You can repeat this process for the *Y Label* and the *Title* of the plot. Other options available in the *Insert* menu are *TextBox, Text Arrow, Arrow*, and others. When you click any one of these options, a crosshair will appear, and you can then move the item to the location where you want it to appear, then left-click the mouse to fix the location. You can then type in the desired text. To remove the outlines of a *TextBox*, place the cursor in the *TextBox* and right-click the mouse. This will bring up a drop-down menu, then select *Line Style*, and then left-click on *none*. This will remove the lines from the *TextBox*.

■ **Saving plots**
To save a plot, click on the *File* in the plot window and select the *Save* option from the drop-down menu. This produces a window where you can enter a file name. The disadvantage of this method is that if you decide to rerun the script, the items that you manually inserted will not be saved. If you wish to copy the figure into a report, you can click on *Edit* in the plot window and then select *Copy Figure* from the drop-down menu. You can then paste the figure into your report. If you need a monochrome version of your plot (for best reproduction

on a photocopier), you can make all of your curves black by choosing *File* from the task bar menu, then selecting *Export Setup* from the drop-down menu. This will open a window in which you need to click on *Rendering* and change the *Colorspace* to *black* and *white*.

There are many more options available in the plot window; however, we leave it up to the student to explore further.

Review 2.7

1. What is the name of the function that will allow you to plot several graphs on one page?
2. How does one enter Greek symbols into a plot?
3. What are the commands that will allow you to enter text onto a plot once the plot has been created?

2.11 Debugging a Program

It is common when writing a program to make typographical errors such as omitting a parenthesis, forgetting a comma in a 2-D array, etc. This type of bug is called a *syntax error*. When this occurs, MATLAB will provide an error message pointing out the line in which the error has occurred. However, there are cases where there are no syntax errors, but the program still fails to run or gives an obvious incorrect answer. When this occurs, you can utilize the debug feature in MATLAB. The debug feature allows you to set breakpoints in your program. The program will run up to, but not including, the line containing the breakpoint. To set a breakpoint, left-click the mouse in the narrow column next to the line number that you wish to be a breakpoint or click on the *Breakpoint* option in the Toolstrip and select *Set/Clear* in the drop down menu. A breakpoint will be set at the current line. A small red circle will appear next to the breakpoint line as shown in Figure 2.27 at line 22. When you click on the *Run* option in the Toolstrip, the script will run up to, but not including the line containing the breakpoint, as shown in Figure 2.28 (no plot appeared). The K>> prompt indicates that you are in the debug mode. Once you run the program up to the breakpoint, additional items will appear in the Toolstrip, such as, *Continue, Step, Step In, Step Out, Run to Cursor* and *Quit Debugging*. You can then click on the *Continue* option in the Toolstrip and the program will run to the next breakpoint, if one exists, or to the end of the program if no additional breakpoints exist. You can also execute one line at a time by clicking on the *Step* item in the Toolstrip. If your script contains a self written function, you can execute one line at a time in the function by clicking on the *Step In* item in the Toolstrip. To return to the calling program, click on the *Step Out*

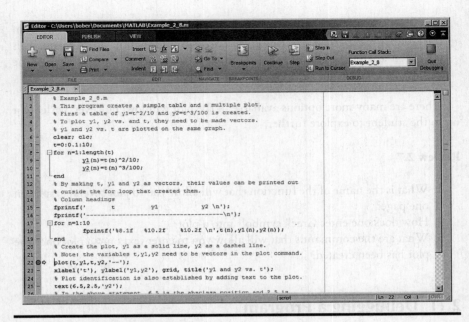

Figure 2.27 Debugging breakpoint.

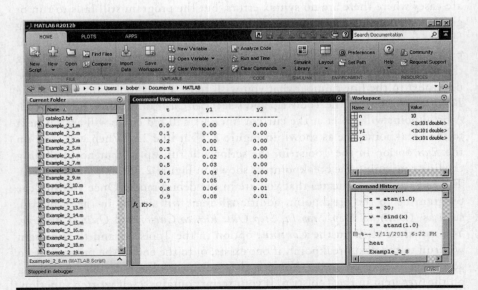

Figure 2.28 Program executed up to breakpoint.

option in the Toolstrip. If you wish to remove all breakpoints, click on the *Breakpoints* item in the Toolstrip and a drop down menu will appear giving you the option to *Clear all* breakpoints. If you wish to exit the debug mode, click on the *Quit Debugging* option in the Toolstrip.

Projects

P2.1 Though atmospheric conditions vary from day to day, it is convenient for design purposes to have a model for atmospheric properties with altitude. The U.S. Standard Atmosphere, modified in 1976, is such a model. The model consists of two types of regions: one in which the temperature varies linearly with altitude and the other is a region where the temperature is a constant (see Figure P2.1).

The temperature and approximate pressure relations are as follows:

a. For a region where the temperature varies linearly,

$$p = p_i \left(1 - \frac{\lambda_i (z - z_i)}{T_i} \right)^{g_i / \lambda_i R} \tag{P2.1a}$$

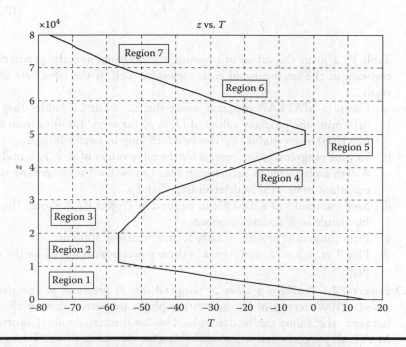

Figure P2.1 Temperature profile of the U.S. Standard Atmosphere.

$$T = T_i - \lambda_i(z - z_i) \tag{P2.1b}$$

$$\rho = \frac{p}{RT} \tag{P2.1c}$$

where

z is the altitude

z_i is the altitude at the beginning of the region of interest

(p_i, T_i) is the pressure and temperature at the beginning of the region of interest

λ_i is the lapse rate in the region

R is the air gas constant = 287.0 N m/kg K

g_i is the gravitational constant that varies slightly with altitude. The expression for p in Equation P2.1a assumes that within the region of interest g_i is constant; otherwise, the expression for p would be a lot more complicated than the one shown

ρ is the air density

b. For a region where the temperature is constant,

$$p = p_i \exp\left(-\frac{g(z - z_i)}{RT_i}\right)$$
$$T = T_i \tag{P2.1d}$$

Table P2.1 gives the values of pressure, temperature, and the gravitational constant at the beginning of each region, as well as the lapse rate of the region.

a. Create a MATLAB program using the `if-elseif` ladder that will determine the property values of (T, p, ρ) for every 1,000 m from $z = 0$ (sea level) to $z = 80,000$ m. Use one `for` loop in establishing z.

b. In your program, create vectors for the table values of z, T, p, λ, and g.

c. Assign a character to each region that can be used to determine which equations to be used in determining T and p.

d. Save the results as a data file to be used in Project P2.2. Note: The data file should only contain numbers.

e. Print the results to a file in table format for every 1000 m.

f. Plot T vs. z, p vs. z, and ρ vs. z as three separate plots but all on the same page.

P2.2 Project P2.1 provided a table of temperature, T, pressure, p, and density ρ, every 1000 meters of altitude. Atmospheric property values at altitudes between table values can be determined by linear interpolation. Construct a MATLAB program that will load the data file created in Project P2.1 giving (T, p, ρ) every 1000 m. The program is to determine by linear interpolation

Table P2.1 U.S. Standard Atmospheric Property Table

Regional Properties of U.S. Standard Atmosphere					
Region	z_i (km)	T_i (K)	p_i (Pa)	λ_i (K/m)	g_i (m/s²)
	0	288.15	101,325		9.810
1				0.0065	
	11.0	216.65	22,632.1		9.776
2				0.0000	
	20.0	216.65	5,474.9		9.749
3				−0.001	
	32.0	228.65	868.02		9.712
4				−0.0028	
	47.0	270.65	110.91		9.666
5				0.0000	
	51.0	270.65	66.94		9.654
6				0.0028	
	71.0	214.65	3.956		9.594
7				0.0020	
	84.9	186.95	0.373		9.553

(Equation 2.46 and the method described in Example 2.13) the properties of (T, p, ρ) at altitudes not in the loaded table. Make the program interactive; that is, the program is to ask the user if he or she wishes to know the temperature, pressure, and density at a specific altitude. If the answer is "y," the program is to ask the user to enter an altitude from the keyboard. The program then determines (T, p, ρ) at the entered altitude and prints the results to the screen. The program then asks the user if he or she wishes to enter another altitude. The program is to continue as long as the response to the question is "y."

P2.3 The properties of specific volume, v, and pressure, p, as a function of temperature, T, for carbon dioxide based on the Redlich–Kwong equation of state are given in Table P2.3.

Determine v and p at temperatures T_2, where

$$T_2 = [367 \ 634 \ 420 \ 587 \ 742]$$

Table P2.3   Carbon Dioxide Properties

T (K)	v (m³/kmol)	p (bar)
350	0.28	7.65
400	0.32	8.57
450	0.36	9.16
500	0.40	9.55
550	0.44	9.81
600	0.48	10.00
650	0.52	10.14
700	0.56	10.24
750	0.60	10.31

Note: 1 bar = 10^5 N/m².

Use interpolation formula of Equation 2.46 and the method described in Example 2.13.

Write a MATLAB program that will do the following:

a. Construct three separate vectors containing the carbon dioxide properties of T, v, and p.
b. Print Table P2.3 to the screen (with table headings).
c. Determine v and p at temperatures T_2 using Equation 2.46 and the method described in Example 2.13.
d. Print to the screen in a table format (with table headings) values of v and p at temperatures T_2.

P2.4 The positioning of a piston in an internal combustion engine is shown in Figure P2.4a and b.

The piston's position, s, as seen from the crankshaft center can be determined by the law of cosines; that is,

$$b^2 = s^2 + r^2 - 2sr\cos\theta$$

or

$$s^2 - (2r\cos\theta)s + (r^2 - b^2) = 0 \tag{P2.4a}$$

where
b is the length of the piston rod
r is the radius of the crankshaft
θ is the angle that the crankshaft journal makes with the vertical axis

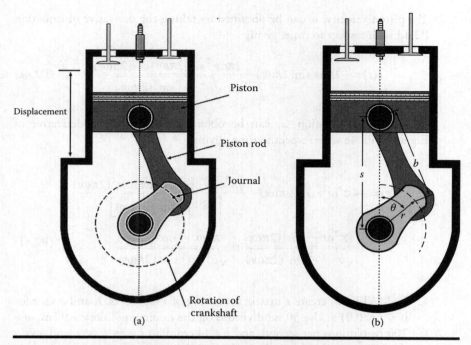

Figure P2.4 (a) Piston configuration. (b) Piston position variables.

Equation P2.4a is a quadratic equation in s, and using the + sign with the square root term gives,

$$s = \frac{1}{2}\left(2r\cos\theta + \sqrt{4r^2\cos^2\theta - 4(r^2 - b^2)}\right) = r\cos\theta + \sqrt{r^2(\cos^2\theta - 1) + b^2}$$

or

$$s = r\cos\theta + \sqrt{b^2 - r^2\sin^2\theta} \qquad \text{(P2.4b)}$$

The piston is constrained to move in the vertical direction, and thus, s varies as the crankshaft rotates. The angle, θ, varies with time, t, and can be expressed in terms of the rotational speed, ω, of the crankshaft. The angle θ is thus given by

$$\theta = 2\pi\omega t \qquad \text{(P2.4c)}$$

where ω is in revolutions per second. Substituting Equation P2.4c into Equation P2.4b gives

$$s(t) = r\cos(2\pi\omega t) + \sqrt{b^2 - r^2\sin^2(2\pi\omega t)} \qquad \text{(P2.4d)}$$

The piston velocity, v, can be obtained by taking the derivative of Equation P2.4d with respect to time, giving

$$v(t) = -2\pi\omega r \sin(2\pi\omega t) - \frac{2\pi\omega r^2 \sin(2\pi\omega t)\cos(2\pi\omega t)}{\sqrt{b^2 - r^2 \sin^2(2\pi\omega t)}} \qquad \text{(P2.4e)}$$

The piston acceleration, a, can be obtained by taking the derivative of Equation P2.4e with respect to time, giving

$$a(t) = -4\pi^2\omega^2 r \cos(2\pi\omega t) - \frac{4\pi^2\omega^2 r^4 \sin^2(2\pi\omega t)\cos^2(2\pi\omega t)}{\left[b^2 - r^2 \sin^2(2\pi\omega t)\right]^{3/2}}$$

$$- \frac{4\pi^2\omega^2 r^2 \cos^2(2\pi\omega t)}{\sqrt{b^2 - r^2 \sin^2(2\pi\omega t)}} + \frac{4\pi^2\omega^2 r^2 \sin^2(2\pi\omega t)}{\sqrt{b^2 - r^2 \sin^2(2\pi\omega t)}} \qquad \text{(P2.4f)}$$

a. In MATLAB, create a matrix consisting of s vs. t, v vs. t, and a vs. t for $0 \le t \le 0.01$ s. Use 50 subdivisions on the t domain. Take $r = 9$ cm, $\omega = 100$ revolutions per second, and $b = 14$ cm. Plot s vs. t, v vs. t, and a vs. t on three separate plots, all on the same page.

b. Using MATLAB's max function and the matrix obtained in part (a), determine the approximate maximum velocity and maximum acceleration, and print out those values to the screen.

c. Plot on a single graph s vs. t for $\omega = [50\ 100\ 150\ 200]$ revolutions per second.

P2.5 This project involves plotting the various motions that can occur with a mass–spring–dashpot system. A sketch of such a system is shown in Figure P2.5.

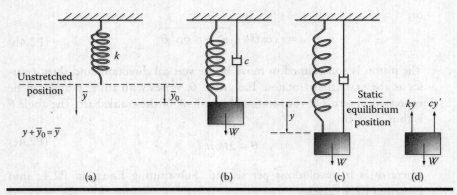

Figure P2.5 **Spring–dashpot system. (a) Unloaded spring, (b) loaded spring in equilibrium position, (c) loaded spring in nonequilibrium position, and (d) forces acting on the mass.**

Development of the governing equation describing the motion of the mass follows [4]:

Equilibrium state

$$W - k\bar{y}_0 = 0 \tag{P2.5a}$$

Nonequilibrium state

$$W - k\bar{y} - c\bar{y}' = m\bar{y}'' \tag{P2.5b}$$

where
  k is the spring constant
  c is the damping factor
  m is the object mass
  $\bar{y}$ is the mass displacement from unstretched position
  $\bar{y}', \bar{y}''$ are the velocity and acceleration of the mass, respectively

Let y be the mass displacement from the equilibrium position, then

$$\bar{y} = \bar{y}_0 + y, \quad \bar{y}' = y' \quad \text{and} \quad \bar{y}'' = y''$$

Substituting these values into Equation P2.5b gives

$$W - k(\bar{y}_0 + y) - cy' = my'' \tag{P2.5c}$$

Since $W - k\bar{y}_0 = 0$, Equation P2.5c reduces to

$$y'' + \frac{c}{m}y' + \frac{k}{m}y = 0 \tag{P2.5d}$$

We seek a function that satisfies this differential equation. Such a function is one in which its derivatives reproduce the function multiplied by a constant. A function that satisfies this condition is e^{pt}. Assume that $y = ae^{pt}$ and then

$$y' = pae^{pt}, \quad \text{and} \quad y'' = p^2 ae^{pt}$$

Substituting these terms into the differential equation P2.5d gives

$$\left(p^2 + \frac{c}{m}p + \frac{k}{m} \right)ae^{pt} = 0 \tag{P2.5e}$$

Now $e^{pt} \neq 0$; therefore,

$$\left(p^2 + \frac{c}{m}p + \frac{k}{m} \right) = 0 \tag{P2.5f}$$

Thus,

$$p = -\frac{1}{2}\frac{c}{m} \pm \frac{1}{2}\sqrt{\left(\frac{c}{m}\right)^2 - 4\frac{k}{m}} = -\frac{1}{2}\frac{c}{m} \pm \sqrt{\left(\frac{c}{2m}\right)^2 - \frac{k}{m}} \qquad \text{(P2.5g)}$$

We see that there are two solutions that satisfy the differential equation. It can be shown that the sum of the two solutions is also a solution to the differential equation. The general solution is

$$y = A\exp\left(-\frac{c}{2m}t + \sqrt{\left(\frac{c}{2m}\right)^2 - \frac{k}{m}}\,t\right) + B\exp\left(-\frac{c}{2m}t - \sqrt{\left(\frac{c}{2m}\right)^2 - \frac{k}{m}}\,t\right)$$

or

$$y = \exp\left(-\frac{c}{2m}t\right)\left\{A\exp\left(\sqrt{\left(\frac{c}{2m}\right)^2 - \frac{k}{m}}\,t\right) + B\exp\left(-\sqrt{\left(\frac{c}{2m}\right)^2 - \frac{k}{m}}\,t\right)\right\}$$

$$\text{(P2.5h)}$$

where A and B are constants and $\exp(x) = e^x$. The type of motion depends on the variables k, c, and m.

If $(c/2m)^2 < k/m$, then Equation P2.5h is the one to use. The system is said to be *overdamped*.

If $(c/2m)^2 > k/m$, then the square root term becomes imaginary, and the system is said to be *underdamped*. Noting that

$$e^{ix} = \cos x + i\sin x \quad \text{and} \quad e^{-ix} = \cos x - i\sin x$$

Equation P2.5h reduces to

$$y = \exp\left(-\frac{c}{2m}t\right)\left\{A\cos\left(\sqrt{\frac{k}{m} - \left(\frac{c}{2m}\right)^2}\,t\right) + B\sin\left(\sqrt{\frac{k}{m} - \left(\frac{c}{2m}\right)^2}\,t\right)\right\}$$

$$\text{(P2.5i)}$$

If $(c/2m)^2 = k/m$, then the square root term is zero, and the system is said to be *critically damped*. For this case, the solution is

$$y = (A + Bt)\exp\left(-\frac{c}{2m}t\right) \qquad \text{(P2.5j)}$$

a. Given the following parameters,

$$m = 75\,\text{kg}, \quad k = 87.5\,\text{N/m}, \quad c = 875\,\text{N-s/m}$$

Develop a computer program to determine the coefficients, A and B, for the following cases:
1. $y(0) = 0.5$ m, $y'(0) = 1.0$ m/s
2. $y(0) = 0.5$ m, $y'(0) = -1.0$ m/s
3. $y(0) = 0.5$ m, $y'(0) = 0$ m/s
For each case,
1. Determine $y(t)$ for $0 \le t \le 10$ s in steps of 0.1 s.
2. Print out a table of y vs. t every 1 s.
3. Plot y vs. t for all three cases on the same graph. Label each curve with the value of y'.
b. Given the following parameters,

$$m = 25\,\text{kg}, \quad k = 200\,\text{N/m}, \quad c = 5\,\text{N-s/m}$$

$$y(0) = 5\,\text{m}, \quad y'(0) = 0\,\text{m/s}$$

Determine $y(t)$ for $0 \le t \le 40$ s in steps of 0.1 s.
The envelope of the solution graph for this case is given by

$$y = \pm A \exp\left(-\frac{c}{2m}t\right)$$

Plot y vs. t for both the oscillating function and its envelope on the same graph.

P2.6 In this project we use a mass-spring-dashpot system subjected to an oscillatory driving force to demonstrate the concept of resonance. Resonance is a phenomenon in which the response amplitude at a particular frequency is much larger than at other frequencies. Suppose the mass-spring-dashpot system described in Project P2.5 to be subjected to a driving force $F_o \sin \omega t$. The governing equation then becomes

$$y'' + \frac{c}{m}y' + \frac{k}{m}y = \frac{F_0}{m}\sin \omega t \qquad (\text{P2.6a})$$

The solution can be obtained by assuming the solution is the sum of the complementary solution plus a particular solution. The complementary solution is the solution to the homogeneous equation, which was obtained in Project P2.5. To obtain the particular solution, y_p, assume

$$y_p = a \sin \omega t + b \cos \omega t \tag{P2.6b}$$

where a and b are constants, then

$$y_p' = \omega(a \cos \omega t - b \sin \omega t)$$

and

$$y_p'' = \omega^2(-a \sin \omega t - b \cos \omega t)$$

Substituting these expressions into Equation 2.6a gives

$$\left(-a\omega^2 - \frac{c}{m}\omega b + \frac{k}{m}a\right)\sin \omega t + \left(-b\omega^2 + \frac{c}{m}\omega a + \frac{k}{m}b\right)\cos \omega t = \frac{F_0}{m}\sin \omega t \tag{P2.6c}$$

Collecting coefficients of the sin and cos terms on the left side of Equation P2.6c and equating them to the sin and cos coefficients on the right side of that equation give

$$\left(\frac{k}{m} - \omega^2\right)a - \frac{c\omega}{m}b = \frac{F_0}{m} \tag{P2.6d}$$

$$\frac{c\omega}{m}a + \left(\frac{k}{m} - \omega^2\right)b = 0 \tag{P2.6e}$$

Solving the above two equations for a and b, gives

$$a = \frac{F_0}{m} \times \frac{\dfrac{k}{m} - \omega^2}{\left(\dfrac{c\omega}{m}\right)^2 + \left(\dfrac{k}{m} - \omega^2\right)^2} \tag{P2.6f}$$

$$b = -\frac{F_0}{m} \times \frac{\dfrac{c\omega}{m}}{\left(\dfrac{c\omega}{m}\right)^2 + \left(\dfrac{k}{m} - \omega^2\right)^2} \tag{P2.6g}$$

Substituting Equations P2.6f and P2.6g into Equation P2.6b gives

$$y_p = \frac{F_0}{m} \times \frac{1}{\left(\frac{c\,\omega}{m}\right)^2 + \left(\frac{k}{m} - \omega^2\right)^2} \left\{ \left(\frac{k}{m} - \omega^2\right) \sin\omega t - \frac{c\,\omega}{m} \cos\omega t \right\} \quad \text{(P2.6h)}$$

We can rewrite Equation P2.6h using the trig identity

$$\alpha \sin\omega t + \beta \cos\omega t = \gamma \sin(\omega t - \phi)$$

where

$$\gamma = \sqrt{\alpha^2 + \beta^2}$$

and

$$\phi = \tan^{-1}\frac{\beta}{\alpha}$$

Applying these relations to Equation 2.6h gives

$$y_p = \frac{F_0}{m} \times \frac{1}{\sqrt{\left(\frac{k}{m} - \omega^2\right)^2 + \left(\frac{c\,\omega}{m}\right)^2}} \sin(\omega t - \phi) \quad \text{(P2.6i)}$$

For a mass-spring-dashpot system with no driving force and no damping, the system would oscillate at a frequency, $\omega_n = \sqrt{\frac{k}{m}}$ (see Equation P2.5h). It is convenient to introduce the ratio $\zeta = \frac{c}{c_c}$, where $c_c = 2m\omega_n$, which is the value of c that would make the system with no driving force critically damped. After some algebraic manipulation, Equation P2.6i can be put into the form

$$y_p = \frac{F_0}{k} \times \frac{1}{\sqrt{\left(1 - \frac{\omega^2}{\omega_n^2}\right)^2 + \left(2\zeta\frac{\omega}{\omega_n}\right)^2}} \sin(\omega t - \varphi) \quad \text{(P2.6j)}$$

The system will oscillate with an amplitude equal to

$$\frac{F_0}{k}\frac{1}{\sqrt{\left(1-\left(\dfrac{\omega}{\omega_n}\right)^2\right)^2+\left(2\zeta\dfrac{\omega}{\omega_n}\right)^2}}$$

Let $ampl = \dfrac{1}{\sqrt{\left(1-\left(\dfrac{\omega}{\omega_n}\right)^2\right)^2+\left(2\zeta\dfrac{\omega}{\omega_n}\right)^2}}$

For a given, F_0/k, the larger the *ampl* the larger is the oscillation.

Construct a MATLAB program to create a plot of *ampl* vs. ω/ω_n for values of $\zeta = 1.0, 0.5, 0.25, 0.10, 0.05$, and $0 < \omega/\omega_n < 2$ in steps of 0.01. What happens as $\omega \to \omega_n$?

P2.7 The RLC circuit is a fundamental problem in electrical engineering. In this project, we consider the parallel RLC circuit shown in Figure P2.7 and derive the governing equation for the voltage across each circuit component when at $t = 0$ the switch is opened. We start the analysis with the constituent voltage–current relations for resistors, inductors, and capacitors:

Resistor (Ohm's law)

$$v_R = Ri_R \tag{P2.7a}$$

Inductor

$$v_L = L\frac{di_L}{dt} \tag{P2.7b}$$

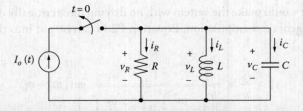

Figure P2.7　Parallel RLC circuit.

Capacitor

$$i_C = C \frac{dv_C}{dt} \tag{P2.7c}$$

where R, L, and C are the component values for the resistor (in ohms), inductor (in henries), and capacitor (in farads) and the voltages v and currents i are as defined in Figure P2.7. In addition, we assume that the stateful circuit elements L and C have known initial conditions $i_L(0)$ and $v_C(0)$ at the moment that the switch is opened.

Kirchhoff's current law (KCL) states that the sum of currents at any circuit node is zero. Applying KCL at the top node in the circuit gives

$$i_R + i_L + i_C = 0 \tag{P2.7d}$$

Also, the parallel topology of the circuit gives

$$v_R = v_L = v_C \equiv v \tag{P2.7e}$$

Differentiating Equation P2.7d gives

$$\frac{di_R}{dt} + \frac{di_L}{dt} + \frac{di_c}{dt} = 0 \tag{P2.7f}$$

Substituting Equation P2.7e into Equation P2.7a and differentiating give

$$\frac{di_R}{dt} = \frac{1}{R} \frac{dv}{dt} \tag{P2.7g}$$

Substituting Equation P2.7e into Equation P2.7b gives

$$\frac{di_L}{dt} = \frac{1}{L} v \tag{P2.7h}$$

Substituting Equation P2.7e into Equation P2.7c gives

$$\frac{dv}{dt} = \frac{i_c}{C} \tag{P2.7i}$$

Differentiating Equation P2.7i and rearranging gives

$$\frac{di_c}{dt} = C\frac{d^2v}{dt^2} \tag{P2.7j}$$

Substituting Equations P2.7g through P2.7j into Equation P2.7f gives

$$\frac{d^2v}{dt^2} + \frac{1}{RC}\frac{dv}{dt} + \frac{1}{LC}v = 0 \tag{P2.7k}$$

To solve this homogeneous differential equation, we seek a function $v(t)$ such that the derivatives $\frac{dv}{dt}(t)$ and $\frac{d^2v}{dt^2}(t)$ reproduce the form of $v(t)$. A function that satisfies this condition is $Ae^{\alpha t}$, where A and α are arbitrary constants. If we assume that $v = Ae^{\alpha t}$, then

$$\frac{dv}{dt} = \alpha Ae^{\alpha t} \quad \text{and} \quad \frac{d^2v}{dt^2} = \alpha^2 Ae^{\alpha t}$$

Substituting these terms into the differential equation P2.7k gives

$$\left(\alpha^2 + \frac{1}{RC}\alpha + \frac{1}{LC}\right)Ae^{\alpha t} = 0 \tag{P2.7l}$$

Since $e^{\alpha t} \neq 0$, then

$$\left(\alpha^2 + \frac{1}{RC}\alpha + \frac{1}{LC}\right) = 0 \tag{P2.7m}$$

The solutions are

$$\alpha = -\frac{1}{2RC} \pm \sqrt{\left(\frac{1}{2RC}\right)^2 - \frac{1}{LC}} \tag{P2.7n}$$

Thus, there are two solutions for α, and there are two solutions that satisfy the differential equation P2.7k. The general solution to Equation P2.7k is the sum of all known solutions:

$$v = A\exp\left(-\frac{1}{2RC}t + \sqrt{\left(\frac{1}{2RC}\right)^2 - \frac{1}{LC}}\,t\right) + B\exp\left(-\frac{1}{2RC}t - \sqrt{\left(\frac{1}{2RC}\right)^2 - \frac{1}{LC}}\,t\right)$$

or

$$v = \exp\left(-\frac{1}{2RC}t\right)\left\{A\exp\left(\sqrt{\left(\frac{1}{2RC}\right)^2 - \frac{1}{LC}}\,t\right) + B\exp\left(-\sqrt{\left(\frac{1}{2RC}\right)^2 - \frac{1}{LC}}\,t\right)\right\}$$

(P2.7o)

where $\exp(x) = e^x$ and A and B are constants that depend on the initial conditions of the circuit.

Note that this solution has three regions of interest:

1. *Overdamped*: If $(1/2RC)^2 > 1/LC$, then the solutions are decaying exponentials over time.
2. *Underdamped*: If $(1/2RC)^2 < 1/LC$, then the solutions are decaying sinusoids over time.

   For the underdamped case, we can show the sinusoidal behavior by applying the identities $e^{jx} = \cos x + j\sin x$ and $e^{-jx} = \cos x - j\sin x$ to Equation P2.7o, giving

$$v = \exp\left(-\frac{1}{2RC}t\right)\left\{A\cos\left(\sqrt{\frac{1}{LC} - \left(\frac{1}{2RC}\right)^2}\,t\right) + B\sin\left(\sqrt{\frac{1}{LC} - \left(\frac{1}{2RC}\right)^2}\,t\right)\right\}$$

(P2.7p)

where the coefficients A and B are to be determined by initial conditions.

3. If $(1/2RC)^2 = 1/LC$, then the square root term is zero, and the system is said to be *critically damped*. For this case, the solution is

$$v = (A + Bt)\exp\left(-\frac{1}{2RC}t\right)$$

(P2.7q)

We now consider the underdamped case (Equation P2.7p) where A and B are constants to be determined by initial conditions and have the units of volts. For the component values $R = 100\ \Omega$, $L = 1$ mH, and $C = 1\ \mu$F and initial conditions $i_L(0) = 0$ and $v_C(0) = 6$ V,

a. Solve for the coefficients A and B.
b. Create a MATLAB program that will calculate $v(t)$ for $0 \le t \le 500\ \mu$s in steps of 5 μs.
c. Plot $v(t)$ vs. t for $0 \le t \le 500\ \mu$s in steps of 5 μs.

P2.8 Using Equations P2.7a through P2.7e, show that the governing differential equation for the inductor current, i_L, in the parallel RLC circuit of Figure P2.7 is given by

$$\frac{d^2 i_L}{dt^2} + \frac{1}{RC}\frac{di_L}{dt} + \frac{1}{LC}i_L = 0 \qquad \text{(P2.8a)}$$

Comparing Equations P2.8a and P2.7k, we see that the equations are alike, with i_L replacing v, and therefore, they have similar solutions. For the underdamped case,

$$i_L = \exp\left(-\frac{1}{2RC}t\right)\left\{\alpha\cos\left(\sqrt{\frac{1}{LC} - \left(\frac{1}{2RC}\right)^2}\,t\right) + \beta\sin\left(\sqrt{\frac{1}{LC} - \left(\frac{1}{2RC}\right)^2}\,t\right)\right\}$$

$$\text{(P2.8b)}$$

For the component values $R = 100\ \Omega$, $L = 1$ mH, and $C = 1\ \mu$F and initial conditions $i_L(0) = 0$ and $v_C(0) = 6$ V,

a. Solve for the coefficients α and β.
b. Create a MATLAB program that will calculate $i_L(t)$ for $0 \le t \le 500\ \mu$s in steps of 5 μs.
c. If you did Project P2.7, plot $v(t)$ vs. t and $i_L(t)$ vs. t on the same graph; otherwise, just plot $i_L(t)$ vs. t. For both cases, take $0 \le t \le 500\ \mu$s in steps of 5 μs.

P2.9 The classical form for a second-order differential equation is

$$\frac{d^2 y}{dt^2} + 2\zeta\omega_n\frac{dy}{dt} + \omega_n^2 y = 0 \qquad \text{(P2.9a)}$$

where
  ζ is the *damping factor*
  ω_n is the *natural frequency*

We can match the terms of Equation P2.9a with Equation P2.8a to find the damping factor and natural frequency for the parallel RLC circuit. Thus,

$$\omega_n = \frac{1}{\sqrt{LC}} \quad \text{and} \quad \zeta = \frac{1}{2RC\omega_n} = \frac{\sqrt{LC}}{2RC} \tag{P2.9b}$$

Note that for $\zeta < 1$, the circuit is underdamped; for $\zeta > 1$, the circuit is overdamped; and for $\zeta = 1$, it is critically damped.

For the component values $L = 1$ mH and $C = 1$ μF and initial conditions $i_L(0) = 0.25$ A and $v_C(0) = 6$ V,

a. Determine the resistor value R_{crit} that makes the circuit critically damped.
b. Plot the inductor current i_L vs. time for the two values of R: $R = R_{crit}$ and $R = 5R_{crit}$. Assume the interval $0 \le t \le 500$ μs in steps of 1/2 μs. Plot all of the waveforms on one graph.

P2.10 Using Equations P2.7c, P2.7d, P2.8b, and P2.7n, we can show that for underdamped case in the parallel RLC circuit,

$$i_R(t) = -\exp\left(-\frac{1}{2RC}t\right)\left[\alpha\cos\left(\sqrt{\frac{1}{LC}-\left(\frac{1}{2RC}\right)^2}\,t\right)+\beta\sin\left(\sqrt{\frac{1}{LC}-\left(\frac{1}{2RC}\right)^2}\,t\right)\right]$$

$$+\frac{1}{2R}\exp\left(-\frac{1}{2RC}t\right)\left[A\cos\left(\sqrt{\frac{1}{LC}-\left(\frac{1}{2RC}\right)^2}\,t\right)+B\sin\left(\sqrt{\frac{1}{LC}-\left(\frac{1}{2RC}\right)^2}\,t\right)\right]$$

$$-C\exp\left(-\frac{1}{2RC}t\right)\left[-A\sin\left(\sqrt{\frac{1}{LC}-\left(\frac{1}{2RC}\right)^2}\,t\right)+B\cos\left(\sqrt{\frac{1}{LC}-\left(\frac{1}{2RC}\right)^2}\,t\right)\right]$$

$$\times\sqrt{\frac{1}{LC}-\left(\frac{1}{2RC}\right)^2} \tag{P2.10a}$$

Take the following component values to be $R = 100$ Ω, $L = 1$ mH, and $C = 1$ μF and initial conditions $i_L(0) = 0$ and $v(0) = 6$ V:

a. If you have completed Project P2.7, use the values for A and B obtained in that program in Equation P2.10a; otherwise, do part (a) of Project P2.7.
b. If you have completed Project P2.8, use the values for α and β obtained in that program in Equation P2.10a; otherwise, do part (a) of Project P2.8.

c. Construct a MATLAB program to calculate $i_R(t)$ for $0 \le t \le 500$ μs in steps of 5 μs.

d. Plot $i_R(t)$ vs. t for $0 \le t \le 500$ μs in steps of 5 μs.

P2.11 This project involves a series RLC circuit as shown in Figure P2.11.

The resistor, inductor, and capacitor voltage–current relations using their respective constituent relations are

$$v_R = i_R R \tag{P2.11a}$$

$$v_L = L \frac{di_L}{dt} \tag{P2.11b}$$

$$i_C = C \frac{dv_C}{dt} \tag{P2.11c}$$

where

v_R, v_L, and v_C are the voltages across the resistor, inductor, and capacitor, respectively

i_R, i_L, and i_C are their respective currents

R, L, and C are their respective component values in ohms, farads, and henries

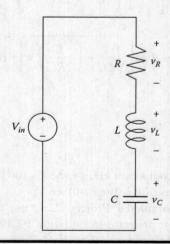

Figure P2.11 Series RLC circuit.

Applying Kirchhoff's voltage law around the RLC loop gives

$$V_{in} - v_R - v_L - v_C = 0 \qquad \text{(P2.11d)}$$

In the series case, all of the components have the identical current, and thus, we define

$$i = i_R = i_L = i_C \qquad \text{(P2.11e)}$$

a. Using Equations P2.11a through P2.11e, derive the governing differential equation for the capacitor voltage v_C.

b. Using Project P2.7 as a guide, derive the three possible solutions to the governing differential equation for the voltage v_C. Assume that the voltage V_{in} becomes zero for $t > 0$.

c. The obtained solutions in part (b) should involve two arbitrary coefficients A and B. Determine the values for A and B for the following two cases:

   1. $R = 100\ \Omega$, $L = 1$ mH, and $C = 1\ \mu$F and initial conditions $i(0) = 0$ and $V_{in}(0) = 6$ V

   2. $R = 100\ \Omega$, $L = 1$ mH, and $C = 0.1\ \mu$F and initial conditions $i(0) = 0$ and $V_{in}(0) = 6$ V

d. For the two cases of part (c), plot i vs. t, for $0 \le t \le 10^{-4}$ s in steps of 10^{-6} s on the same page.

P2.12 This project is a variation of Project P2.11. Instead of solving for the voltage v_C in the series RLC circuit of Figure P2.11, we can alternatively solve for the current i through the circuit elements.

a. Use Equations P2.11a through P2.11e to show that for the case when $V_{in} = 0$, the governing differential equation for i is:

$$\frac{d^2 i}{dt^2} + \frac{R}{L}\frac{di}{dt} + \frac{i}{LC} = 0 \qquad \text{(P2.12a)}$$

b. Show that the general solution to Equation P2.12a is:

$$i = \exp\left(-\frac{R}{2L}t\right)\left\{\alpha \cos\left(\sqrt{\frac{1}{LC} - \left(\frac{R}{2L}\right)^2}\ t\right) + \beta \sin\left(\sqrt{\frac{1}{LC} - \left(\frac{R}{2L}\right)^2}\ t\right)\right\}$$

$$\text{(P2.12b)}$$

c. For the following component and initial conditions, determine α and β:
$R = 10\ \Omega$, $L = 1$ mH, and $C = 2\ \mu$F and initial conditions $v_C(0) = 10$ V
and $\dfrac{dv_C}{dt}(0) = \dfrac{6}{\sqrt{LC}}$.

d. Plot i vs. t for $0 \leq t \leq 4\ \mu$s in steps of 0.1 μs.

P2.13 A common problem in serial data communication is that digital bitstreams
often contain long sequences of ones or zeros that can cause the receiver to
become unsynchronized. One solution is to use *Manchester coding* in order
to ensure that the bitstream contains many transitions so that it is easier to
recover the clock.

10BaseT Ethernet uses Manchester coding as depicted in Figure P2.13a,
where a logical "one" is encoded as a low-to-high transition and a logical
"zero" is encoded as a high-to-low transition. By ensuring that each bit contains a transition, the receiver can extract (or "recover") a clock signal from
the transmitted waveform by using a phase-locked loop circuit.

a. Using the `if-else` ladder in MATLAB, generate the waveform of
Figure P2.13b for the bit sequence "1101" and plot. Assume the follow-
ing parameters:
"High" voltage: +2.5 V
"Low" voltage: –2.5 V
Bit rate: 10 Mbps (i.e., each bit is 0.1 μs in duration)
When modeling your waveform in MATLAB, generate values from
zero to 0.4 μs in steps of 0.01 μs. Plot using MATLAB's `plot` command.

b. What is one disadvantage of Manchester coding?

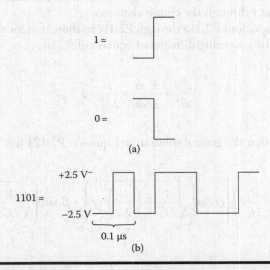

(a)

(b)

Figure P2.13 **(a) Manchester coding. (b) Bit sequence "1101."**

References

1. Kernighen, B.W. and Ritchie, D.M., *The C Programming Language*, 2nd edn., Prentice Hall, Englewood Cliffs, NJ, 1988.
2. Holman, J.P., *Heat Transfer*, 9th edn., McGraw-Hill, New York, 2002.
3. Bober, W., The use of the Swamee-Jain formula in pipe network problems, *Journal of Pipelines*, 4, 315–317, 1984.
4. Thomson, W.T., *Theory of Vibration with Applications*, Prentice Hall, Englewood Cliffs, NJ, 1972.
5. Bober, W. and Stevens, A., *Numerical and Analytical Methods with MATLAB for Electrical Engineers*, CRC Press, Boca Raton, FL, 2012.

Chapter 3

Taylor Series, Self-Written Functions and MATLAB®'s `interp1` Function

3.1 Introduction

In this chapter, we introduce the concept of expressing various functions as a series. In particular, we cover the Taylor series expansions of several functions. Numerical methods for solving various types of engineering problems are based on the use of a few terms in the Taylor series expansions. The chapter contains several exercises and projects to demonstrate the equivalence of the Taylor series expansion of several functions and MATLAB's evaluation of the function by the use of an arithmetic statement.

The next topic in this chapter is the self-written function. The self-written function is another building block that can be used in constructing a manageable computer program. The use of many MATLAB® functions that solve various types of mathematical problems requires the user to write a self-written function. Self-written functions are usually saved as separate *.m* files. However, if the self-written function involves a single statement, an anonymous function can be used which is contained within the calling program and thus avoids an additional *.m* file.

The next topic covered in the chapter is MATLAB's `interp1` function. Many types of engineering problems involve material properties that are tabulated. If a particular problem involves properties that lie between tabulated values, one needs to interpolate for the property of interest. MATLAB's `interp1` function provides the means to do this.

Lastly, the use of characters and strings in MATLAB programs is discussed.

3.2 Functions Expressed as a Series

▪ Taylor series expansion

Any rational function can be evaluated by a power series, such as a Taylor series. The function $f(x)$ can be approximated by the n-term Taylor series, which is

$$f(x) = f(a) + f'(a)(x-a) + f''(a)\frac{(x-a)^2}{2!}$$

$$+ f'''(a)\frac{(x-a)^3}{3!} + \cdots + f^{(n)}(a)\frac{(x-a)^n}{n!} \tag{3.1}$$

where

$f'(a)$ is the first derivative of $f(x)$ evaluated at $x = a$
$f''(a)$ is the second derivative of $f(x)$ evaluated at $x = a$
$\vdots$
$f^{(n)}(a)$ is the nth derivative of $f(x)$ evaluated at $x = a$

Since a number of the sample programs involve determining e^x by a Taylor series expansion, we start with a discussion of the series expansion of e^x.

We wish to determine the Taylor series expansion of e^x about $x = 0$:

$$f(x) = e^x, \quad f(0) = e^0 = 1$$

$$f' = e^x, \quad f'(0) = 1$$

$$f'' = e^x, \quad f''(0) = 1$$

$$f''' = e^x, \quad f'''(0) = 1$$

etc.

Substituting into Equation 3.1, the Taylor series expansion of e^x to n terms is

$$e^x = 1 + x + \frac{x^2}{2!} + \frac{x^3}{3!} + \frac{x^4}{4!} + \cdots + \frac{x^n}{n!} \tag{3.2}$$

There are several possible programming algorithms to calculate this series. The simplest method is to start with `ex = 1` and then evaluate each term using exponentiation and MATLAB®'s `factorial` function and add the obtained term to `ex`. An alternative approach would be to evaluate each term individually, save them in an array, and then use MATLAB's `sum` function to add them all up. However, for some series expressions, it is best to compute each term in the series based on the value of the preceding term. For example, in Equation 3.2, we can see that the third term in the series can be obtained from the second term by multiplying the second term by x and dividing 3; that is, term 3 = term $2 \cdot x/3$. In general, term (n) = term $(n - 1) \cdot x/n$. We can set up a term index to identify each term in the series:

Term index:	1	2	3	4	⋯	n

$$e^x = 1 + x + \frac{x^2}{2!} + \frac{x^3}{3!} + \frac{x^4}{4!} + \cdots + \frac{x^n}{n!}$$

This concept is used in several of the following sample programs.

The following example determines e^x by using MATLAB's factorial function.

Example 3.1

```
% Example_3_1.m
% This program calculates e^x by series and by MATLAB's
% exp() function
% e^x=1 + x + x^2/2! + x^3/3! + x^4/4! +···
% The 'for' loop ends when the term only affects the seventh
% significant figure.
clear; clc;
x=5.0;
ex1=1.0;
for n=1:100
    term=x^n/factorial(n);
    ex1=ex1+term;
    if(abs(term)<=ex1*1.0e-7)
        break;
    end
end
ex2 = exp(x);
fprintf('x=%3.1f    ex1=%7.2f    ex2=%7.2f \n',x,ex1,ex2);
```

Program results

```
x=5.0    ex1=148.41    ex2=148.41
>>
```

The following program illustrates the use of nested loops to calculate e^x both by series and by MATLAB's `exp()` function for $-0.5 < x < 0.5$.

Example 3.2

```
% Example_3_2.m
% This program illustrates the use of nested loops; that is,
% an inner 'for' loop inside an outer 'for' loop. The
% program calculates e^x by both an arithmetic statement
```

```
% (ex2), and by a Taylor series expansion (ex1), where
% -0.5<x<0.5. The outer 'for' loop is used to determine
% the x values. The inner loop is used to determine the
% Taylor series method for evaluating e^x. In this example
% term(n+1) is obtained by multiplying term(n) by x/n.
% The variable, term, is established as a vector so that
% MATLAB's built in'sum' function can be used to sum all the
% terms calculated in the Taylor series method. A maximum
% of fifty terms is used in the series.
% e^x=1+x+x^2/2!+x^3/3!+x^4/4!+···
% Program output is both to the screen and to a file.
% By printing the output to a file, one can edit the output
% file, such as lining up column headings, etc. One cannot
% directly edit output to the screen.
clear; clc;
fo = fopen('output.txt','w');
% Table headings
fprintf(' x            ex1          ex2 \n');
fprintf('------------------------------------\n');
fprintf(fo,' x          ex1           ex2 \n');
fprintf(fo,'------------------------------------\n');
for x=-0.5:0.1:0.5
    ex2=exp(x);
    term(1)=1.0;
    for n=1:49
        term(n+1)=term(n)*x/n;
        if abs(term(n+1))<=1.0e-7
            break;
        end
    end
    ex1=sum(term);
    fprintf('%5.2f   %10.5f   %10.5f \n',x,ex1,ex2);
    fprintf(fo,'%5.2f   %10.5f   %10.5f \n',x,ex1,ex2);
end
fclose(fo);
```

Program results

```
   x       ex1        ex2
---------------------------
-0.50    0.60653    0.60653
-0.40    0.67032    0.67032
-0.30    0.74082    0.74082
-0.20    0.81873    0.81873
-0.10    0.90484    0.90484
 0.00    1.00484    1.00000
```

```
0.10    1.10517    1.10517
0.20    1.22140    1.22140
0.30    1.34986    1.34986
0.40    1.49182    1.49182
0.50    1.64872    1.64872
>>
```

The following program uses a while loop to determine e^x by series. The program also determines e^x by MATLAB's exp() function. The program is made interactive by asking the user to enter an x value from the keyboard.

Example 3.3

```
% Example_3_3.m
% Calculation of e^x by both MATLAB's exp function and the
% Taylor series expansion of e^x. The input() function is
% used to establish the exponent x.
% A 'while loop' is used in determining the series solution.
% A break statement is used to end the loop if the number of
% terms becomes greater than 50. The display statement is
% used to display the values of x, ex and ex2 in the
% command window.
% e^x = 1+x+x^2/2!+x^3/3!+x^4/4!+...
clear; clc;
x = input('Enter a value for the exponent x \n');
n = 1; ex = 1.0; term = 1.0;
while abs(term) > ex*1.0e-6
    term = x^n/factorial(n);
    ex = ex+term;
    n = n+1;
    if n > 50
        break;
    end
end
ex2 = exp(x);
disp('x = '); disp(x); disp('e^x = '); disp(ex);
```

Program results

```
Enter a value for the exponent x
5
x =
     5
e^x =
  148.4131
```

Exercises

E3.1 Use a Taylor series expansion of $\log(1 + x)$ about $x = 0$ to show that

$$\log(1+x) = x - \frac{1}{2}x^2 + \frac{1}{3}x^3 - \frac{1}{4}x^4 + \frac{1}{5}x^5 - \cdots + \cdots$$

This formula is valid for $x^2 < 1$. Create a MATLAB program to determine $\log(1 + x)$ for $-0.9 \leq x \leq 0.9$ in steps of 0.1 by both series and MATLAB's built-in $\log()$ function. Create a table similar to Table 3.1 and print it to a file. Print the results to 5 decimal places.

E3.2 Determine the Taylor series expansion of $\cos(x)$ about $x = 0$ and develop a computer program in MATLAB that will compute the series from $-\pi \leq x \leq \pi$ in steps of 0.1π by determining each term in the series according to the equation term $(k + 1) =$ term $(k) \times$ a multiplication factor, where the multiplication factor needs to be determined. Use as many as 50 terms; however, stop adding terms when the last term only affects the 6th decimal place in the answer. Also compute the value of $\cos(x)$ by using MATLAB's built-in $\cos()$ function over the same interval and step size. Create a table similar to Table 3.2 and print it to a file. Print the results to 5 decimal places.

E3.3 Determine the Taylor Series expansion of $\sin(x)$ about $x = 0$ and develop a computer program in MATLAB that will evaluate $\sin(x)$ from $-\pi \leq x \leq \pi$ in steps of 0.1π. Compute the series by using MATLAB's $factorial$ function and using as many as 50 terms. However, stop adding terms when the last term only affects the 6th decimal place in the answer.

Also compute the value of $\sin(x)$ by using MATLAB's built-in $\sin()$ function over the same interval and step size. Create a table similar to the table shown in Exercise E3.2 except replace $\cos(x)$ with $\sin(x)$.

Table 3.1   Table Format for Exercise E3.1

x	$\log(1 + x)$ by $\log()$	$\log(1 + x)$ by Series	Terms in the Series
−0.9	−	−	−
−0.8	−	−	−
⋮	⋮	⋮	⋮
0.9	−	−	−

Table 3.2   Table Format for Exercise E3.2

x	cos x by cos ()	cos x by Series	Terms in the Series
−1.0π	—	—	—
−0.9π	—	—	—
⋮	⋮	⋮	⋮
0.9π	—	—	—
1.0π	—	—	—

E3.4 Using the Taylor series expansions for e^x, $\cos(x)$, and $\sin(x)$, show that

$$e^{ix} = \cos(x) + i\sin(x).$$

Also show that

$$e^{-ix} = \cos(x) - i\sin(x)$$

where $i = \sqrt{-1}$

3.3 Self-Written Functions

Self-written functions are useful if you have a complicated program and wish to break it down into smaller segments. Also, if a series of statements is to be used many times, it is convenient to place them in a function. Many MATLAB functions (such as `fzero`, `quad`, `ode45`) require a self-written function to define the problem of interest. Self written functions are equivalent to subroutines in most programming languages, but in MATLAB they are usually stored in a separate file and are saved as a *.m* file (though small functions can be defined in the same file as your main script, as described in the next section). The first executable statement in the function file must be `function` and be similar to the form shown in the MATLAB template for writing a function (see Figure 3.1). As can be seen from Figure 3.1, the function template is of the form:

`function[` output arguments `] =` `function_name(` input arguments `)`

Some example function definitions are shown in Table 3.3.

If the function has more than one output value, then the output variables must be in brackets. If there is only one output value, then no brackets are necessary. If there are no output values, use empty brackets.

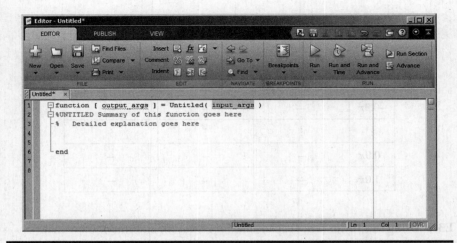

Figure 3.1   Function template.

Table 3.3   Example of Function Usage

Function Definition Line	Function File Name
`function [P,V] = power(i,v)`	*power.m*
`function ex = exf (x)`	*exf.m*
`function[] = output(x,y)`	*output.m*

■ It should be noted that the function output arguments may be passed to the calling program or to another function or to the Command window. Variables defined and manipulated inside the function are local to the function. **This means that the only communication between the calling program and the function is through the input and output arguments of the function.** The exception to this is when a global statement is contained in both the calling program and the function.

■ Many of MATLAB's "built-in" functions are actually implemented as *.m* files. For example, MATLAB's `factorial` function is implemented in the file *factorial.m*. You can find *factorial.m* by typing **which factorial** in the Command window.

In the following example, e^x is determined using a self-written function.

Example 3.4

```
% exf1.m
% This code defines a function, 'exf1', that evaluates e^x.
% This function takes 'x' as an input argument and is called
% either from the Command Window or from a script. The
```

```
% resulting output, 'ex', is available to be used in another
% program or in the Command Window.
% In this example, term(n+1) is obtained from term(n) by
% multiplying term(n) by x/n.
% e^x = 1+x+x^2/2!+x^3/3!+x^4/4!+...
function ex = exf1(x)
term(1) = 1.0;
for n = 1:99
    term(n+1) = term(n)*x/n;
    if abs(term(n+1)) <= 1.0e-7
        break;
    end
end
ex = sum(term);
```

To test out this function, run it from the Command Window. Some examples:

```
>> exf1(1.0)
ans =
        2.7183
>> y = exf1(5.0)
y =
    148.4132
>>
```

In the following example, function exf1 created in Example 3.4 is used in two different arithmetic statements.

Example 3.5

```
% Example_3_5.m
% This program uses the function exf1 (defined in Example_3_4)
% in two different arithmetic statements, i.e., it is used
% to calculate w, y and z.
% In this program, the output is sent to a file.
clear; clc;
fo = fopen('output.txt','w');
fprintf(fo,' x              y              w              z \n');
fprintf(fo,'----------------------------------------------\n');
for x = 0.1:0.1:2.0
    y = exf1(x);
    w = 5*x*exf1(x);
    z = 10*x/y;
    fprintf(fo,'%6.2f   %10.3f   %10.3f   %10.3f \n',x,y,w,z);
end
fclose(fo);
```

Program results

x	y	w	z
0.10	1.105	0.553	0.905
0.20	1.221	1.221	1.637
0.30	1.350	2.025	2.222
0.40	1.492	2.984	2.681
0.50	1.649	4.122	3.033
⋮	⋮	⋮	⋮
1.60	4.953	39.624	3.230
1.70	5.474	46.529	3.106
1.80	6.050	54.447	2.975
1.90	6.686	63.516	2.842
2.00	7.389	73.891	2.707

The following two examples demonstrate that the names of the arguments in the calling program need not be the same as those in the function. It is only the argument list in the calling program that needs to be in the same order as the argument list defined in the function. This feature is useful when a function is to be used with several different scripts, each script having different variable names, but each of the variable names corresponds to variables in the function. This concept is used in all of MATLAB's built-in functions.

Example 3.6

```
% Example_3_6.m
% This program uses the function exf2 to calculate q=e^w/z.
% This calling program uses w and z as input variables to
% function exf2.
% The function exf2 calculates q and returns it to this
% program.
% The function exf2 names its input variables as (x,y) which
% is not the same names used in the calling program, which
% is (w,z). We see that the calling program need not use
% the same names for the input variables as those used in
% the function.
% For one-to-one correspondence, the argument positions in
% the calling program have to be the same as the argument
% positions in the function.
% Note: function exf2 (defined below) must be created before
% this program is executed because the function exf2
% requires input from this program.
clear; clc;
z = 5;
```

```
for w=0.2:0.2:1.0
    q=exf2(w,z);
    fprintf('printout of w,z and q from calling program \n');
    fprintf('w=%5.2f z=%2.0f q=%10.5f \n', w, z, q);
end
```

```
% exf2.m
% This script is used in Example_3_6 and creates a function
% 'exf2' that evaluates e^x/y. In this function, x and y are
% input variables that need to be defined in another
% program, but need not have the same names. The output
% variable in the function exf2 is sent back to the calling
% program.
% This example demonstrates that the names of the arguments
% in the calling program need not be the same as those in
% the function. It is the order list in the calling program
% that needs to be the same as the order list in the function.
% In this example, the next term is obtained from the
% previous term by multiplying the previous term by x and
% dividing by index n. The program does its own summing of
% terms.
% e^x=1+x+x^2/2!+x^3/3!+x^4/4!+···
function exy=exf2(x,y)
s=1.0; term=1.0;
for n=1:100
        term=term*x/n;
        s=s+term;
        if abs(term)<=s*1.0e-6
            break;
        end
end
exy=s/y;
fprintf('printout of x,y and exy from exf2 \n');
fprintf('x=%5.2f    y=%2.0f    exy=%10.5f \n', x, y, exy);
% Compare the printout of x,y and exy from function exf2
% with the printout of w,z and q from the calling program,
% Example_3_6.m. Observe that (w,z) = (x,y) and q = exy.
```

Program results

```
printout of x,y and exy from exf2
x=0.20    y=5    exy=0.24428
printout of w,z and q from calling program
w=0.20    z=5    q=0.24428
printout of x,y and exy from exf2
x=0.40    y=5    exy=0.29836
printout of w,z and q from calling program
w=0.40    z=5    q=0.29836
  ⋮        ⋮        ⋮
```

Review 3.1

1. What is the series expansion for e^x?
2. What are two different approaches for evaluating the terms in the series for e^x?
3. When does it seem appropriate to write a self-written function?
4. A self-written function usually has both an input and an output. Where does the input come from? Where does the output go to?
5. If a self-written function has more than one output, how must the output be presented?
6. How does a self-written function communicate with the calling program?
7. What can be said about variables within a self-written function that are not in the input or output arguments of the function?
8. Do the variable names in the input and output arguments between the calling program and the function have to be the same?

3.4 Anonymous Functions

Sometimes, it is more convenient to define a function inside your script rather than in a separate file. For example, if a function is brief (perhaps a single line) and unlikely to be used in other scripts, then the *anonymous* form of a function can be used. This will save you from having to create another *.m* file. An example is

$$fh = @(x,y)(y*sin(x)+x*cos(y));$$

MATLAB defines the @ sign as a *function handle*. A function handle is equivalent to accessing a function by calling its function name, which in this case is fh. The (x,y) defines the input arguments to the function, and (y*sin(x) + x*(cos(y)) is the function. Anonymous functions may be used in a script or in the Command window.

Example: In the Command window, type in the following two lines:

```
>> fh=@(x,y)(y*sin(x)+x*cos(y));
>> w=fh(pi,2*pi)
     w =
         3.1416
```

Additional information on anonymous functions can be obtained by typing **help function _ handle** in the Command window.

The following example uses an anonymous function and is a modification of Example 2.12. The script loads the data file *atm_properties.txt* (shown at the end of the script). The first column is the altitude, the second column is the temperature, the third column is the pressure, and the fourth column is density. Note: The data file should *only* contain numbers. The script is interactive; it asks the user to enter an altitude from the keyboard.

Example 3.7

```
% Example_3_7.m
% This program interpolates for atmospheric properties T,p and rho
% at an altitudes entered from the keyboard.
% Atmospheric table data is loaded into the program from the
% data file atm_properties. The data table includes properties
% every 1000 m from z=0 to z=5000 m. Properties of
% temperature, pressure and density at the specified altitudes
% are determined by interpolation and printed to the screen.
% An anonymous function (avoids creating an extra .m file)is
% used to do the interpolation:
clear; clc;
% anonymous function
yf = @(z,z1,z2,y1,y2) (y1+(z-z1)*(y2-y1)/(z2-z1));
load 'atm_properties.txt'
zt=atm_properties(:,1);
Tt=atm_properties(:,2);
pt=atm_properties(:,3);
rhot=atm_properties(:,4);
fprintf('Do you wish to determine atmospheric \n');
fprintf('properties at a particular altitude \n');
char=input('Enter Y for yes or any other letter key for no \n','s');
% None of the commands within the while loop will be executed
% unless the user enters Y from the key board.
while char=='Y'
    fprintf('Enter the altitude at which atmospheric properties ');
    fprintf('are to be determined. ');
    z=input('Altitude range is from 0 to 5000 m \n');
    if z>5000
        fprintf('You entered an altitude > 5000 m.\n');
        fprintf('Restart the program.\n');
        break;
    end
    for i=1:length(zt)-1
        if z >= zt(i) && z < zt(i+1)
            z1=zt(i); z2=zt(i+1); T1=Tt(i); T2=Tt(i+1);
            T=yf(z,z1,z2,T1,T2);
            p1=pt(i); p2=pt(i+1); p=yf(z,z1,z2,p1,p2);
            rho1=rhot(i); rho2=rhot(i+1);
            rho=yf(z,z1,z2,rho1,rho2);
            fprintf('z=%.1f(m) T=%.2f(C) p=%.5e(Pa) ',z,T,p);
            fprintf('rho=%12.5e(kg/m^3) \n',rho);
        end
    end
    fprintf('\n Do you wish to enter another altitude.');
    fprintf('Enter Y for yes or');
    char=input(any other letter key for no \n','s');
end
fprintf('The program will now close. Have a good day \n');
```

The following *atm_properties.txt* file needs to be created before running Example 3.7.

0	288.15	1.0133e+005	1.2252
1000	281.65	8.9869e+004	1.1118
2000	275.15	7.9485e+004	1.0065
3000	268.65	7.0095e+004	0.9091
4000	262.15	6.1624e+004	0.8191
5000	255.65	5.4002e+004	0.7360

Program results

```
Enter the altitude at which atmospheric
properties are to be determined.
Altitude range is from 0 to 5000 m
2340
z =2340.0(m)  T=275.15(C)
p=7.94850e+04(Pa)  rho=1.00650(kg/m^3)
Do you wish to enter another altitude
enter Y for yes or N for no
Y
Enter the altitude at which atmospheric
properties are to be determined.
Altitude range is from 0 to 5000 m
4580
z=4580.0(m)  T=262.15(C)
p=6.16240e+04(Pa)  rho=0.81910(kg/m^3)
Enter the altitude at which atmospheric
properties are to be determined.
Altitude range is from 0 to 5000 m
N
>>
```

3.5 MATLAB®'s `interp1` Function

MATLAB has a function named `interp1` that performs interpolation.
The syntax for `interp1` is

```
Yi = interp1(X,Y,Xi)
```

where

X and Y are a set of known (x, y) data points

Xi is the set of x values at which the set of y values, Yi, are to be determined by linear interpolation

Arrays X and Y must be of the same length. Note: If Xi is a vector, then Yi will also be a vector. The function `interp1` can also be used for interpolation methods other than linear interpolation, and this is covered in Chapter 9 on Curve Fitting. The next example demonstrates the use of `interp1` function for interpolating for internal energy of refrigerant R134a [1] at temperatures specified in vector T2.

Example 3.8

```
% Example_3_8.m
% This program uses MATLAB's function interp1 to interpolate
% for the internal energy, u, as a function of temperature,
% T, of refrigerant 134a at a pressure of 0.6 bar.
% Measured values of the internal energy (kJ/kg) vs.
% temperature (C) are specified in vectors ut and Tt
% respectively.
% The temperatures at which the internal energy is to be
% determined are specified in vector T2.
% The program also creates a plot of u vs. T and includes
% points of u at temperature T2.
clear; clc;
Tt = -20:10:90;
ut = [217.86 224.97 232.24 239.69 247.32 255.12 263.10 ...
      271.25 279.58 288.08 296.75 305.58];
fprintf('This program interpolates for the internal energy, \n');
fprintf('u at a specified temperature T. \n');
fprintf('The allowable temperature range is');
fprintf('-20 to +90 C \n\n');
T2 = [-12 6 24 32 64 82];
u = interp1(Tt,ut,T2);
fprintf(' T2(K)     u(kJ/kg)      \n');
fprintf('----------------------\n');
for i = 1:6
    fprintf('%6.1f     %8.3f \n',T2(i),u(i));
end
plot(Tt,ut,T2,u,'o');
xlabel('T'), ylabel('u'), title('u vs. T'), grid;
Legend ('u vs. T', 'u2 vs. T2');
```

Program results

```
This program interpolates for the internal energy,
u at a specified temperature T.
The allowable temperature range is -20 to +90 C.

  T2(K)    u(kJ/kg)
------------------
  -12.0    223.548
    6.0    236.710
   24.0    250.440
   32.0    256.716
   64.0    282.980
   82.0    298.516
>>
```

See Figure 3.2.

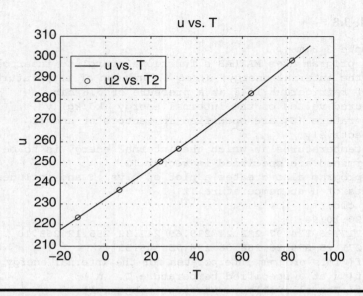

Figure 3.2 Plot of *u* vs. *T*.

Exercises

E3.5 Modify Example 3.7 by removing the anonymous function `yf` and doing the interpolation by MATLAB's `interp1` function.

E3.6 The mobility of electrons, μ, in silicon is dependent on temperature, T, and is often presented as a table of laboratory measurements for a few distinct temperatures [2]. If we wish to determine the electron mobility at a temperature that is not in the table, we need to interpolate. Table values for electron mobility, Mu (cm^2/V-s), at temperatures, T (K), are specified in the vectors Mut and Tt, respectively:

```
Tt = [100 150 200 250 300 350 400 450];
Mut = [19650 7427 3723 2180 1407 972 705.5 531.8];
```

We wish to determine Mu at temperatures T2 where

$$T2 = [130\ 425\ 280\ 335]$$

Create a MATLAB program that evaluates Mu at temperatures T2 using MATLAB's `interp1` function.
a. Print the results to a file in table format.
b. Create a plot of Mu vs. T

Review 3.2

1. If a programmer wishes to write a self-written function, but does not wish to create an additional *.m* file, what can the programmer do and what is the constraint?
2. What is the name of MATLAB's function that does interpolation?
3. What are the inputs to MATLAB's interpolation function?
4. What are the outputs from MATLAB's interpolation function?

3.6 Working with Characters and Strings

In this section, we demonstrate several programs that use characters and strings. Usually characters and strings need to be enclosed with single quotation marks.

Example 3.9

```
% Example_3_9.m
% Sometimes you might wish to print out a string of
% characters from a loop. This can be done by declaring a
% 2-D character string matrix as shown in this example.
% Note that all row character strings must have the same
% number of columns and that character strings must be
% enclosed by single quotation marks.
clear; clc;
% Assign a string matrix.
% Each row in the matrix must have the same number of
% columns.
parts = ['Internal modem  '
         'Graphics adapter'
         'CD drive        '
         'DVD drive       '
         'Floppy drive    '
         'Hard disk drive '];
for i = 1:6
   fprintf('%16s \n',parts(i,1:16));
end
```

Program results

```
Internal modem
Graphics adapter
CD drive
DVD drive
Floppy drive
Hard disk drive
>>
```

Example 3.10

This example is an interactive program. The user has to input whether to print the string matrix to the screen or to a file.

```
% Example_3_10.m
% This example is a modification of Example 3.9. The program
% asks the user if he/she wishes to have the output go to the
% screen or to a file.
% This example also illustrates the use of the switch statement.
clear; clc;
char = ['Internal modem          '
        'External modem          '
        'Graphics circuit board'
        'CD drive                '
        'Hard disk drive        '];
fprintf('Choose whether to send the output to the\n');
fprintf('screen or to a file named ''output.txt''. \n\n');
% Note that in order to get MATLAB to permit quotation marks
% surrounding output.txt, we had to use double quotation marks.
fprintf('Enter ''s'' for screen or ''f'' for ');
var=input('file (without quotes)\n', 's');
switch(var)
    case 's'
        for i = 1:5
            fprintf('%22s \n',char(i,1:22));
        end
    case 'f'
        fo = fopen('output.txt','w');
        for i = 1:5
            fprintf(fo,'%22s \n',char(i,1:22));
        end
        fclose(fo);
    otherwise
        fprintf('you did not enter an s or a f, try again \n');
        exit;
end
```

Program results

```
Choose whether to send the output to the
screen or to a file named 'output.txt'.

Enter 's' for screen or 'f' for file (without quotes)
s
Internal modem
External modem
Graphics circuit board
CD drive
Hard disk drive
>>
```

Example 3.11

When a program requires the reading in of a data file that contains both character strings and numbers, you need to use data containers called cells, where each cell can contain any type of data. To access the contents of a cell, use curly braces, { }. Also, when reading in items in the data file, it is best to use the textscan function, which will place the data into cells. For example, suppose we have a data file as follows named inv2.

```
10    Internal modem        36.50
11    Graphics adapter      74.20
12    CD drive             125.00
13    DVD drive            175.80
```

First, we would need to open the file for reading with the following command:

```
fo = fopen('inv2','r')
```

Then we can use the textscan command to place the formatted data into cells. The syntax for the textscan command follows:

```
C = textscan(fo,'format',N).
```

The computer will read N rows of the data in inv2.

```
% Example_3_11.m
% This program demonstrates the use of the textscan command
% which places formatted data into cells. Note that in the
% textscan command, you need to use %16c for text info while
% in the print command you need to use %16s for text info.
clear; clc;
fo=fopen('inv2.txt','r');
C=textscan(fo,'%d    %16c  %f',4);
Ln=C{1}; parts=C{2}; cost=C{3};
for i=1:4
    fprintf('%d    %16s  %8.2f \n',Ln(i),parts(i,1:16),cos t(i));
end
```

Program results

```
10    Internal modem        36.50
11    Graphics adapter      74.20
12    CD drive             125.00
13    DVD drive            175.80
>>
```

Example 3.12

In this interactive example, a `for` loop is used to establish the interval containing the grade.

```
% Example_3_12.m
% The program determines a letter grade depending on the
% score the user enters from the keyboard.
% This version uses a loop to determine the correct interval of
```

```
% interest. For a large number of intervals, this method is
% more efficient (fewer statements) than the method in
% Example 3.11
clear; clc;
gradearray = ['A'; 'B'; 'C'; 'D'; 'F'];
sarray = [100 90 80 70 60 0];
score = input('Enter your test score: ');
% The following 2 statements are needed for the case when
% score = 100.
if score == 100
   grade = gradearray(1);
else
   for i = 1:5
       if (score >= sarray(i+1) && score < sarray(i))
           grade = gradearray(i);
           break;
       end
   end
end
fprintf('score is: %i, grade is: %c \n', score, grade);
```

Program results

```
Enter your test score: 82
score is: 82, grade is: B
```

The names used in the following examples do not represent real people.

Example 3.13

This example combines the use of a string matrix and the establishment of a grade.

```
% Example_3_13.m
% This program determines the letter grades of several students.
% Student's names and their test scores are entered in the
% program. This example uses nested 'for' loops and an 'if'
% statement to determine the correct letter grade for each
% student.
clear; clc;
gradearray = ['A'; 'B'; 'C'; 'D'; 'F'];
sarray = [100 90 80 70 60 0];
Lname = [ 'Smith        '
          'Lambert      '
          'Kurtz        '
          'Jones        '
          'Hutchinson   '
          'Diaz         ']; 
Fname = [ 'Joe          '
          'Jane         '
          'Howard       '
          'Mary         '
```

```
                  'Peter        '
                  'Carlos       '];
score = [84;  86;  67;  92;  81;  75];
% The score = 100 is treated separately.
for j = 1:6
   if score(j) == 100
       grade(j) = gradearray(1);
   else
       for i = 1:5
           if (score(j) >= sarray(i+1) && score(j) < sarray(i))
               grade(j) = gradearray(i);
           end
       end
   end
end
fprintf('last name    first name   grade \n');
fprintf('--------------------------------------\n');
for j = 1:6
    fprintf ('%11s  %11s  %c \n', ...
        Lname(j,1:11),Fname(j,1:11),grade(j));
end
```

Program results

```
last name        first name     grade
---------------------------------

Smith            Joe            B
Lambert          Jane           B
Kurtz            Howard         D
Jones            Mary           A
Hutchinson       Peter          B
Diaz             Carlos         C
>>
```

Example 3.14

This example involves a self-written function and string arrays.

An important building block in program development is the self-written function. In the following text, we construct the function `func _ grade` to determine the grade in the previous example. The input to `func _ grade` is the score and the output is the corresponding grade. Remember that a user-defined function is saved in a separate *.m* file.

```
% Example_3_14.m
% This program uses func_grade.m (defined below) to
% determine the grade of a student. The input to the
% function is score and the function returns the grade to
% the calling program.
clear; clc;
```

```
Lname = ['Smith        '
         'Lambert      '
         'Kurtz        '
         'Jones        '
         'Hutchinson   '
         'Diaz         '];
Fname = ['Joe          '
         'Jane         '
         'Howard       '
         'Mary         '
         'Peter        '
         'Carlos       '];
% The scores of the students named above are given in the
% vector score in the order listed in Lname.
score = [84; 72; 93; 64; 81; 75];
fprintf('last name    first name    grade \n');
fprintf('------------------------------------\n');
for j = 1:6
    grade(j) = func_grade(score(j));
    fprintf('%11s     %11s      %c    \n',...
            Lname(j,1:11),Fname(j,1:11),grade(j));
end
```

```
% func_grade.m
% This function works with Example_3_14.m
function grade = func_grade(score)
gradearray = ['A'; 'B'; 'C'; 'D'; 'F'];
sarray = [100 90 80 70 60 0];
if score == 100
   grade = gradearray(1);
else
   for i = 1:5
       if (score >= sarray(i+1) && score < sarray(i))
           grade = gradearray(i);
       end
   end
end
```

Program results

```
last name   first name   grade
----------------------------

Smith       Joe          B
Lambert     Jane         B
Kurtz       Howard       D
Jones       Mary         A
Hutchinson  Peter        B
Diaz        Carlos       C
>>
```

Review 3.3

1. Suppose you wish to define a matrix consisting of string elements. What are the rules that need to be followed in setting up this matrix?
2. Suppose our independent variable is x, and the x domain is subdivided into small intervals and we wish to determine which interval contains an item of interest. What is the most efficient way to determine the interval that contains our item of interest?

Projects

P3.1 Develop a computer program in MATLAB that will evaluate the following functions for $-0.9 \leq x \leq 0.9$ in steps of 0.1 by
  a. An arithmetic statement.
  b. A series allowing for as many as 50 terms. However, stop adding terms when the last term only affects the 6th decimal place in the answer.
  c. Print to a file a table similar to Table P3.1 with $f(x)$ to 6 decimal places.

The functions and their series expansions are

1. $f(x) = (1+x)^{1/2} = 1 + \dfrac{1}{2}x - \dfrac{1 \cdot 1}{2 \cdot 4}x^2 + \dfrac{1 \cdot 1 \cdot 3}{2 \cdot 4 \cdot 6}x^3 - \dfrac{1 \cdot 1 \cdot 3 \cdot 5}{2 \cdot 4 \cdot 6 \cdot 8}x^4 + \cdots$

2. $f(x) = (1+x)^{-1/2} = 1 - \dfrac{1}{2}x + \dfrac{1 \cdot 3}{2 \cdot 4}x^2 - \dfrac{1 \cdot 3 \cdot 5}{2 \cdot 4 \cdot 6}x^3 + \dfrac{1 \cdot 3 \cdot 5 \cdot 7}{2 \cdot 4 \cdot 6 \cdot 8}x^4 - \cdots$

Table P3.1   Template for Output Table

x	f(x) (By Arithmetic Statement)	f(x) (By Series)	# of Terms Used in the Series
−0.9	—	—	—
−0.8	—	—	—
−0.7	—	—	—
⋮	⋮	⋮	⋮
0.7	—	—	—
0.8	—	—	—
0.9	—	—	—

3. $f(x) = (1+x^2)^{-1/2} = 1 - \dfrac{1}{2}x^2 + \dfrac{1\cdot3}{2\cdot4}x^4 - \dfrac{1\cdot3\cdot5}{2\cdot4\cdot6}x^6 + \dfrac{1\cdot3\cdot5\cdot7}{2\cdot4\cdot6\cdot8}x^8 - \cdots$

4. $f(x) = (1+x)^{1/3} = 1 + \dfrac{1}{3}x - \dfrac{1\cdot2}{3\cdot6}x^2 + \dfrac{1\cdot2\cdot5}{3\cdot6\cdot9}x^3 - \dfrac{1\cdot2\cdot5\cdot8}{3\cdot6\cdot9\cdot12}x^4 + \cdots$

Hint: $\text{term}(k+1) = \text{term}(k) \times \left(\dfrac{1}{3} - (k-1)\right) \times \dfrac{x}{k}$, for $k = 1,2,3,\ldots$

P3.2 The binomial expansion for $(1 + x)^n$, where n is an integer, is as follows:

$$(1+x)^n = 1 + nx + \frac{n(n-1)x^2}{2!} + \frac{n(n-1)(n-2)x^3}{3!} + \cdots$$

$$+ \frac{n(n-1)(n-2)\cdots(n-r+2)x^{r-1}}{(r-1)!} + \cdots + x^n \tag{P3.2}$$

Construct a MATLAB program that will evaluate $(1 + x)^n$ by both the series and by an arithmetic statement for $n = 10$ and $1.0 \le x \le 10.0$ in steps of 0.5. Print out the results in a table as shown in Table P3.2.

P3.3 This project is a modification of Project P2.2. Instead of making the program interactive, enter the following altitudes, z2, at which the atmospheric properties of (T, p, ρ) are to be determined by linear interpolation using MATLAB's interp1 function:

```
z2 = [23500 8570 14320 34865 78790 56820 64780]
```

Print the results to a file in a table format.

Table P3.2   Template for Output Table

x	$(1 + x)^n$ (by Arithmetic Statement)	$(1 + x)^n$ (by Series)
1.0	—	—
1.5	—	—
2.0	—	—
2.5	—	—
⋮	⋮	⋮
10.0	—	—

P3.4 The properties of specific volume, v, and pressure, p, as a function of temperature, T, for carbon dioxide based on the Redlich–Kwong equation of state are given in Table P3.4.

Write a MATLAB program that will do the following:
a. Construct three separate vectors containing the carbon dioxide properties of T, v, and p.
b. Print Table P3.4 to the screen (with table headings).
c. Determine v and p at temperatures T2 using MATLAB's `interp1` function where

$$T2 = [367 \ 634 \ 420 \ 587 \ 742]$$

d. Print to the screen in a table format (with table headings) values of v and p at temperatures T2.

P3.5 Mathematician Joseph Fourier is credited with the theorem that any periodic waveform may be expressed as a summation of pure sines and cosines. For example, the square wave of Figure P3.5a can be written as a sum of sines:

$$v(t) = \frac{4}{\pi} \sin\frac{2\pi t}{T} + \frac{4}{3\pi} \sin\frac{6\pi t}{T} + \frac{4}{5\pi} \sin\frac{10\pi t}{T} + \cdots$$

$$= \sum_{\substack{k=1 \\ k \text{ odd}}}^{\infty} \frac{4}{\pi k} \sin\frac{2\pi k t}{T}$$

Table P3.4   Carbon Dioxide Properties

T (K)	v (m³/kmol)	p (bar)
350	0.28	7.65
400	0.32	8.57
450	0.36	9.16
500	0.40	9.55
550	0.44	9.81
600	0.48	10.00
650	0.52	10.14
700	0.56	10.24
750	0.60	10.31

Note: 1 bar = 10^5 N/m².

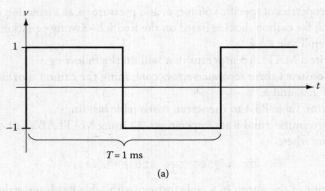

(a)

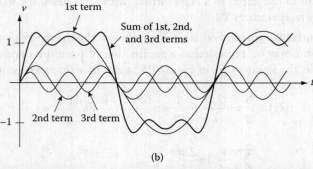

(b)

Figure P3.5 **(a) Square waveform. (b) Three-term Fourier series of square waveform.**

Figure P3.5b shows the first three terms of the series and their summation.

a. Write a MATLAB script that implements the self-written function sqwave(n,T,i) that takes the following arguments:

n is the number of terms of the Fourier series.

T is the period of the square wave in seconds.

i is the number samples per period.

The function should return two arrays, t and v, each containing i elements, where

    t = an array of i time points

    v = an array of i computed values of the nth-degree approximated square wave

b. Run your sqwave(n,T,i) function and plot the results for the following arguments:

$$T = 1 \text{ ms}, \quad i = 1001, \quad n = 1, 3, 10, 100$$

c. For $n = 100$, create a plot of v vs. t.

P3.6 Bob's Hardware Store wishes to create an online program to sell items in its inventory. You are to create an interactive MATLAB program for this purpose. The program is to contain the following items:

Data file
a. A data file that contains a description of the inventory items for sale, their cost, and the quantity available for sale. The list should contain at least 10 items. This data file will be read into the main script. Whenever an item is purchased, the inventory available for sale should be updated in the data file.

Main script
b. The script is to print to the screen the items for sale, their cost, and the quantity available for purchase.
c. The script is to ask the user if he or she wishes to make a purchase. If yes, the script is to ask the user the line number of the item he or she wishes to purchase and the quantity.
d. If the user does not wish to make a purchase, exit the program. If the user does not respond correctly to the query, print an error message to the screen and inform the user to restart the program. Then exit the program.
e. If the user does make a purchase, the program is to continue asking the user if he or she wishes to make another purchase. If yes, print to the screen the list of items, their cost, and the updated items available for purchase. Then, the script is to ask the user the line number of the item he or she wishes to purchase and the quantity. If the answer is no, exit the program.

Billing function
f. When the user is finished making all his or her purchases, the script is to call a function that will print out a bill for the items purchased. The bill should contain the name and address of the store, the user's first and last names, the user's address, bill headings, a list of all the items purchased and their unit prices, the total price for each item purchased, and, finally, the total price of all the items purchased.

P3.7 Create a MATLAB script that does the following:
1. Create a function named atmf with input argument z and output arguments T, p, ρ, where z is the altitude, T is the temperature, p is the pressure and ρ is the density.
2. Create the *atm_properties.txt* file used in Example 3.7 (column 1 is z, column 2 is T, column 3 is p, and column 4 is ρ).
3. The function atmf is to load the *atm_properties.txt*.
4. The function atmf is to use MATLAB's function `interp1` to interpolate for T, p, and ρ at the z values specified in the calling program.

5. The calling program is to define vector z = [635 1376 3642 4783].
6. The calling program is to open a file for printing and print table headings into that file.
7. The calling program is to create a loop that calls the function atmf at each of the altitudes listed in vector z.
8. The calling program then prints to the output file values of z, T, p, and ρ at the altitudes defined in vector z.

References

1. Moran, M.J., Shapiro, H.N., Boettner, D.D., and Bailey, M.B., *Fundamentals of Engineering Thermodynamics*, John Wiley & sons, Hoboken, NJ, 2011.
2. Sze, S.M., *Physics of Semiconductor Devices*, 2nd ed., Wiley, New York, 1981.

Chapter 4

Matrices

4.1 Introduction

In engineering, we frequently deal with matrices or problems involving a set of linear equations. In addition, the basic element in MATLAB® is a matrix and it is therefore appropriate to have a chapter on matrices. First, we discuss some basic matrix concepts and operations and the treatment of matrices in MATLAB, including several sample MATLAB programs involving matrices. This is followed by a discussion of MATLAB's `inv` function for solving a system of linear equations. Problems in statics and in resistive circuits, which involve a system of linear equations, are covered. Next, the Gauss elimination and the Gauss–Jordan methods are described, including examples on these methods. A method for obtaining the inverse of a matrix is also discussed. These latter three topics do not involve MATLAB and may be skipped without affecting the remaining chapters in the book. Finally, the matrix eigenvalue problem is discussed, including applications in vibration and resonant circuit problems.

4.2 Matrix Operations

- A rectangular array of numbers of the form shown as follows is called a matrix:

$$\mathbf{A} = \begin{bmatrix} a_{11} & a_{12} & \cdots & a_{1n} \\ a_{21} & a_{22} & \cdots & a_{2n} \\ \vdots & \vdots & \ddots & \vdots \\ a_{m1} & a_{m2} & \cdots & a_{mn} \end{bmatrix}$$

The numbers a_{ij} in the array are called the elements of the $m \times n$ matrix $\mathbf{A}$.

- A matrix of m rows and one column is called a column vector.
- A matrix of one row and n columns is called a row vector.
- Matrices obey certain rules of addition, subtraction, and multiplication.
- Addition and subtraction: If matrices **A** and **B** have the same number of rows and columns, then

$$\mathbf{C} = \mathbf{A} + \mathbf{B} = \begin{bmatrix} (a_{11}+b_{11}) & (a_{12}+b_{12}) & \cdots & (a_{1n}+b_{1n}) \\ (a_{21}+b_{21}) & (a_{22}+b_{22}) & \cdots & (a_{2n}+b_{2n}) \\ \vdots & \vdots & \ddots & \vdots \\ (a_{m1}+b_{m1}) & (a_{m2}+b_{m2}) & \cdots & (a_{mn}+b_{mn}) \end{bmatrix}$$

$$\mathbf{C} = \mathbf{A} - \mathbf{B} = \begin{bmatrix} (a_{11}-b_{11}) & (a_{12}-b_{12}) & \cdots & (a_{1n}-b_{1n}) \\ (a_{21}-b_{21}) & (a_{22}-b_{22}) & \cdots & (a_{2n}-b_{2n}) \\ \vdots & \vdots & \ddots & \vdots \\ (a_{m1}-b_{m1}) & (a_{m2}-b_{m2}) & \cdots & (a_{mn}-b_{mn}) \end{bmatrix}$$

- Addition and subtraction of the matrices **A** and **B** are only defined if **A** and **B** have the same number of rows and columns.
- Multiplication: The product **AB** is only defined if the number of columns in **A** equals the number of rows in **B**. If $\mathbf{C} = \mathbf{AB}$, where

$$\mathbf{A} = \begin{bmatrix} a_{11} & a_{12} & a_{13} \\ a_{21} & a_{22} & a_{23} \end{bmatrix} \quad \text{and} \quad \mathbf{B} = \begin{bmatrix} b_{11} & b_{12} \\ b_{21} & b_{22} \\ b_{31} & b_{32} \end{bmatrix}$$

then

$$\mathbf{C} = \begin{bmatrix} (a_{11}b_{11}+a_{12}b_{21}+a_{13}b_{31}) & (a_{11}b_{12}+a_{12}b_{22}+a_{13}b_{32}) \\ (a_{21}b_{11}+a_{22}b_{21}+a_{23}b_{31}) & (a_{21}b_{12}+a_{22}b_{22}+a_{23}b_{32}) \end{bmatrix}$$

- If **A** has M rows and **B** has K columns, then $\mathbf{C} = \mathbf{AB}$ will have M rows and K columns. A general expression for the element c_{ij} is

$$c_{ij} = \sum_{k=1}^{K} a_{ik}b_{kj}$$

In MATLAB, the multiplication of matrices **A** and **B** is entered as **A*B**.

■ The transpose of a matrix **A** is an operation in which rows become columns and columns become rows. In matrix algebra, the transpose of matrix **A** is written as $\mathbf{A}^{\mathrm{T}}$. If

$$\mathbf{A} = \begin{bmatrix} 2 & 5 & 1 \\ 7 & 3 & 8 \\ 4 & 5 & 21 \\ 16 & 3 & 10 \end{bmatrix}$$

then

$$\mathbf{A}^{\mathrm{T}} = \begin{bmatrix} 2 & 7 & 4 & 16 \\ 5 & 3 & 5 & 3 \\ 1 & 8 & 21 & 10 \end{bmatrix}$$

In MATLAB, the transpose of matrix **A** is entered as **A'**.
■ Summing the elements of a vector or the columns of a matrix:
If $\mathbf{A} = \begin{bmatrix} a_1 & a_2 & a_3 \end{bmatrix}$, then sum(**A**) = $a_1 + a_2 + a_3$.
   If

$$\mathbf{B} = \begin{bmatrix} b_{11} & b_{12} & b_{13} \\ b_{21} & b_{22} & b_{23} \\ b_{31} & b_{32} & b_{33} \end{bmatrix}$$

then

$$\text{sum}(\mathbf{B}) = \begin{bmatrix} (b_{11} + b_{21} + b_{31}) & (b_{12} + b_{22} + b_{32}) & (b_{13} + b_{23} + b_{33}) \end{bmatrix}$$

■ Dot product: In vector notation, the dot product of two vectors is defined as

$$\mathbf{A} \cdot \mathbf{B} = \sum a_i b_i$$

For example, if $\mathbf{A} = \begin{bmatrix} 4 & -1 & 3 \end{bmatrix}$ and $\mathbf{B} = \begin{bmatrix} -2 & 5 & 2 \end{bmatrix}$, then

$$\mathbf{A} \cdot \mathbf{B} = (4)(-2) + (-1)(5) + (3)(2) = -7$$

The result of a dot product is always a scalar (i.e., a single number).
   In MATLAB, the dot product, $\mathbf{A} \cdot \mathbf{B}$, is entered as **dot(A,B)**.

■ Identity matrix: The identity matrix, **I**, is a matrix where the main diagonal elements are all one and all other elements are zero.

$$IA = AI = A$$

For example, for the case of a three-by-three matrix **A**,

$$AI = \begin{bmatrix} a_{11} & a_{12} & a_{13} \\ a_{21} & a_{22} & a_{23} \\ a_{31} & a_{32} & a_{33} \end{bmatrix} \begin{bmatrix} 1 & 0 & 0 \\ 0 & 1 & 0 \\ 0 & 0 & 1 \end{bmatrix}$$

$$= \begin{bmatrix} (a_{11} + 0 + 0) & (0 + a_{12} + 0) & (0 + 0 + a_{13}) \\ (a_{21} + 0 + 0) & (0 + a_{22} + 0) & (0 + 0 + a_{23}) \\ (a_{31} + 0 + 0) & (0 + a_{32} + 0) & (0 + 0 + a_{33}) \end{bmatrix}$$

In MATLAB, you can generate an $n \times n$ identity matrix with `eye(n)`.
■ Inverse of a matrix: The inverse of matrix **A**, denoted $\mathbf{A}^{-1}$, is a matrix such that

$$\mathbf{A}^{-1}\mathbf{A} = \mathbf{A}\mathbf{A}^{-1} = \mathbf{I} = \begin{bmatrix} 1 & 0 & 0 \\ 0 & 1 & 0 \\ 0 & 0 & 1 \end{bmatrix} \quad \text{(for a } 3 \times 3 \text{ matrix)}$$

In MATLAB, the matrix inverse $\mathbf{A}^{-1}$ is entered as `inv(A)`.
■ In MATLAB, the determinant of the matrix **A** is entered as `det(A)`.
Before the use of computers, determinants were developed as a method for obtaining a solution to a system of linear equations. Computationally, it is only practical for a system involving just a few equations [1]. However, determinants do have an application in the eigenvalue problem that is discussed at the end of this chapter.

The determinant of a 2×2 matrix is

$$D = \begin{vmatrix} a_{11} & a_{12} \\ a_{21} & a_{22} \end{vmatrix} = a_{11}a_{22} - a_{12}a_{21}$$

Note that the determinant of a matrix is written with two straight lines that enclose the matrix elements.

The determinant of a 3 × 3 matrix is

$$D = \begin{vmatrix} a_{11} & a_{12} & a_{13} \\ a_{21} & a_{22} & a_{23} \\ a_{31} & a_{32} & a_{33} \end{vmatrix} = a_{11}\begin{vmatrix} a_{22} & a_{23} \\ a_{32} & a_{33} \end{vmatrix} - a_{12}\begin{vmatrix} a_{21} & a_{23} \\ a_{31} & a_{33} \end{vmatrix} + a_{13}\begin{vmatrix} a_{21} & a_{22} \\ a_{31} & a_{32} \end{vmatrix}$$

■ You can obtain the size of matrix **A** by the command **size(A)**. This command is useful when you run a script and you get an error message like "Index exceeds matrix dimensions." Entering the size() command in the script will help you determine the problem.

■ Element-by-element operations: Given two vectors of the same dimensions, we can perform element-by-element multiplication and division in MATLAB with the .* and ./ operators.

Given $\mathbf{A} = \begin{bmatrix} a_1 & a_2 & a_3 \end{bmatrix}$ and $\mathbf{B} = \begin{bmatrix} b_1 & b_2 & b_3 \end{bmatrix}$, then

$$\mathbf{C} = \mathbf{A}.*\mathbf{B} = \begin{bmatrix} a_1 b_1 & a_2 b_2 & a_3 b_3 \end{bmatrix}$$

$$\mathbf{D} = \mathbf{A}./\mathbf{B} = \begin{bmatrix} \dfrac{a_1}{b_1} & \dfrac{a_2}{b_2} & \dfrac{a_3}{b_3} \end{bmatrix}$$

We see that the element-by-element operation results in a vector that is the same dimension as the vectors that are involved in the operation. Also, note the dot product can be expressed as a combination of an element-by-element multiplication and a sum:

$$\mathbf{A} \cdot \mathbf{B} = \text{sum}(\mathbf{A}.*\mathbf{B}) = \text{sum}\left(\begin{bmatrix} a_1 b_1 & a_2 b_2 & a_3 b_3 \end{bmatrix}\right) = a_1 b_1 + a_2 b_2 + a_3 b_3$$

■ Product of functions of two vectors: As described in Chapter 2, MATLAB allows taking a function of a vector and that the result is also a vector. If a script involves a mathematical operation of two vector functions (such as a product of two vector functions), then the operation will require an element-by-element operation. In Example 4.2 that follows, we compute the product of two vector functions, both directly and indirectly by using a for loop and multiplying the elements of each vector.

Example 4.1

```
% Example_4_1.m
% This program demonstrates matrix algebra in MATLAB
clear; clc;
a = [1 5 9]
b = [2 6 12]
c = a+b
d = dot(a,b)
e = a.*b
f = a./b
g = sum(a.*b)
h = a*b'
```

Program results

```
a =
     1     5     9
b =
     2     6    12
c =
     3    11    21
d =
   140
e =
     2    30   108
f =
    0.5000    0.8333    0.7500
g =
   140
h =
   140
>>
```

Example 4.2

```
% Example_4_2.m
% This example illustrates element-by element operation
% of vector functions
clear; clc;
x = 0:30:180;
% y1 is the product of two vector functions
y1 = sind(x) .* cosd(x);
fprintf('    x        y1        y2\n');
fprintf('--------------------------\n');
for n = 1:length(x);
    % y2(n) is the product of the elements of the two functions.
    y2(n) = sind(x(n)) * cosd(x(n));
    fprintf('%5.1f    %8.5f    %8.5f \n',x(n),y1(n),y2(n));
end
```

Program results

x	y1	y2
0.0	0.00000	0.00000
30.0	0.43301	0.43301
60.0	0.43301	0.43301
90.0	0.00000	0.00000
120.0	−0.43301	−0.43301
150.0	−0.43301	−0.43301
180.0	−0.00000	−0.00000

```
>>
```

We see that the two different methods for computing y1 and y2 give the same answer.

Review 4.1

1. If matrix $C = A + B$, what must be true about matrices A and B?
2. If $C = AB$, what must be true about matrices A and B?
3. What command in MATLAB is used to obtain the inverse of a matrix?
4. What symbol is used in MATLAB to transpose a matrix?
5. Is the dot product of two vectors a scalar or a vector?
6. Is the element-by-element multiplication of two vectors a scalar or a vector?
7. Does the use of MATLAB's sum command on a vector produce a scalar or a vector?
8. Does the use of MATLAB's sum command on a matrix of two or more columns produce a scalar or a vector?

4.3 System of Linear Equations

The set of equations

$$a_{11}x_1 + a_{12}x_2 + a_{13}x_3 + \cdots + a_{1n}x_n = c_1$$

$$a_{21}x_1 + a_{22}x_2 + a_{23}x_3 + \cdots + a_{2n}x_n = c_2$$

$$\vdots$$

$$a_{n1}x_1 + a_{n2}x_2 + a_{n3}x_3 + \cdots + a_{nn}x_n = c_n$$

(4.1)

can be represented by the matrix equation

$$AX = C$$

(4.2)

where

$$
\mathbf{A} = \begin{bmatrix} a_{11} & a_{12} & \cdots & a_{1n} \\ a_{21} & a_{22} & \cdots & a_{2n} \\ \vdots & \vdots & \ddots & \vdots \\ a_{n1} & a_{n2} & \cdots & a_{nn} \end{bmatrix}, \quad \mathbf{X} = \begin{bmatrix} x_1 \\ x_2 \\ \vdots \\ x_n \end{bmatrix}, \quad \mathbf{C} = \begin{bmatrix} c_1 \\ c_2 \\ \vdots \\ c_n \end{bmatrix}
$$

Matrix $\mathbf{A}$ has n rows and n columns.
Matrix $\mathbf{X}$ has n rows and 1 column.
Matrix $\mathbf{C}$ has n rows and 1 column.

Note: The number of columns in $\mathbf{A}$ must equal the number of rows in $\mathbf{X}$; otherwise, matrix multiplication is not defined.

In matrix algebra, $\mathbf{X}$ can be obtained by multiplying both sides by $\mathbf{A}^{-1}$, that is,

$$
\underbrace{\mathbf{A}^{-1}\mathbf{A}}_{\mathbf{I}} \mathbf{X} = \mathbf{A}^{-1}\mathbf{C} \Rightarrow \mathbf{IX} = \mathbf{X} = \mathbf{A}^{-1}\mathbf{C}
$$

MATLAB offers the following functions for solving a set of linear equations:

■ `inv` Function:
To solve a system of linear equations in MATLAB, you can use

```
X = inv(A) * C
```

The method of solving a system of linear equations by using the `inv` function is more computationally complicated than a method called Gauss elimination, which is discussed in the following text.

■ Gauss Elimination Function:
MATLAB represents the Gauss elimination method as matrix division; that is, if $\mathbf{AX} = \mathbf{C}$, then

```
X = A\C
```

Note the use of MATLAB's backslash operator to solve for X by Gauss elimination.

Example 4.3

The following example solves a third-order system of linear equations and writes the results to a file:

$$
\begin{aligned}
3x_1 + 2x_2 - x_3 &= 10 \\
-x_1 + 3x_2 + 2x_3 &= 5 \\
x_1 - x_2 - x_3 &= -1
\end{aligned}
$$

```
% Example_4_3.m
% This program solves a simple linear system of equations by
% both the use of MATLAB's inverse matrix function and by
% the Gauss elimination method.
clc; clear;
A = [3 2 -1; -1 3 2; 1 -1 -1]
C = [10 5 -1]'
X1 = inv(A)*C    % X1 is the solution using matrix inverse
X2 = A\C         % X2 is the solution using Gauss elimination
% check solution:
A*X1
% Use the size() command to determine the number of rows and
% the number of columns in matrix A.
[A_rows A_cols] = size(A)
% Print matrix A, matrix C and the solution of the linear
% system of equations to the file output.txt:
fid = fopen('output.txt','w');
fprintf(fid,'The A matrix is:\n');
for i = 1:A_rows
        for j = 1:A_cols
                fprintf(fid,'%6.1f',A(i,j));
        end
        fprintf(fid,'\n');
end
fprintf(fid,'The C vector is:\n');
for i = 1:length(C)
        fprintf(fid,'%6.1f', C(i));
end
fprintf(fid,'\n');
fprintf(fid,'The solution X1 (using inverse matrix) is:\n');
for i = 1:length(X1)
        fprintf(fid,'%6.1f', X1(i));
end
fprintf(fid,'\n');
fprintf(fid,'The solution X2 (using Gauss elimination) is:\n');
for i = 1:length(X2)
        fprintf(fid,'%6.1f',X2(i));
end
fprintf(fid,'\n');
fclose(fid);
```

Program results

```
The A matrix is:
   3.0    2.0    -1.0
  -1.0    3.0     2.0
   1.0   -1.0    -1.0
The C vector is:
  10.0    5.0    -1.0
```

```
The solution X1 (using inverse matrix) is:
   -2.0    5.0    -6.0
The solution X2 (using Gauss elimination) is:
   -2.0    5.0    -6.0
```

We see that the Gauss elimination method produces the same answer as using the inverse of the coefficient matrix method.

Review 4.2

1. Given a set of linear equations in the form $\mathbf{AX} = \mathbf{C}$, where $\mathbf{A}$ is the coefficient matrix and $\mathbf{X}$ and $\mathbf{C}$ are column vectors, what are the two ways for solving for $\mathbf{X}$ in MATLAB?

Exercises

E4.1 Solve the following set of linear equations by the method of determinants:

$$2x_1 - x_2 = 12$$
$$4x_1 + 3x_2 = -8$$

E4.2 Solve the following set of linear equations by MATLAB's inv function:

a. $2x_1 + 3x_2 - x_3 = 20$
   $4x_1 - x_2 + 3x_3 = -14$
   $x_1 + 5x_2 + x_3 = 21$

b. $4x_1 + 8x_2 + x_3 = 8$
   $-2x_1 - 3x_2 + 2x_3 = 14$
   $x_1 + 3x_2 + 4x_3 = 30$

4.4 Statics Truss Problem

An engineering example involving a large system of linear equations can be found in the field of statics. The problem is to solve for the internal forces in the structural members of a truss. In truss analysis, the structural members are considered to be connected by pins at the joints. A pin can exert a single force with components in both the horizontal and vertical directions, but cannot exert a moment. It is assumed that external loads act in a plane, and thus, the truss may be treated as a 2-D structure. It is also assumed that external loads are only applied at the ends of the structural members and therefore, the members may be considered as a two-force body. For a two-force body, the force exerted on the member by the pin must be in the member's axial direction. To demonstrate that this is true, isolate a structural member, and take the moment about one end of the member.

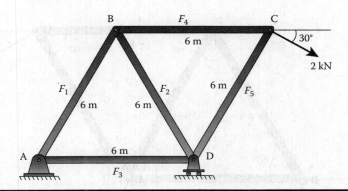

Figure 4.1   Simple truss.

For the body to be in equilibrium, the moment about any point on the body must be zero. Thus, the force at the other end must be along the longitudinal axis of the member resulting in a zero moment. In the analysis, we consider making an imaginary cut in a structural member and removing one part of the member. We replace the effect that the removed part had on the remaining part by a force that is considered as the internal force of the member.

For illustration purposes, a truss consisting of five structural members is considered (see Figure 4.1). Since there are five structural members, there are five unknown internal forces (reactions, which are the forces that the supports exert on the structure, can be obtained from the equilibrium equations involving only the external forces and are thus considered to be known). Also note that there are four joints. We can write two scalar equations at each joint; that is,

$$\sum_{n}(F_x)_n = 0 \quad \text{and} \quad \sum_{n}(F_y)_n = 0 \qquad (4.3)$$

However, we can only use five independent equations. We will write two scalar equations at joints A and B and one scalar equation (in the x direction) at joint C. The system of equations can then be represented by the matrix equation:

$$\mathbf{AF} = \mathbf{P} \qquad (4.4)$$

where
  A is the coefficient matrix
  F is the column matrix of the unknown internal forces
  P is the column matrix containing given external forces

The coefficient matrix **A** will be made up of elements a_{ij} where the first index is the equation number and the second index represents the force member associated with that element. In writing the equations, take all unknown internal forces to be in tension.

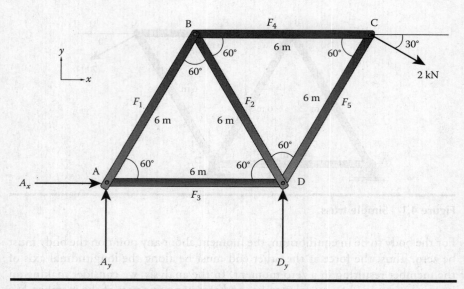

Figure 4.2   Reactions for truss shown in Figure 4.1.

If F_i comes out negative, then the internal force is in compression. First, we solve for the reactions at points A and D (see Figure 4.2). Unknown reactions (A_x, A_y and D_y) are to be assumed pointing in the positive (x,y) directions. Forces are in kN:

Taking the sum of the moments about point A gives

$$\sum M_A = 0 = 6D_y - 2\cos(30°) \times 6\sin(60°) - 2\sin(30°) \times 6(1 + \cos(60°))$$

Solving for D_y gives

$$D_y = \frac{1}{6} \times [2\cos(30°) \times 6\sin(60°) + 2\sin(30°) \times 6*(1 + \cos(60°))]\, kN$$

Taking the sum of the external forces in the x direction gives

$$\sum F_x = 0 = A_x + 2\cos(30°) \rightarrow A_x = -2\cos(30°)\, kN$$

Taking the sum of the external forces in the y direction gives

$$\sum F_y = 0 = A_y + D_y - 2\sin(30°) \rightarrow A_y = -D_y + 2\sin(30°)\, kN$$

We now determine the non-zero elements of the coefficient matrix, **A**
Internal forces at joint A (see Figure 4.3)

$$\sum F_x = 0 = F_1\cos(60°) + F_3 + A_x \tag{1}$$

$$a_{11} = \cos(60°), \quad a_{13} = 1, \quad P_1 = -A_x$$

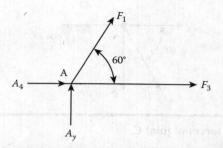

Figure 4.3 Internal forces at joint A.

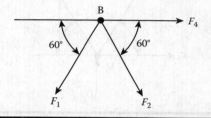

Figure 4.4 Internal forces at joint B.

$$\sum F_y = 0 = F_1 \sin(60°) + A_y$$

$$a_{21} = \sin(60°), \quad P_2 = -A_y \tag{2}$$

Internal forces at joint B (see Figure 4.4)

$$\sum F_x = 0 = F_4 + F_2 \cos(60°) - F_1 \cos(60°)$$

$$a_{32} = \cos(60°), \quad a_{31} = -\cos(60°), \quad a_{34} = 1, \quad P_3 = 0 \tag{3}$$

$$\sum F_y = 0 = -F_1 \sin(60°) - F_2 \sin(60°)$$

$$a_{41} = -\sin(60°), \quad a_{42} = -\sin(60°), \quad P_4 = 0 \tag{4}$$

Internal forces at joint C (see Figure 4.5)

$$\sum F_x = 0 = -F_4 - F_5 \cos(60°) + 2\cos(30°)$$

$$a_{54} = -1, \quad a_{55} = -\cos(60°), \quad P_5 = -2\cos(30°) \tag{5}$$

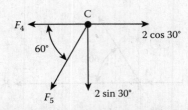

Figure 4.5   Internal forces at joint C.

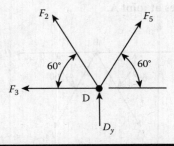

Figure 4.6   Internal forces at joint D.

Since joint D has not been used in establishing the coefficient matrix, it can be used as a check on the obtained solution; that is, the sum of all the forces acting at joint D, both in the x and y directions, must be zero.

Forces at joint D (see Figure 4.6)

$$CHDx = -F_3 - F_2 \cos(60°) + F_5 \cos(60°) \tag{4.5}$$

$$CHDy = F_2 \sin(60°) + F_5 \sin(60°) + D_y \tag{4.6}$$

where $CHDx$ and $CHDy$ represent a check on the sum of the forces at joint D in the x and y directions respectively.

The following example program solves for the internal forces in the truss.

Example 4.4

```
% Example_4_4.m
% This program solves for the internal forces in a truss by
% the matrix method
% The a and p matrices are sparse, so
% Set all a and p elements to zero and overwrite
% the elements that are not zero.
```

```
clear; clc;
th60 = pi/3; th30 = pi/6; ie = 5; je = 5;
a = zeros(5); p = zeros(5,1);
sth60 = sin(th60); cth60 = cos(th60); sth30 = sin(th30);
cth30 = cos(th30);
% Calculation of reactions:
% Sum of moments about joint A gives the vertical component of
% the reaction at joint
Dy = (2*cth30*6*sth60+2*sth30*6*(1+cth60))/6;
% Sum of external forces in the x direction gives the
% horizontal component of the reaction at joint A.
Ax = -2*cth30;
% the sum of external forces in vertical direction gives
% the vertical component of the reaction at joint A.
Ay = 2*sth30-Dy;
fo = fopen('output.txt','w');
fprintf(fo,' Example_4_4.m \n');
fprintf(fo,' statics_problem.m \n');
fprintf(fo,'Program solves for the internal forces ');
fprintf(fo,'of a truss \n\n');
fprintf(fo,' Reactions in kN \n');
fprintf(fo,' Ax=%8.3f \t\t Ay=%8.3f \t\t Dy=%8.1f \n\n',...
          Ax,Ay,dy);
% Overwrite the non-zero elements of matrix a and matrix p.
a(1,1) = cth60; a(1,3) = 1.0; p(1) = -Ax;
a(2,1) = sth60; p(2) = -Ay;
a(3,1) = -cth60; a(3,2) = cth60; a(3,4) = 1.0;
a(4,1) = -sth60; a(4,2) = -sth60;
a(5,4) = -1.0; a(5,5) = -cth60; p(5) = -2*cth30;
fprintf(fo,' A matrix \n\n');
jindex = 1:je;
fprintf(fo,'   ');
for i = 1:ie
   fprintf(fo,'     %5i',jindex(i));
end
fprintf(fo,'\n');
fprintf(fo,'---------------------------------------\n');
for i = 1:ie
   fprintf(fo,'%4i ',i);
   for j = 1:je
      fprintf(fo,'%8.4f',a(i,j));
   end
fprintf(fo,'\n');
end
F = a\p;
fprintf(fo,'Internal forces, F');
fprintf(fo,'and external forces p, in kN \n\n');
fprintf(fo,' Member # F(kN) Equation # p(kN) \n');
fprintf(fo,' ======================================= \n');
for i = 1:ie
```

```
    fprintf(fo,'  %3.0f  %9.4f    %3.0f      %5.3f \n',...
            i,F(i),i,p(i));
end
% check if the sum of the horizontal components at joint D
% is zero:
fprintf(fo,'\n check if sum of the x components at');
fprintf(fo,'joint D is zero \n');
CHDx = -F(2)*cth60+F(5)*cth60-F(3);
fprintf(fo,' CHDx =%12.4e \n',CHDx);
% check if the sum of the vertical components at joint D
% is zero:
fprintf(fo,'\n check if sum of the y components at ');
fprintf(fo,'joint D is zero \n');
CHDy = Dy+F(2)*sth60+F(5)*sth60;
fprintf(fo,' CHDy=%12.4e \n',CHDy);
fclose(fo);
```

Program results

```
Example_4_4.m
statics_problem.m
Program solves for the internal forces of a truss
  Reactions in kN
  Ax = -1.732      Ay = -2.000      Dy = 3.0

  A matrix

              1          2          3          4          5
----------------------------------------------------------------

  1      0.5000     0.0000     1.0000     0.0000     0.0000
  2      0.8660     0.0000     0.0000     0.0000     0.0000
  3     -0.5000     0.5000     0.0000     1.0000     0.0000
  4     -0.8660    -0.8660     0.0000     0.0000     0.0000
  5      0.0000     0.0000     0.0000    -1.0000    -0.5000


Internal forces,F and external forces p,in kN

Member #     F(kN)    Eq. #    p(kN)

===============================================================

  1          2.3094     1      1.732
  2         -2.3094     2      2.000
  3          0.5774     3      0.000
  4          2.3094     4      0.000
  5         -1.1547     5     -1.732

check if sum of the x components at joint D is zero
  CHDx = 0.0000e+00
check if sum of the y components at joint D is zero
  CHDy = -8.8818e-16
```

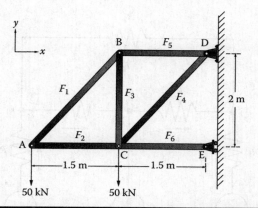

Figure 4.7   Truss for Exercise E4.3.

Exercise

E4.3 For the truss structure shown in Figure 4.7, write a MATLAB program that will determine the internal forces in the structural members by the method described in Section 4.4.

Write two equations at joints A, B, C; that is,

$$\sum (F_x)_i = 0 \quad \text{and} \quad \sum (F_y)_i = 0$$

That should give six equations in six unknowns. Print out the reactions, the coefficient matrix, the member's internal force, and a check on the solution at joint D.

4.5 Resistive Circuit Problem

A typical example involving a system of linear equations can be found in the resistive circuit of Figure 4.8. The goal is to solve for the node voltages v_1, v_2, and v_3 as functions of the input voltages V_1 and V_2 and the input current I_1.

Using Ohm's law ($v = iR$) and Kirchhoff's current law ("KCL"), we can write expressions for the sum of currents at each node of interest. KCL states that the total current entering any circuit node must sum to zero. In Figure 4.8, the resistor currents are written as i_n and are arbitrarily defined in the directions as drawn. Writing three KCL equations for the currents *entering* each node (and noting that currents drawn as *leaving* the node are written mathematically as negative numbers) gives

At node ①,

$$i_1 + i_2 + i_3 = 0 \tag{4.7}$$

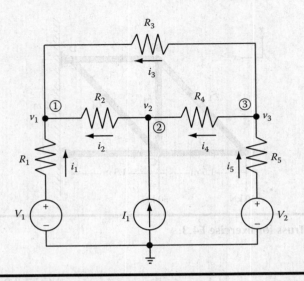

Figure 4.8   Resistive circuit for Example 4.5.

At node ②,

$$I_1 - i_2 + i_4 = 0 \qquad (4.8)$$

At node ③,

$$-i_3 - i_4 + i_5 = 0 \qquad (4.9)$$

We will now rewrite these three equations in terms of voltages using Ohm's law. Note that in these types of problems, it is often easier to write the equations in terms of conductances, G_n, instead of resistances, R_n (where $G_n = 1/R_n$), and Ohm's law may be rewritten as $i = vG$. Thus,

$$i_1 = \frac{V_1 - v_1}{R_1} \quad \rightarrow \quad i_1 = (V_1 - v_1)G_1 \qquad (4.10)$$

$$i_2 = \frac{v_2 - v_1}{R_2} \quad \rightarrow \quad i_2 = (v_2 - v_1)G_2 \qquad (4.11)$$

$$i_3 = \frac{v_3 - v_1}{R_3} \quad \rightarrow \quad i_3 = (v_3 - v_1)G_3 \qquad (4.12)$$

$$i_4 = \frac{v_3 - v_2}{R_4} \quad \rightarrow \quad i_4 = (v_3 - v_2)G_4 \tag{4.13}$$

$$i_5 = \frac{V_2 - v_3}{R_5} \quad \rightarrow \quad i_5 = (V_2 - v_3)G_5 \tag{4.14}$$

Substituting these expressions for i_1 through i_5 into Equations 4.7 through 4.9 gives

$$(V_1 - v_1)G_1 + (v_2 - v_1)G_2 + (v_3 - v_1)G_3 = 0 \tag{4.15}$$

$$I_1 - (v_2 - v_1)G_2 + (v_3 - v_2)G_4 = 0 \tag{4.16}$$

$$-(v_3 - v_1)G_3 - (v_3 - v_2)G_4 + (V_2 - v_3)G_5 = 0 \tag{4.17}$$

These expressions can be rearranged into a system of three equations with three unknowns:

$$\begin{bmatrix} (G_1 + G_2 + G_3) & -G_2 & -G_3 \\ -G_2 & (G_2 + G_4) & -G_4 \\ -G_3 & -G_4 & (G_3 + G_4 + G_5) \end{bmatrix} \begin{bmatrix} v_1 \\ v_2 \\ v_3 \end{bmatrix} = \begin{bmatrix} V_1 G_1 \\ I_1 \\ V_2 G_5 \end{bmatrix} \tag{4.18}$$

This equation has the form $\mathbf{AX} = \mathbf{C}$ and may be solved by the use of MATLAB's inv function or by MATLAB'S Gauss elimination method as shown in Example 4.2.

The following script solves for the node voltages for the following circuit values:

$$R_1 = 2200\,\Omega, \quad R_2 = 10\,\mathrm{k}\Omega, \quad R_3 = 6900\,\Omega, \quad R_4 = 9100\,\Omega, \quad R_5 = 3300\,\Omega$$

$$V_1 = 12\,\mathrm{V}, \quad V_2 = 3.3\,\mathrm{V}, \quad I_1 = 2\,\mathrm{mA}$$

Example 4.5

```
% Example_4_5.m
% Resistive Circuit Problem
% This program solves for the internal node voltages for the
% circuit shown in Figure 4.8.
% The conductances G are in units of siemens
% The node voltage V are in units of volts
% The currents I are in units of amps
clear; clc;
g1 = 1/2200; g2 = 1/10000; g3 = 1/6900; g4 = 1/9100;
g5= 1/3300; V1 = 12; V2 = 3.3; I1 =.002;
% Put the problem into standard form AX = C:
A = [g1+g2+g3 -g2 -g3; -g2 g2+g4 -g4; -g3 -g4 g3+g4+g5];
C = [V1*g1; I1; V2*g5];
X = A\C;
% print the results
fprintf('V1=%5.1f V  V2=%5.1f V  I1=%5.1e A\n',V1,V2,I1);
fprintf('g1=%8.5f S  g2=%8.5f S  g3=%8.5f S\n',g1,g2,g3);
fprintf('g4=%8.5f S  g5=%8.5f S\n',g4,g5);
fprintf('\n\n');
fprintf(' Node #     v(volts) \n');
fprintf('---------------------\n');
for n = 1:length(C)
    fprintf(' %3i      %9.1f\n', n,X(n));
end
```

Program results

```
V1 = 12.0 V V2 = 3.3 V I1 = 2.0e-03 A
g1 = 0.00045 S g2 = 0.00010 S g3 = 0.00014 S
g4 = 0.00011 S g5 = 0.00030 S
Node #   v (volts)
---------------------
  1          12.6
  2          20.3
  3           9.0
>>
```

Exercise

E4.5 For the circuit of Figure 4.8, solve for the five unknown currents instead of the three unknown voltages. Develop the system of five linear equations using Equations 4.7 through 4.9 and by using Kirchhoff's voltage law ("KVL") to write two additional equations. KVL states that the sum of the voltage drops around any loop in the circuit must be zero. Write KVL equations for these two circuit loops: $V_1 \rightarrow R_1 \rightarrow R_2 \rightarrow R_4 \rightarrow R_5 \rightarrow V_2$ and $R_3 \rightarrow R_4 \rightarrow R_2$. You should wind up with a system of five equations and five unknowns. Solve with MATLAB using the inverse matrix technique and confirm that your answer matches the results found in Example 4.4.

The following sections cover (a) the Gauss elimination and Gauss–Jordon methods for solving a system of linear algebraic equations, (b) the number of solutions that exists in a set of linear algebraic equations, (c) the theoretical methods for obtaining the inverse matrix, and (d) the eigenvalue problem. Except for the eigenvalue problem, the listed sections do not involve MATLAB and may be skipped without effecting the understanding of the remaining chapters in the textbook.

4.6 Gauss Elimination

As previously discussed, the Gauss elimination method is computationally more efficient than the inverse matrix method or the method of determinants with Cramer's rule. We wish to use the Gauss elimination method to solve the system of linear equations described by Equation 4.1 and represented in matrix form by Equation 4.2. In the Gauss elimination method, the original system is reduced to an equivalent triangular set that can readily be solved by back substitution. The reduced *equivalent* set would appear like the following set of equations:

$$
\begin{aligned}
\tilde{a}_{11}x_1 + \tilde{a}_{12}x_2 + \tilde{a}_{13}x_3 + \quad \cdots \quad + \tilde{a}_{1n}x_n &= \tilde{c}_1 \\
\tilde{a}_{22}x_2 + \tilde{a}_{23}x_3 + \quad \cdots \quad + \tilde{a}_{2n}x_n &= \tilde{c}_2 \\
\tilde{a}_{33}x_3 + \quad \cdots \quad + \tilde{a}_{3n}x_n &= \tilde{c}_3 \\
\ddots \qquad \vdots \quad &= \vdots \\
\tilde{a}_{n-1,n-1}x_{n-1} + \tilde{a}_{n-1,n}x_n &= \tilde{c}_{n-1} \\
\tilde{a}_{n,n}x_n &= \tilde{c}_n
\end{aligned}
\tag{4.19}
$$

where the tilde ($\sim$) variables are a new set of coefficients (to be determined) and where the new coefficient matrix $\tilde{\mathbf{A}}$ is *diagonal* (i.e., all of the coefficients left of the main diagonal are zero). Then,

$$
x_n = \frac{\tilde{c}_n}{\tilde{a}_{n,n}}
$$

$$
x_{n-1} = \frac{1}{\tilde{a}_{n-1,n-1}}\left(\tilde{c}_{n-1} - \tilde{a}_{n-1,n}x_n\right)
$$

$$
x_{n-2} = \frac{1}{\tilde{a}_{n-2,n-2}}\left(\tilde{c}_{n-2} - \tilde{a}_{n-2,n-1}x_{n-1} - \tilde{a}_{n-2,n}x_n\right)
$$

$$
\vdots
$$

etc.

To determine the reduced equivalent sets $\tilde{\mathbf{A}}$ and $\tilde{\mathbf{C}}$, it is convenient to augment the original coefficient matrix $\mathbf{A}$ with the $\mathbf{C}$ matrix as shown in Equation 4.20:

$$
\mathbf{A}_{aug} =
\begin{bmatrix}
a_{11} & a_{12} & \cdots & a_{1n} & c_1 \\
a_{21} & a_{22} & \cdots & a_{2n} & c_2 \\
\vdots & \vdots & \ddots & \vdots & \vdots \\
a_{n1} & a_{n2} & \cdots & a_{nn} & c_n
\end{bmatrix}
\tag{4.20}
$$

The following procedure is used to obtain the reduced equivalent set:

1. Multiply the first row of Equation 4.1 by a_{21}/a_{11} and subtract from the second row, giving

$$
a'_{21} = a_{21} - \frac{a_{21}}{a_{11}} \times a_{11}, \quad a'_{22} = a_{22} - \frac{a_{21}}{a_{11}} \times a_{12}, \quad a'_{23} = a_{23} - \frac{a_{21}}{a_{11}} \times a_{13}, \ldots \text{ etc.}
$$

and

$$
c'_2 = c_2 - \frac{a_{21}}{a_{11}} \times c_1
$$

This gives

$$
a'_{21} = 0
$$

2. For the third row, multiply the first row of Equation 4.1 by a_{31}/a_{11} and subtract from row 3, giving

$$
a'_{31} = a_{31} - \frac{a_{31}}{a_{11}} \times a_{11}, \quad a'_{32} = a_{32} - \frac{a_{31}}{a_{11}} \times a_{12}, \quad a'_{33} = a_{33} - \frac{a_{31}}{a_{11}} \times a_{13}, \ldots \text{ etc.}
$$

and

$$
c'_3 = c_3 - \frac{a_{31}}{a_{11}} \times c_1
$$

This gives

$$
a'_{31} = 0.
$$

3. This process is repeated for the remaining rows 4, 5, 6, ..., n. The original row 1 is kept in its original form. All other rows have been modified, and the new coefficients are designated by a (′). Except for the first row, the resulting set will not contain x_1.

4. For the preceding steps, the first row of Equation 4.1 was used as the *pivot row*, and a_{11} was the *pivot element*. We now use the new row 2 as the pivot row and repeat steps 1–3 for the remaining rows below row 2. Thus, multiply the new row 2 by a'_{32}/a'_{22} and subtract from row 3, giving

$$a''_{32} = a'_{32} - \frac{a'_{32}}{a'_{22}} \times a'_{22}, \quad a''_{33} = a'_{33} - \frac{a'_{32}}{a'_{22}} \times a'_{23}, \quad a''_{34} = a'_{34} - \frac{a'_{32}}{a'_{22}} \times a'_{24}, \ldots \text{ etc.}$$

and

$$c''_3 = c'_3 - \frac{a'_{32}}{a'_{22}} \times c'_2$$

This gives

$$a''_{32} = 0$$

Similarly, multiply the new row 2 by a'_{42}/a'_{22} and subtract from row 4, giving

$$a''_{42} = a'_{42} - \frac{a'_{42}}{a'_{22}} \times a'_{22}, \quad a''_{43} = a'_{43} - \frac{a'_{42}}{a'_{22}} \times a'_{23}, \quad a''_{44} = a'_{43} - \frac{a'_{42}}{a'_{22}} \times a'_{24}, \ldots \text{ etc.}$$

and

$$c''_4 = c'_4 - \frac{a'_{42}}{a'_{22}} \times c'_2$$

This gives

$$a''_{42} = 0$$

This process is continued for the remaining rows 5, 6, …, n. Except for the new row 2, the set does not contain x_2.

5. The next row is now used as a pivot row and the process is continued until the $(n-1)$th row is used as the pivot row. When this is complete, the new system is triangular and can be solved by back substitution.

The general expression for the new coefficients is

$$a'_{ij} = a_{ij} - \frac{a_{ik}}{a_{kk}} \times a_{kj} \quad \text{for } i = k+1, k+2, \ldots, n \quad \text{and} \quad j = k+1, k+2, \ldots, n$$

where k is the pivot row.

These operations only affect the a_{ij} and c_j. Thus, we need only to operate on the **A** and **C** matrices.

Example 4.6

Given

$$x_1 - 3x_2 + x_3 = -4$$
$$-3x_1 + 4x_2 + 3x_3 = -10$$
$$2x_1 + 3x_2 - 2x_3 = 18$$

the augmented coefficient matrix is

$$\mathbf{A}_{aug} = \begin{bmatrix} 1 & -3 & 1 & -4 \\ -3 & 4 & 3 & -10 \\ 2 & 3 & -2 & 18 \end{bmatrix}; \text{ multiply row 1 by } \frac{a_{21}}{a_{11}} = \frac{-3}{1} \text{ and subtract from row 2.}$$

Row 2 becomes $(-3 + 3)$, $(4 - 9)$, $(3 + 3)$, $(-10 - 12) = (0, -5, 6, -22)$.

The new matrix is

$$\mathbf{A}_{aug,Equiv} = \begin{bmatrix} 1 & -3 & 1 & -4 \\ 0 & -5 & 6 & -22 \\ 2 & 3 & -2 & 18 \end{bmatrix}; \text{ multiply row 1 by } \frac{a_{31}}{a_{11}} = \frac{2}{1} \text{ and subtract from row 3.}$$

Row 3 becomes $(2 - 2)$, $(3 + 6)$, $(-2 - 2)$, $(18 + 8) = (0, 9, -4, 26)$.

The new matrix is

$$\mathbf{A}_{aug,Equiv} = \begin{bmatrix} 1 & -3 & 1 & -4 \\ 0 & -5 & 6 & -22 \\ 0 & 9 & -4 & 26 \end{bmatrix}$$

We now use row 2 as the pivot row.

$$\mathbf{A}_{aug,Equiv} = \begin{bmatrix} 1 & -3 & 1 & -4 \\ 0 & -5 & 6 & -22 \\ 0 & 9 & -4 & 26 \end{bmatrix}; \text{multiply row 2 by } \frac{a_{32}}{a_{22}} = -\frac{9}{5} \text{ and subtract}$$

from row 3.

Row 3 becomes

$$(0-0), \left(9 - \frac{9}{5} \times 5\right), \left(-4 + \frac{9}{5} \times 6\right), \left(26 - \frac{9}{5} \times 22\right) = \left(0, 0, \frac{34}{5}, -\frac{68}{5}\right)$$

The new matrix is

$$\mathbf{A}_{aug,Equiv} = \begin{bmatrix} 1 & -3 & 1 & -4 \\ 0 & -5 & 6 & -22 \\ 0 & 0 & \dfrac{34}{5} & -\dfrac{68}{5} \end{bmatrix}$$

The system is now triangular and the system can be rewritten as

$$\mathbf{A}_{Equiv}\mathbf{X} = \mathbf{C}_{Equiv}$$

which gives

$$x_1 - 3x_2 + x_3 = -4$$

$$-5x_2 + 6x_3 = -22$$

$$\frac{34}{5}x_3 = -\frac{68}{5}$$

Solving

$$\frac{34}{5}x_3 = -\frac{68}{5} \quad \rightarrow \quad x_3 = -2$$

$$-5x_2 + 6(-2) = -22 \quad \rightarrow \quad x_2 = 2$$

$$x_1 - 3(2) + (-2) = -4 \quad \rightarrow \quad x_1 = 4$$

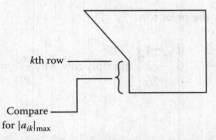

kth row

Compare
for $|a_{ik}|_{\max}$

Figure 4.9   Row interchange.

Two important considerations:

1. If a_{kk} is zero, where k is the pivot row, then the process cannot be carried out.
2. Greater accuracy in the solution is obtained if the pivot element is the absolute maximum available from the set. That is, if the pivot row is k, one compares the a_{ik}'s for $i = k + 1, k + 2, \ldots, n$ (see Figure 4.9). If $|a_{ik}|_{\max} \neq |a_{kk}|$, then the row containing the $|a_{ik}|_{\max}$ is interchanged with the kth row. This only affects the ordering of the equations and does not affect the solution.

If after row interchange is carried out and one of the a_{kk}'s remains zero, then the system is singular and no solution can be obtained.

One last consideration: It can be shown that if the magnitude of the pivot element is much smaller than other elements in the matrix, the use of the small pivot element will cause a decrease in the accuracy of the solution. To check if this is the case, you can first scale the equations; that is, divide each equation by the absolute maximum coefficient in that equation. This makes the absolute maximum coefficient in that equation equal to 1.0. If $|a_{kk}| \ll 1$, then the solution is likely to be inaccurate.

4.7 Gauss–Jordan Method

The Gauss–Jordan method is a modification of the Gauss elimination method and can be used to solve a system of linear equations of the form $\mathbf{AX} = \mathbf{C}$. In this method, the objective is to obtain an equivalent coefficient matrix such that, except for the main diagonal, all elements are zero. The method starts out, as in the Gauss elimination method, by finding an equivalent matrix that is triangular. It then continues, assuming that $\mathbf{A}$ is an $n \times n$ matrix, using the nth row as the pivot row, multiplying the nth row by $a_{n-1,n}/a_{n,n}$, and subtracting the result from row $n - 1$, thus making the new $a_{n-1,n} = 0$. The process is repeated for pivot rows $n - 2, n - 3, \ldots, 2$.

Example 4.7

As an example, the Gauss–Jordan method is applied as follows. Starting with the triangular equivalent matrix of Example 4.5,

$$
\mathbf{A}_{aug,Equiv} =
\begin{bmatrix}
1 & -3 & 1 & -4 \\
0 & -5 & 6 & -22 \\
0 & 0 & \dfrac{34}{5} & -\dfrac{68}{5}
\end{bmatrix}
$$

Multiply row 3 by $a_{23}/a_{33} = 6 \times (5/34)$ and subtract from row 2. Row 2 becomes

$$
\left(0 - \frac{30}{34} \times 0\right), \left(-5 - \frac{30}{34} \times 0\right), \left(6 - \frac{30}{34} \times \frac{34}{5}\right), \left(-22 + \frac{30}{34} \times \frac{68}{5}\right) = (0, -5, 0, -10).
$$

Next, multiply row 3 by $a_{13}/a_{33} = 5/34$ and subtract from row 1. Row 1 becomes

$$
\left(1 - \frac{5}{34} \times 0\right), \left(-3 - \frac{5}{34} \times 0\right), \left(1 - \frac{5}{34} \times \frac{34}{5}\right), \left(-4 + \frac{5}{34} \times \frac{68}{5}\right) = (1, -3, 0, -2).
$$

The new $\mathbf{A}_{aug,\,Equiv}$ becomes

$$
\mathbf{A}_{aug,Equiv} =
\begin{bmatrix}
1 & -3 & 0 & -2 \\
0 & -5 & 0 & -10 \\
0 & 0 & \dfrac{34}{5} & -\dfrac{68}{5}
\end{bmatrix}
$$

Now, row 2 is used as the pivot row. Multiply row 2 by $a_{12}/a_{22} = 3/5$ and subtract from row 1. Row 1 then becomes

$$
\left(1 - \frac{3}{5} \times 0\right), \left(-3 + \frac{3}{5} \times 5\right), \left(0 - \frac{3}{5} \times 0\right), \left(-2 + \frac{3}{5} \times 10\right) = (1, 0, 0, 4)
$$

The new $\mathbf{A}_{aug, Equiv}$ becomes

$$\mathbf{A}_{aug,Equiv} = \begin{bmatrix} 1 & 0 & 0 & 4 \\ 0 & -5 & 0 & -10 \\ 0 & 0 & \dfrac{34}{5} & -\dfrac{68}{5} \end{bmatrix}$$

Thus, the equivalent set of equations becomes

$$x_1 = 4$$

$$-5x_2 = -10$$

$$\frac{34}{5}x_3 = -\frac{68}{5}$$

And thus, $x_1 = 4$, $x_2 = 2$, and $x_3 = -2$, which is the same answer that was obtained earlier by the Gauss elimination method.

4.8 Number of Solutions

Suppose a Gauss elimination program is carried out and the following results are obtained:

$$a_{11}x_1 + a_{12}x_2 + a_{13}x_3 + \quad \cdots \quad + a_{1n}x_n = c_1$$

$$a_{22}x_2 + a_{23}x_3 + \quad \cdots \quad + a_{2n}x_n = c_2$$

$$a_{33}x_3 + \quad \cdots \quad + a_{3n}x_n = c_3$$

$$\ddots \qquad \vdots \qquad \vdots$$

$$a_{rr}x_r + \cdots + a_{rn}x_n = c_r \qquad (4.21)$$

$$0 = c_{r+1}$$

$$0 = c_{r+2}$$

$$\vdots$$

$$0 = c_n$$

where $r < n$ and $a_{11}, a_{22}, \ldots, a_{rr}$ are not zero. There are two possible cases:

1. No solution exists if any one of the c_{r+1} through c_n is not zero.
2. Infinitely many solutions exist if c_{r+1} through c_n are all zero.

If $r = n$ and $a_{11}, a_{22}, \ldots, a_{nn}$ are not zero, then the system would appear as follows:

$$a_{11}x_1 + a_{12}x_2 + a_{13}x_3 + \cdots + a_{1n}x_n = c_1$$

$$a_{22}x_2 + a_{23}x_3 + \cdots + a_{2n}x_n = c_2$$

$$a_{33}x_3 + \cdots + a_{3n}x_n = c_3 \qquad (4.22)$$

$$\ddots \quad \vdots \quad \vdots$$

$$a_{nn}x_n = c_n$$

For this case, there is only one solution.

Exercise

E4.6 Obtain a solution to the following linear equations using the Gauss elimination method.

a.
$$2x_1 + 3x_2 - x_3 = 20$$
$$4x_1 - x_2 + 3x_3 = -14$$
$$x_1 + 5x_2 + x_3 = 21$$

b.
$$4x_1 + 8x_2 + x_3 = 8$$
$$-2x_1 - 3x_2 + 2x_3 = 14$$
$$x_1 + 3x_2 + 4x_3 = 30$$

c.
$$2x_1 + 3x_2 + x_3 - x_4 = 1$$
$$5x_1 - 2x_2 + 5x_3 - 4x_4 = 5$$
$$x_1 - 2x_2 + 3x_3 - 3x_4 = 3$$
$$3x_1 + 8x_2 - x_3 + x_4 = -1$$

4.9 Inverse Matrix

Given a matrix system of equations of the general form

$$\mathbf{AX} = \mathbf{C}$$

we can solve by front-multiplying both sides of the equation by the matrix inverse $\mathbf{A}^{-1}$:

$$\mathbf{A}^{-1}\mathbf{AX} = \mathbf{IX} = \mathbf{X} = \mathbf{A}^{-1}\mathbf{C}$$

MATLAB's method of solution is

$$X = inv(A)*C \qquad \text{(solves by } A^{-1})$$

or

$$X = A \backslash C \qquad \text{(solves by Gauss elimination)}$$

Let us see what is involved in determining A^{-1}.

Let $B = A^{-1}$. Then, $BA = I$. We will demonstrate for a 3×3 matrix:

$$\begin{bmatrix} b_{11} & b_{12} & b_{13} \\ b_{21} & b_{22} & b_{23} \\ b_{31} & b_{32} & b_{33} \end{bmatrix} \begin{bmatrix} a_{11} & a_{12} & a_{13} \\ a_{21} & a_{22} & a_{23} \\ a_{31} & a_{32} & a_{33} \end{bmatrix} = \begin{bmatrix} 1 & 0 & 0 \\ 0 & 1 & 0 \\ 0 & 0 & 1 \end{bmatrix} \qquad (4.23)$$

First Row of BA

Element (1,1): $b_{11}a_{11} + b_{12}a_{21} + b_{13}a_{31} = 1$

Element (1,2): $b_{11}a_{12} + b_{12}a_{22} + b_{13}a_{32} = 0$

Element (1,3): $b_{11}a_{13} + b_{12}a_{23} + b_{13}a_{33} = 0$

Here, b_{11}, b_{12}, and b_{13} are the unknowns. Also note the transpose matrix

$$A^{T} = \begin{bmatrix} a_{11} & a_{21} & a_{31} \\ a_{12} & a_{22} & a_{32} \\ a_{13} & a_{23} & a_{33} \end{bmatrix}$$

Let $B_1 = \begin{bmatrix} b_{11} \\ b_{12} \\ b_{13} \end{bmatrix}$, then, $A^{T}B_1 = \begin{bmatrix} 1 \\ 0 \\ 0 \end{bmatrix}$.

Solve for b_{11}, b_{12}, and b_{13} by Gauss elimination.

Second Row of BA

Element (2,1): $b_{21}a_{11} + b_{22}a_{21} + b_{23}a_{31} = 0$

Element (2,2): $b_{21}a_{12} + b_{22}a_{22} + b_{23}a_{32} = 1$

Element (2,3): $b_{21}a_{13} + b_{22}a_{23} + b_{23}a_{33} = 0$

Here, b_{21}, b_{22}, and b_{23} are the unknowns.

Let $B_2 = \begin{bmatrix} b_{21} \\ b_{22} \\ b_{23} \end{bmatrix}$, then, $A^{T}B_2 = \begin{bmatrix} 0 \\ 1 \\ 0 \end{bmatrix}$.

Solve for b_{21}, b_{22}, and b_{23} by Gauss elimination.

Third Row of **BA**

$$\text{Element (3,1): } b_{31}a_{11} + b_{32}a_{21} + b_{33}a_{31} = 0$$
$$\text{Element (3,2): } b_{31}a_{12} + b_{32}a_{22} + b_{33}a_{32} = 0$$
$$\text{Element (3,3): } b_{31}a_{13} + b_{32}a_{23} + b_{33}a_{33} = 1$$

Here, b_{31}, b_{32}, and b_{33} are the unknowns.

$$\text{Let } \mathbf{B}_3 = \begin{bmatrix} b_{31} \\ b_{32} \\ b_{33} \end{bmatrix}, \text{ then, } \mathbf{A}^T \mathbf{B}_3 = \begin{bmatrix} 0 \\ 0 \\ 1 \end{bmatrix}.$$

Solve for b_{31}, b_{32}, and b_{33} by Gauss elimination.

An alternative [1] to the previous method is to augment the coefficient matrix with the identity matrix, then apply the Gauss–Jordan method making the coefficient matrix the identity matrix. The original identity matrix then becomes $\mathbf{A}^{-1}$. The starting augmented matrix for a 3×3 coefficient matrix is shown as follows:

$$\begin{bmatrix} a_{11} & a_{12} & a_{13} & | & 1 & 0 & 0 \\ a_{21} & a_{22} & a_{23} & | & 0 & 1 & 0 \\ a_{31} & a_{32} & a_{33} & | & 0 & 0 & 1 \end{bmatrix} \tag{4.24}$$

Example 4.8

This method is illustrated by the following example:

$$x_1 - 3x_2 + x_3 = -4$$
$$-3x_1 + 4x_2 + 3x_3 = -10$$
$$2x_1 + 3x_2 - 2x_3 = 18$$

Writing the augmented matrix,

$$\mathbf{A}_{aug} = \begin{bmatrix} 1 & -3 & 1 & | & 1 & 0 & 0 \\ -3 & 4 & 3 & | & 0 & 1 & 0 \\ 2 & 3 & -2 & | & 0 & 0 & 1 \end{bmatrix}$$

Multiply row 1 by –3/1 and subtract from row 2, giving

$$(-3 + 3 \times 1), (4 - 3 \times 3), (3 + 3 \times 1), (0 + 3 \times 1), (1 + 3 \times 0), (0 + 3 \times 0) = (0, -5, 6, 3, 1, 0).$$

Multiply row 1 by (2/1) = 2 and subtract from row 3, giving

$(2 - 2 \times 1), (3 - 2 \times (-6)), (-2 - 2 \times 1), (0 - 2 \times 1), (0 - 2 \times 0), (1 - 2 \times 0) = (0, 9, -4, -2, 0, 1)$

$$\mathbf{A}_{aug, Equiv} = \begin{bmatrix} 1 & -3 & 1 & 1 & 0 & 0 \\ 0 & -5 & 6 & 3 & 1 & 0 \\ 0 & 9 & -4 & -2 & 0 & 1 \end{bmatrix}$$

Multiply row 2 by –9/5 and subtract from row 3, giving

$$\left(0 + \frac{9}{5} \times 0\right), \left(9 + \frac{9}{5} \times (-5)\right), \left(-4 + \frac{9}{5} \times 6\right), \left(-2 + \frac{9}{5} \times 3\right), \left(0 + \frac{9}{5} \times 1\right), \left(1 + \frac{9}{5} \times 0\right)$$

$$= \left(0, 0, \frac{34}{5}, \frac{17}{5}, \frac{9}{5}, 1\right)$$

$$\mathbf{A}_{aug, Equiv} = \begin{bmatrix} 1 & -3 & 1 & 1 & 0 & 0 \\ 0 & -5 & 6 & 3 & 1 & 0 \\ 0 & 0 & \dfrac{34}{5} & \dfrac{17}{5} & \dfrac{9}{5} & 1 \end{bmatrix}$$

Multiply row 3 by 30/34 and subtract from row 2, giving

$$\left(0 - \frac{30}{34} \times 0\right), \left(-5 - \frac{30}{34} \times 0\right), \left(6 - \frac{30}{34} \times \frac{34}{5}\right), \left(3 - \frac{30}{34} \times \frac{17}{5}\right),$$

$$\left(1 - \frac{30}{34} \times \frac{9}{5}\right), \left(0 - \frac{30}{34} \times 1\right) = \left(0, -5, 0, 0, -\frac{20}{34}, -\frac{30}{34}\right)$$

Multiply row 3 by 5/34 and subtract from row 1, giving

$$\left(1 - \frac{5}{34} \times 0\right), \left(-3 - \frac{5}{34} \times 0\right), \left(1 - \frac{5}{34} \times \frac{34}{5}\right), \left(1 - \frac{5}{34} \times \frac{17}{5}\right),$$

$$\left(0 - \frac{5}{34} \times \frac{9}{5}\right), \left(0 - \frac{5}{34} \times 1\right) = \left(1, -3, 0, \frac{1}{2}, -\frac{9}{34}, -\frac{5}{34}\right)$$

$$\mathbf{A}_{aug,Equiv} = \begin{bmatrix} 1 & -3 & 0 & \Big| & \dfrac{1}{2} & -\dfrac{9}{34} & -\dfrac{5}{34} \\[3mm] 0 & -5 & 0 & \Big| & 0 & -\dfrac{20}{34} & -\dfrac{30}{34} \\[3mm] 0 & 0 & \dfrac{34}{5} & \Big| & \dfrac{17}{5} & \dfrac{9}{5} & 1 \end{bmatrix}$$

Multiply row 2 by 3/5 and subtract from row 1, giving

$$\left(1 - \frac{3}{5} \times 0\right), \left(-3 - \frac{3}{5} \times (-5)\right), \left(0 - \frac{3}{5} \times 0\right), \left(\frac{1}{2} - \frac{3}{5} \times 0\right),$$

$$\left(-\frac{9}{34} + \frac{3}{5} \times \frac{20}{34}\right), \left(-\frac{5}{34} + \frac{3}{5} \times \frac{30}{34}\right) = \left(1, 0, 0, \frac{1}{2}, \frac{3}{34}, \frac{13}{34}\right)$$

$$\mathbf{A}_{aug,Equiv} = \begin{bmatrix} 1 & 0 & 0 & \Big| & \dfrac{1}{2} & \dfrac{3}{34} & \dfrac{13}{34} \\[3mm] 0 & -5 & 0 & \Big| & 0 & -\dfrac{20}{34} & -\dfrac{30}{34} \\[3mm] 0 & 0 & \dfrac{34}{5} & \Big| & \dfrac{17}{5} & \dfrac{9}{5} & 1 \end{bmatrix}$$

Divide row 2 by –5 and row 3 by 34/5, giving

$$\mathbf{A}_{aug,Equiv} = \begin{bmatrix} 1 & 0 & 0 & \Big| & \dfrac{1}{2} & \dfrac{3}{34} & \dfrac{13}{34} \\[3mm] 0 & 1 & 0 & \Big| & 0 & \dfrac{4}{34} & \dfrac{6}{34} \\[3mm] 0 & 0 & 1 & \Big| & \dfrac{1}{2} & \dfrac{9}{34} & \dfrac{5}{34} \end{bmatrix}$$

Thus,

$$\mathbf{A}^{-1} = \begin{bmatrix} \dfrac{1}{2} & \dfrac{3}{34} & \dfrac{13}{34} \\[3mm] 0 & \dfrac{4}{34} & \dfrac{6}{34} \\[3mm] \dfrac{1}{2} & \dfrac{9}{34} & \dfrac{5}{34} \end{bmatrix}$$

Use MATLAB to show that $\mathbf{A}\mathbf{A}^{-1} = \mathbf{I}$.

4.10 Eigenvalue Problem in Mechanical Vibrations

One very important application of the eigenvalue problem is in the theory of vibrations. Consider the two degrees of freedom problem shown in Figure 4.10.

The governing differential equations describing the motion of the two masses m_1 and m_2 are

$$m_1\ddot{x}_1 = -k_2(x_1 - x_2) - k_1 x_1 \tag{4.25}$$

$$m_2\ddot{x}_2 = k_2(x_1 - x_2) - k_3 x_2 \tag{4.26}$$

where k_1 and k_2 are spring constants and x_1 and x_2 are the positions of the masses.

We wish to determine the modes of oscillation such that each mass undergoes harmonic motion at the same frequency ω. To obtain such a solution, set

$$x_1 = A_1 \exp(i\omega t) \tag{4.27}$$

$$x_2 = A_2 \exp(i\omega t) \tag{4.28}$$

where A_1 and A_2 are constants to be determined and $i = \sqrt{-1}$.

Substituting Equations 4.27 and 4.28 into Equations 4.25 and 4.26 gives

$$\left(\frac{k_1 + k_2}{m_1} - \omega^2\right)A_1 - \frac{k_2}{m_1}A_2 = 0 \tag{4.29}$$

$$-\frac{k_2}{m_2}A_1 + \left(\frac{k_2 + k_3}{m_2} - \omega^2\right)A_2 = 0 \tag{4.30}$$

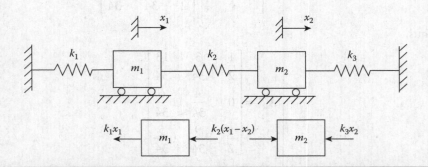

Figure 4.10 Two degrees of freedom vibration system.

Equations 4.29 and 4.30 are two homogeneous linear algebraic equations in two unknowns. There is a theorem in linear algebra that states that the only way for two homogeneous linear algebraic equations in two unknowns to have a nontrivial solution is for the determinant of the coefficient matrix to be zero:

$$\begin{vmatrix} \left(\dfrac{k_1 + k_2}{m_1} - \omega^2 \right) & -\dfrac{k_2}{m_1} \\ -\dfrac{k_2}{m_2} & \left(\dfrac{k_2 + k_3}{m_2} - \omega^2 \right) \end{vmatrix} = 0 \tag{4.31}$$

Letting

$$\mathbf{A} = \begin{bmatrix} \dfrac{k_1 + k_2}{m_1} & -\dfrac{k_2}{m_{11}} \\ -\dfrac{k_2}{m_2} & \dfrac{k_2 + k_3}{m_2} \end{bmatrix} \tag{4.32}$$

$$a_{11} = \frac{k_1 + k_2}{m_1}, \quad a_{12} = -\frac{k_2}{m_1}, \quad a_{21} = -\frac{k_2}{m_2}, \quad a_{22} = \frac{k_2 + k_3}{m_2}, \quad \text{and} \quad \omega^2 = \lambda$$

where λ represents the square of the frequency.

The equation for λ becomes

$$\lambda^2 - (a_{11} + a_{22})\lambda + (a_{11}a_{22} - a_{12}a_{21}) = 0 \tag{4.33}$$

The solution of Equation 4.33 gives the eigenvalues, λ_1 and λ_2, which are the square of the two natural frequencies of oscillations for this system. The ratio of the amplitudes of the oscillation of the two masses can be obtained by substituting the values of λ into Equation 4.29 or 4.30; that is,

$$\frac{A_2}{A_1} = \frac{-(a_{11} - \lambda_1)}{a_{12}} = \frac{m_1}{k_2} \left(\frac{k_1 + k_2}{m_1} - \lambda_1 \right) \quad \text{for the first mode}$$

and

$$\frac{A_2}{A_1} = \frac{-(a_{11} - \lambda_2)}{a_{12}} = \frac{m_1}{k_2} \left(\frac{k_1 + k_2}{m_1} - \lambda_2 \right) \quad \text{for the second mode}$$

The eigenvector, $\mathbf{V}_1$, associated with λ_1 is $\begin{bmatrix} A_1 \\ -A_1(a_{11} - \lambda_1)/a_{12} \end{bmatrix}$ and the eigenvector,

$\mathbf{V}_2$, associated with λ_2 is $\begin{bmatrix} A_1 \\ -A_1(a_{11} - \lambda_2)/a_{12} \end{bmatrix}$.

Since A_1 is arbitrary, we can select $A_1 = 1$, then

$$\mathbf{V}_1 = \begin{bmatrix} 1 \\ -(a_{11} - \lambda_1)/a_{12} \end{bmatrix} \tag{4.34}$$

and

$$\mathbf{V}_2 = \begin{bmatrix} 1 \\ -(a_{11} - \lambda_2)/a_{12} \end{bmatrix} \tag{4.35}$$

If $\mathbf{V}$ is an eigenvector of a matrix $\mathbf{A}$ corresponding to an eigenvalue λ, so is $b\mathbf{V}$ with any $b \neq 0$ [1].

MATLAB has a built-in function named eig that finds the eigenvalues of a square matrix. MATLAB's description of the function follows:

```
E = EIG(X) is a vector containing the Eigen values of
a square matrix X.
[V,D] = EIG(X) produces a diagonal matrix D of Eigen
values and a full matrix V whose columns are the
corresponding eigenvectors so that X*V = V*D.
```

For this problem, the matrix $\mathbf{A}$ represents X. Thus, running

```
[V, D] = eig(A)
```

gives the eigenvectors associated with λ_1 and λ_2. V(:,1) corresponds to λ_1, and V(:,2) corresponds to λ_2.

Example 4.9

The following example program will determine

1. The eigenvalues of the system by both Equation 4.33 and by MATLAB's eig function
2. The eigenvectors by both Equations 4.34 and 4.35 and MATLAB's [V, D] output

For the system shown in Figure 4.10, use the following parameters:

$$m_1 = m_2 = 1500 \text{ kg}, \quad k_1 = 3250 \text{ N/m}, \quad k_2 = 3500 \text{ N/m}, \quad k_3 = 3000 \text{ N/m}$$

The program follows:

```
% Example_4_9.m
% Eigenvalues and Eigenvectors.
% E = eig(a) is a vector containing the Eigenvalues of a
% square matrix a.
% [V,D] = eig(a) produces a diagonal matrix D of Eigenvalues
% and a full matrix V whose columns are the
% corresponding eigenvectors so that X*V = V*D.
% Units are in SI units.
clear; clc;
k1 = 3250; k2 = 3500; k3 = 3000; m1 = 1500; m2 = 1500;
a(1,1) = (k1+k2)/m1;
a(1,2) = -k2/m1;
a(2,1) = -k2/m2;
a(2,2) = (k2+k3)/m2;
% Lamda^2-(a(1,1)+a(2,2))*Lamda+(a(1,1)*a(2,2)-a(1,2)*a(2,1)) = 0
b = -(a(1,1)+a(2,2)); c = a(1,1)*a(2,2)-a(1,2)*a(2,1);
Lamda1 = (-b-sqrt(b^2-4*c))/2;
Lamda2 = (-b+sqrt(b^2-4*c))/2;
E = eig(a);
fprintf('Lamda1=%7.5f  Lamda2=%7.5f  \n',Lamda1,Lamda2)
fprintf('E(1)=%7.5f  E(2)=%7.5f \n',E(1),E(2));
v1 = [1;-(a(1,1)-Lamda1)/a(1,2)]
v2 = [1;-(a(1,1)-Lamda2)/a(1,2)]
[V,D] = eig(a);
V1 = V(:,1)
V2 = V(:,2)
D
```

Program results

```
    Lamda1 = 2.08185   Lamda2 = 6.75149
    E(1) = 2.08185   E(2) = 6.75149
    v1 =
         1.0000
         1.0364
    v2 =
         1.0000
        -0.9649
    V1 =
        -0.6944
        -0.7196
    V2 =
        -0.7196
         0.6944
```

```
    D =
        2.0818        0
        0             6.7515
>>
```

Examining the results, it can be seen that `Lamda1` = `E(1)` and `Lamda2` = `E(2)`. Also, `v1` is a scalar multiple of `V1` and `v2` is a scalar multiple of `V2`.

4.11 Eigenvalue Problem in Electrical Circuits

One application of the eigenvalue problem is in resonant circuits. Consider the LC network shown in Figure 4.11.

Using the constituent relations for voltage across an inductor $\left(v = L\dfrac{di}{dt} \right)$ and current through a capacitor $\left(i = C\dfrac{dv}{dt} \right)$, the governing differential equations describing the two node voltages v_1 and v_2 are

$$\frac{d^2 v_1}{dt^2} = -\left(\frac{1}{L_1 C_1} \right) v_1 + \left(\frac{1}{L_1 C_1} \right) v_2 \tag{4.36}$$

$$\frac{d^2 v_2}{dt^2} = \left(\frac{1}{L_1 C_2} \right) v_1 - \left(\frac{1}{L_1 C_2} + \frac{1}{L_2 C_2} \right) v_2 \tag{4.37}$$

We wish to determine the modes of oscillation such that both voltages oscillate at the same frequency. To obtain such a solution, set

$$v_1 = K_1 e^{j\omega t} \tag{4.38}$$

$$v_2 = K_2 e^{j\omega t} \tag{4.39}$$

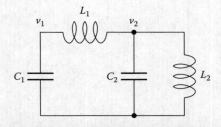

Figure 4.11   Electrical circuit for Example 4.10.

where K_1 and K_2 are constants to be determined and $j = \sqrt{-1}$. Substituting Equations 4.38 and 4.39 into Equations 4.36 and 4.37 gives

$$\left(\frac{1}{L_1 C_1} - \omega^2\right) K_1 - \frac{1}{L_1 C_1} K_2 = 0 \tag{4.40}$$

$$-\frac{1}{L_1 C_2} K_1 + \left(\frac{1}{L_1 C_2} + \frac{1}{L_2 C_2} - \omega^2\right) K_2 = 0 \tag{4.41}$$

Equations 4.40 and 4.41 are two homogeneous linear algebraic equations with two unknowns. Linear algebra tells us that the only way for two homogeneous linear algebraic equations in two unknowns to have a nontrivial solution is for the determinant of the coefficient matrix to be zero. Thus,

$$\begin{vmatrix} \left(\dfrac{1}{L_1 C_1} - \omega^2\right) & -\dfrac{1}{L_1 C_1} \\[2ex] -\dfrac{1}{L_1 C_2} & \left(\dfrac{1}{L_1 C_2} + \dfrac{1}{L_2 C_2} - \omega^2\right) \end{vmatrix} = 0 \tag{4.42}$$

Now, let $\omega^2 = \lambda$ and also let

$$\mathbf{A} = \begin{bmatrix} \dfrac{1}{L_1 C_1} & -\dfrac{1}{L_1 C_1} \\[2ex] -\dfrac{1}{L_1 C_2} & \dfrac{1}{L_1 C_2} + \dfrac{1}{L_2 C_2} \end{bmatrix}$$

giving

$$a_{11} = \frac{1}{L_1 C_1}, \quad a_{12} = -\frac{1}{L_1 C_1}, \quad a_{21} = -\frac{1}{L_1 C_2}, \quad a_{22} = \frac{1}{L_1 C_2} + \frac{1}{L_2 C_2}$$

The equation for λ and the individual elements a_{ij} of $\mathbf{A}$ give

$$\lambda^2 - (a_{11} + a_{22})\lambda + (a_{11}a_{22} - a_{12}a_{21}) = 0 \tag{4.43}$$

The solution of Equation 4.43 gives the eigenvalues λ_1 and λ_2 of matrix **A**, which represent the square of the two natural frequencies for this system. The ratio of the amplitudes of the oscillation of the two voltages can be obtained by substituting the values of λ into Equation 4.40 or 4.41, which gives

$$\frac{K_2}{K_1} = \frac{-(a_{11} - \lambda_1)}{a_{12}} = 1 - L_1 C_1 \lambda_1$$

for the first mode and

$$\frac{K_2}{K_1} = \frac{-(a_{11} - \lambda_2)}{a_{12}} = 1 - L_1 C_1 \lambda_2$$

for the second mode.

The eigenvector $\mathbf{V}_1$ associated with λ_1 is $\begin{bmatrix} K_1 \\ -K_1(a_{11} - \lambda_1)/a_{12} \end{bmatrix}$, and the eigenvector $\mathbf{V}_2$ associated with λ_2 is $\begin{bmatrix} K_1 \\ -K_1(a_{11} - \lambda_2)/a_{12} \end{bmatrix}$.

Since K_1 is arbitrary, we can select $K_1 = 1$, and thus,

$$\mathbf{V}_1 = \begin{bmatrix} 1 \\ -(a_{11} - \lambda_1)/a_{12} \end{bmatrix} \tag{4.44}$$

and

$$\mathbf{V}_2 = \begin{bmatrix} 1 \\ -(a_{11} - \lambda_2)/a_{12} \end{bmatrix} \tag{4.45}$$

If **V** is an eigenvector of a matrix **A** corresponding to an eigenvalue λ, then so is $b\mathbf{V}$, where b is a nonzero constant [1].

Running

```
[V, D] = eig(A)
```

gives the eigenvectors associated with λ_1 and λ_2. V(:,1) corresponds to λ_1, and V(:,2) corresponds to λ_2.

Example 4.10

Suppose in Figure 4.11, the following parameters were given:

$$L_1 = 15 \text{ mH}, \quad L_2 = 25 \text{ mH}, \quad C_1 = 1 \text{ μF}, \quad C_2 = 33 \text{ μF}$$

We wish to create a MATLAB script that will determine

1. The eigenvalues of the system by Equation 4.43 and by MATLAB's eig function
2. The eigenvectors of the system by both Equations 4.44 and 4.45 and MATLAB's [V,D] output

The program follows:

```
% Example_4_10.m
% Eigenvalues and eigenvectors.
% L = eig(A) is a vector containing the eigenvalues of a
% square matrix A.
% [M,D] = eig(A) produces a diagonal matrix D of eigenvalues
% and a full matrix M whose columns are the
% corresponding eigenvectors such that A*L = M*D.
% Units are in SI units.
clear; clc;
L1 = 15e-3; L2 = 25e-3; % henries
C1 = 1e-6; C2 = 33e-6; % farads
A(1,1) = 1/(L1*C1);
A(1,2) = -1/(L1*C1);
A(2,1) = -1/(L1*C2);
A(2,2) = 1/(L1*C2) + 1/(L2*C2);
% Solve for eigenvalues and eigenvectors using derivation
% in text:
% Equation 4.43:
% lambda^2-(A(1,1)+A(2,2))*lambda+(A(1,1)*A(2,2)
%       - A(1,2)*A(2,1)) = 0
% Solve for lambda using quadratic formula:
b = -(A(1,1)+A(2,2)); c = A(1,1)*A(2,2)-A(1,2)*A(2,1);
lambda1 = (-b+sqrt(b^2-4*c))/2;
lambda2 = (-b-sqrt(b^2-4*c))/2;
% From Equations 4.44 and 4.45:
eigenvector1 = [1;-(A(1,1)-lambda1)/A(1,2)];
eigenvector2 = [1;-(A(1,1)-lambda2)/A(1,2)];
fprintf('Eigenvalues solved via Equation 4.43: \n');
fprintf('lambda1=%8.0f   lambda2=%8.0f\n',lambda1, lambda2);
fprintf('Eigenvectors solved via Equation 4.44 and 4.45:\n');
eigenvector1
eigenvector2
% Solve for eigenvalues and eigenvectors using MATLAB's eig
% function.
% Calling eig and returning result to a scalar will give the
% eigenvalues.
L = eig(A);
fprintf('Eigenvalues solved via MATLAB eig function: \n');
fprintf('L(1)=%8.4e   L(2)=%8.4e \n',L(1),L(2));
% Calling eig and returning result to two variables will
% give the eigenvectors and a diagonal matrix of the
% eigenvalues.
```

```
[V D] = eig(A);
fprintf('Eigenvectors solved via MATLAB eig function:\n');
V1 = V(:,1)
V2 = V(:,2)
D
```

Program results

```
Eigenvalues solved via Equation 4.43:
lambda1 = 6.8723e+07 lambda2 = 1.1758e+06
Eigenvectors solved via Equation 4.44 and 4.45:
eigenvector1 =
     1.0000
    -0.0308
eigenvector2 =
     1.0000
     0.9824
Eigenvalues solved via MATLAB eig function:
L(1) = 6.8723e+07 L(2) = 1.1758e+06
Eigenvectors solved via MATLAB eig function:
V1 =
     0.9995
    -0.0308
V2 =
     0.7134
     0.7008
D =
     1.0e+07 *
     6.8723        0
     0             0.1176
>>
```

Examining the results, it can be seen that lambda1 = D(1,1) and lambda2 = D(2,2). Also, eigenvector1 is a scalar multiple of V1, and eigenvector2 is a scalar multiple of V2.

Projects

P4.1 For the truss structures shown in Figure P4.1, create a MATLAB program that will determine the internal forces in the structural members by the method described in Section 4.4. Print out the reactions, the members' internal forces, and a check on the solution.

P4.2 For the truss structure shown in Figure P4.2, create a MATLAB program that will determine the internal forces in the structural members by the method described in Section 4.4. Print out the reactions, the members' internal forces, and a check on the solution.

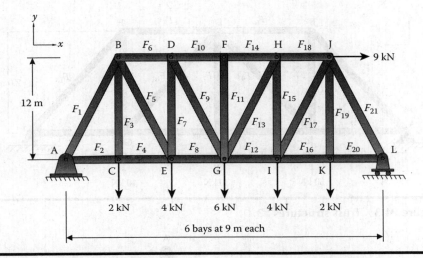

Figure P4.1   Truss structures #1.

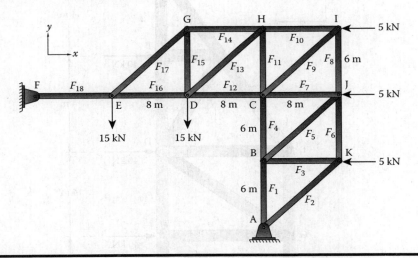

Figure P4.2   Truss structures #2.

P4.3 For the truss structure shown in Figure P4.3, create a MATLAB program that will determine the internal forces in the structural members by the method described in Section 4.4. Print out the reactions, the members' internal forces, and a check on the solution.

P4.4 For the truss structures shown in Figure P4.4, create a MATLAB program that will determine the internal forces in the structural members by the method described in Section 4.4. Print out the reactions, the members' internal forces, and a check on the solution.

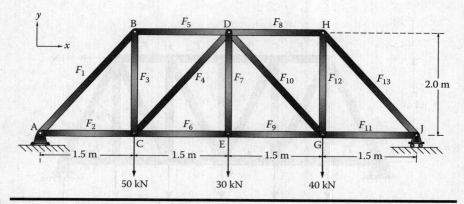

Figure P4.3   Truss structures #3.

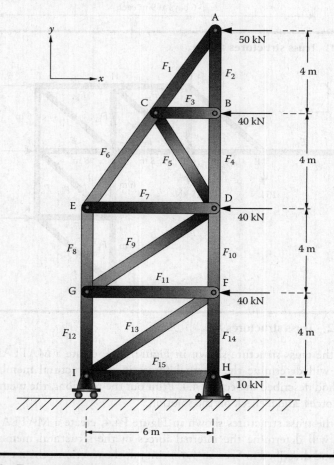

Figure P4.4   Truss structures #4.

P4.5 Figure P4.5 shows a resistive circuit known as a "ladder network."

    a. Using Ohm's law and KCL, write a system of equations for the second-order and third-order networks of Figure P4.5a and b. In addition, write the equations for a fourth-order ladder network.

    b. You should see a pattern emerging from your solutions in part (a). Use this pattern to write the matrix for an eighth-order ladder network. Solve for all circuit voltages using MATLAB for $V_{ref} = 5V$ and the following resistor values:

$$R_{11} = 2200\,\Omega, \quad R_{12} = 2200\,\Omega$$

$$R_{21} = 1200\,\Omega, \quad R_{22} = 6800\,\Omega$$

$$R_{31} = 3900\,\Omega, \quad R_{32} = 2200\,\Omega$$

$$R_{41} = 3300\,\Omega, \quad R_{42} = 5700\,\Omega$$

$$R_{51} = 1800\,\Omega, \quad R_{52} = 5100\,\Omega$$

$$R_{61} = 4700\,\Omega, \quad R_{62} = 3900\,\Omega$$

$$R_{71} = 5700\,\Omega, \quad R_{72} = 6800\,\Omega$$

$$R_{81} = 1200\,\Omega, \quad R_{82} = 2700\,\Omega$$

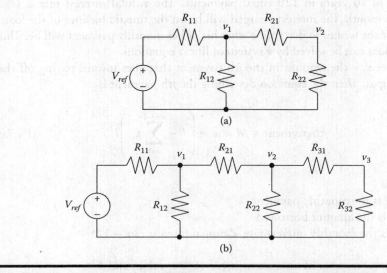

Figure P4.5 **(a) Second-order ladder network. (b) Third-order ladder network.**

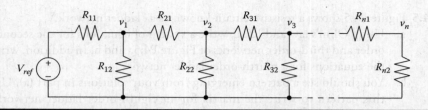

Figure P4.6 *n*th-order ladder network.

P4.6 As a follow-up to Project 4.5, write a MATLAB program to automatically solve the problem of a "ladder network" for an arbitrary n and R_o under the following conditions (see Figure P4.6):

$$R_o = R_{11} = R_{21} = R_{31} = \cdots = R_{(n-1)1} = R_{n1}$$

$$2R_o = R_{12} = R_{22} = R_{32} = \cdots = R_{(n-1)2}$$

Find solutions for the following two cases:

$$n = 16, \quad R_o = 1000\,\Omega, \quad V_{\text{ref}} = 10$$

$$n = 24, \quad R_o = 100\,\Omega, \quad V_{\text{ref}} = 1$$

P4.7 Suppose a manufacturer wishes to purchase a piece of equipment that costs $40,000. He or she plans to borrow the money from a bank and pay off the loan in 10 years in 120 equal payments. The annual interest rate is 6%. Each month, the interest charged will be on the unpaid balance of the loan. He or she wishes to determine what his or her monthly payment will be. This problem can be solved by a system of linear equations.

Let x_j = the amount in the jth payment that goes toward paying off the principal. Then the equation describing the jth payment is

$$j\text{th payment} = M = x_j + \left(P - \sum_{n=1}^{n=j-1} x_n \right) I \tag{P4.7a}$$

where
  M is the monthly payment
  P is the amount borrowed
  I is the monthly interest rate = annual interest rate ÷ 12

The total number of unknowns is 121 (120 x_j values and M).

Applying Equation P4.7a to each month gives 120 equations. One additional equation is

$$P = \sum_{n=1}^{n=120} x_n \qquad \text{(P4.7b)}$$

Develop a computer program that will
1. Ask the user to enter from the keyboard the amount of the loan, P; the annual interest rate, I; and the time period, Y, in years.
2. Set up the system of linear equations, using $A_{n,m}$ as the coefficient matrix of the system of linear equations. The n represents the equation number and m represents the coefficient of x_m in that equation. Set $x_{121} = M$.
3. Solve the system of linear equations in MATLAB.
4. Print out a table consisting of four columns. The first column should be the month number, the second column the monthly payment, the third column the amount of the monthly payment that goes toward paying off the principal, and the fourth column the interest payment for that month.

Reference

1. Kreyszig, E., *Advanced Engineering Mathematics*, 8th ed., Wiley, New York, 1999.

Chapter 5

Roots of Algebraic and Transcendental Equations

5.1 Introduction

In the analysis of various engineering problems, we often are faced with a need to find roots of equations whose solution is not easily found analytically. Given a function $f(x)$, the roots of the function are the values of x that makes $f(x) = 0$. Typical examples include nth degree polynomials and transcendental equations containing trigonometric functions, exponentials, or logarithms. In this chapter, we review several methods for solving such equations numerically. Also included is a section on MATLAB®'s fzero and roots functions, which may be used to obtain the roots of equations of the types just stated.

5.2 The Search Method

In the search method, we seek a small interval that contains a real root. This only gives an approximate value for the real root. We will then follow with one of several other methods to obtain a more accurate value for the root. This method is especially useful if there is more than one real root. The equation whose roots are to be determined should be put into the following standard form:

$$f(x) = 0 \tag{5.1}$$

We proceed as follows: first we subdivide the x domain into N equal subdivisions of width Δx, giving

$$x_1, x_2, x_3, \ldots, x_{n+1} \quad \text{with } x_{i+1} = x_i + \Delta x$$

Then, we determine where $f(x)$ changes sign (see Figure 5.1).

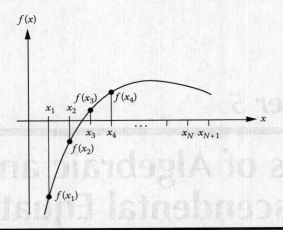

Figure 5.1 **The root of $f(x)$ lies between x_2 and x_3.**

This occurs when the signs of two consecutive values of $f(x)$ are different, that is,

$$f(x_i)f(x_{i+1}) < 0$$

The sign change usually indicates that a real root has been passed. However, it may also indicate a discontinuity in the function. (e.g., $\tan x$ is discontinuous at $x = \pi/2$.)

Once the intervals in which the roots lie have been established, we can use several different methods for obtaining the real roots, as described in the following sections.

5.3 Bisection Method

Suppose it has been established that a root lies between x_i and x_{i+1}. Then, cut the interval in half (see Figure 5.2), and thus,

$$x_{i+1/2} = x_i + \frac{\Delta x}{2}$$

Now compute $f(x_i)f(x_{i+1/2})$:

Case 1: If $f(x_i)f(x_{i+1/2}) < 0$, then the root lies between x_i and $x_{i+1/2}$.

Case 2: If $f(x_i)f(x_{i+1/2}) > 0$, then the root lies between $x_{i+1/2}$ and x_{i+1}.

Case 3: If $f(x_i)f(x_{i+1/2}) = 0$, then x_i or $x_{i+1/2}$ is a real root.

For cases 1 and 2, select the interval containing the root and repeat the process. Continue repeating the process, say r times, then $(\Delta x)_f = \Delta x/2^r$, where Δx is the initial size of the interval containing the root before the start of the bisection

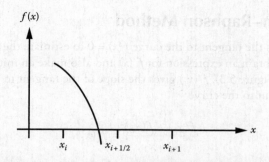

Figure 5.2   Bisecting the interval containing the root.

process ($\Delta x = x_{i+1} - x_i$) and $(\Delta x)_f$ is the size of the interval containing the root after r bisections. If $(\Delta x)_f$ is sufficiently small, then a very good approximation for the root is anywhere within the last bisected interval, say the midpoint of the interval.

For example, for 20 bisections,

$$(\Delta x)_f = \frac{\Delta x}{2^{20}} \approx \Delta x \times 10^{-6}$$

Program method:

$$\text{Set } x_A = x_i$$

$$x_B = x_{i+1}$$

$$x_C = \frac{1}{2}(x_A + x_B)$$

If $f(x_A)f(x_C) < 0$ (indicating that the root lies between x_A and x_C),

$$\text{then set } x_B = x_C$$

$$x_C = \frac{1}{2}(x_A + x_B)$$

and repeat the process.

Otherwise, if $f(x_A)f(x_C) > 0$ (indicating that the root lies between x_B and x_C),

$$\text{then set } x_A = x_C$$

$$x_C = \frac{1}{2}(x_A + x_B)$$

and repeat the process.

If $f(x_A)f(x_C) = 0$, then either x_A or x_C is a root.

5.4 Newton–Raphson Method

This method uses the tangent to the curve $f(x) = 0$ to estimate the root.

We need to obtain an expression for $f'(x)$ and also make an initial guess for the root, say x_1 (see Figure 5.3). $f'(x_1)$ gives the slope of the tangent to the curve at x_1.

On the tangent to the curve

$$\frac{f(x_1) - f(x_2)}{x_1 - x_2} = f'(x_1) \tag{5.2}$$

Our next guess (or iteration) for the root is obtained by setting $f(x_2) = 0$ and solving for x_2; that is,

$$x_2 = x_1 - \frac{f(x_1)}{f'(x_1)} \tag{5.3}$$

Check if $|f(x_2)| < \varepsilon$, where ε is the error tolerance.

If true, then quit. x_2 is the root, so print x_2.

If not true, then set $x_1 = x_2$ and repeat the process.

Continue repeating the process until $|f(x_n)| < \varepsilon$, where n is the number of iterations.

An alternate condition for convergence is

$$\left| \frac{f(x_1)}{f'(x_1)} \right| < \varepsilon \tag{5.4}$$

Usually, ε should be a very small number, such as 10^{-4}, but it can be either larger or smaller depending on how close you wish to get to the real root. You can always test the accuracy of the obtained root, by substituting the obtained root into the function $f(x)$ to see if it is sufficiently close to zero.

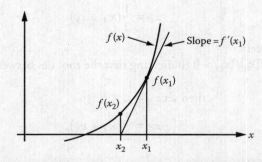

Figure 5.3   Using the tangent to the curve to close in on the root.

The Newton–Raphson method is widely used for its rapid convergence. However, there are cases where convergence will not occur. This can happen if

a. $f'(x)$ changes sign near the root.
b. The initial guess for the root is too far from the true root.

If we combine the Newton–Raphson method with the search method for obtaining a small interval in which the real root lies, then convergence is generally not a problem.

Review 5.1

1. What is meant by the term "root of function $f(x)$?"
2. What is the objective in the search method for determining a root of the equation $f(x) = 0$?
3. Describe the concept in the bisection method for obtaining a more accurate value for a root than what would normally be obtained by the search method alone.
4. Describe the concept in the Newton–Raphson method for obtaining a more accurate value for a root than what would normally be obtained by the search method alone.

5.5 MATLAB®'s Root-Finding Functions

MATLAB has built-in functions to determine the *real* roots of a function of one variable. The `fzero` function should be used for transcendental equations and the `roots` function should be used for polynomial equations. First, we will discuss the `fzero` function.

5.5.1 `fzero` Function

To get started, click in the Command window and type in the following:

```
>> help fzero
```

This gives (from MATLAB, with permission) the following:

```
FZERO Single-variable nonlinear zero finding.
  X = FZERO(FUN,X0) tries to find a zero of the function FUN
  near X0, if X0 is a scalar. It first finds an interval
  containing X0 where the function values of the interval
  endpoints differ in sign, then searches that interval for a
  zero. FUN is a function handle. FUN accepts real scalar input X
  and returns a real scalar function value F, evaluated at X. The
  value X returned by FZERO is near a point where FUN changes
  sign (if FUN is continuous), or NaN if the search fails.

  X = FZERO(FUN,X0), where X0 is a vector of length 2, assumes
  X0 is a finite interval where the sign of FUN(X0(1)) differs
```

```
from the sign of FUN(X0(2)). An error occurs if this is not
true. Calling FZERO with a finite interval guarantees FZERO
will return a value near a point where FUN changes sign.
```

The first usage (X0 is a scalar) is appropriate if you have some idea of where the root lies. This can be accomplished by plotting the function and noting where the function crosses the *x*-axis. The second usage (X0 is a vector of length 2) is more appropriate when there is more than one root and all roots need to be obtained. The second usage should be used in combination with the search method described earlier.

The first usage of the `fzero` function is

```
X = fzero(fun, X0)
```

`fun` is a function handle for the function whose root is to be determined and can be the name of a function file (with the *.m* extension) or can be an anonymous function (as described in Section 3.3). X0 is the initial guess for the root, and X is the solution determined by MATLAB's `fzero` function. Thus, suppose the name of the function file whose root is to be obtained is *myfun.m* and our guess for the root is 3.0. Then, we would write

```
X = fzero('myfun', 3.0)
```

An alternative and equivalent way to run the command is

```
X = fzero(@myfun, 3.0)  (no single quotation marks needed)
```

If no root is found, `fzero` returns NaN ("not a number").

In the second usage of `fzero`, X0 is a vector of length 2 that defines an interval over which `myfun` changes sign, that is, the value of `myfun` at X0(1) is opposite in sign to the value at X0(2). MATLAB gives an error message if this condition is not met.

In some instances, we would like to find the zero of a function of two arguments, say X and P, where P is a parameter and is fixed. In order to solve with the `fzero` function, P must be defined in the calling program. For example, suppose `myfun` is defined in an M-file as a function of two arguments:

```
function f = myfun(X,P)
f = cos(P*X);
```

The `fzero` statement would need to be invoked as follows:

```
P = 1000;
root_example = fzero (@(X) myfun(X, P), X0)
```

where `root` is the zero of function `myfun` when P = 1000. Note that P needs to be defined *before* the `fzero` function is called.

An alternative to adding parameter P as an argument to `myfun` is to use MATLAB's `global` statement to define parameter P in function `myfun` as an externally assigned variable. The `global` statement needs to be used both in the calling program and in the function `myfun`.

Example

Calling program:

```
global P;
P = 1000;
X0 = 10.0;
root_example = fzero(@myfun, X0);
```

The file *myfun.m*:
```
function f = myfun(x)
global P;
f = cos(P*x);
```

Example 5.1

The equation of state for a substance is a relationship between pressure, p; temperature, T; and specific volume, $\bar{v}$. Many gases at low pressures and moderate temperatures behave approximately as an ideal gas. The ideal gas equation of state with p in N/m^2, $\bar{v}$ in m^3/kmol, T in K, and $\bar{R}$ in (N-m)/(K-kmol) is

$$p = \frac{\bar{R}T}{\bar{v}} \tag{5.5}$$

where $\bar{R}$ is the universal gas constant. As temperature decreases and pressure increases, gas behavior deviates from ideal gas behavior. The Redlich–Kwong equation of state is often used to approximate nonideal gas behavior. The Redlich–Kwong equation of state is [1]

$$p = \frac{\bar{R}T}{\bar{v} - b} - \frac{a}{\bar{v}(\bar{v} + b)T^{1/2}}$$

or

$$f(\bar{v}) = \frac{\bar{R}T}{\bar{v} - b} - \frac{a}{\bar{v}(\bar{v} + b)T^{1/2}} - p \tag{5.6}$$

where a and b are constants dependent on the gas.

The values for $\bar{R}$, a, and b for carbon dioxide are tabulated in Table 5.1.

We wish to determine the percent error in the specific volume by using the ideal gas relationship while assuming that the Redlich–Kwong equation of state is the correct equation of state for carbon dioxide. Vary the temperature from 350 to 700 K in steps of 50 K while holding the pressure constant at 1.0132×10^7 N/m^2

Table 5.1 Values of *a*, *b*, and $\bar{R}$ for Carbon Dioxide in the Redlich–Kwong Equation of State

Gas	a (N-m⁴-K¹ᐟ²/kmol²)	b (m³/kmol)	$\bar{R}$ (N-m/K-kmol)
Carbon dioxide	65.43×10^5	0.02963	8314

Source: Moran, M.J. and Shapiro, H.N., *Fundamentals of Thermodynamics*, John Wiley & Sons, Hoboken, NJ, 2004.

Table 5.2 $\bar{V}$ Determined by Redlich–Kwong Equation and by Ideal Gas Law for CO_2

	Ideal Gas	Redlich–Kwong Equation	
T (K)	$\bar{v}$ (m³/kmol)	$\bar{v}$ (m³/kmol)	% Error in $\bar{v}$
350	—	—	—
400	—	—	—
⋮	⋮	⋮	⋮
700	—	—	—

(100 atm). Using the specified temperatures and pressure, determine the specific volumes, $\bar{v}$, by both the ideal gas equation and the Redlich–Kwong equation and determine the percent error in the specific volume resulting from the use of the ideal gas equation. Take the percent error in the specific volume to be

$$\% \text{ error} = \frac{\left| \bar{v}_{ideal\ gas} - \bar{v}_{Redlich-Kwong} \right|}{\bar{v}_{Redlich-Kwong}} \times 100 \qquad (5.7)$$

Write a MATLAB program utilizing the fzero function to calculate the specific volume by the Redlich–Kwong equation. Assume that $\bar{v}$ varies between 0.1 and 2.1 m³/kmol. Plot $f(\bar{v})$ vs. $\bar{v}$ for $T = 350$ and 700 K using 40 subdivisions on the $\bar{v}$ domain. Construct a table as shown in Table 5.2.

```
% Example_5_1.m
% This program compares ideal gas equation of state with
% Redlich-Kwong's equation of state for Carbon Dioxide.
% The pressure, p, is held constant at 100 atmosheres.
% The fzero function is used to obtain the value of v that
% satisfies the Redlich-Kwong equation of state.
% We expect one root so we do not have to use the search method
% for obtaining a small interval containing the root.
% We will use v obtained from the ideal gas law as our guess
% for the root of the Redlich-Kwong equation for v.
clear; clc;
```

```
p=1.0132e+07; R=8314; a=65.43e+5; b=0.02963;
Tt=350:50:700;
fprintf('This program compares ideal vs. real gas \n');
fprintf('behavior. Real gas behavior is determined by the \n');
fprintf('Redlich-Kwong equation of state \n');
fprintf('Pressure is held constant at 100 atmospheres \n\n');
% Table headings
fprintf('   T     v(m^3/kmol)     v(m^3/kmol)    error \n');
fprintf('  (K)       ideal      Redlich_Kwong   percent) \n');
fprintf('----------------------------------------------\n');
for k=1:length(Tt)
    T=Tt(k);
    v_ideal=R*T/p;
    v_RK=fzero(@(v) (R*T/(v-b)-a/(v*(v+b)*T^0.5)-p),v_ideal);
    err=abs((v_RK-v_ideal)/v_RK)*100;
    fprintf('%5.0f   %10.5f  %15.5f  %10.2f    \n',...
      T,v_ideal,v_RK,err);
    fprintf('----------------------------------------------\n');
end
```

Program results

This program compares ideal vs. real gas
behavior. Real gas behavior is determined by the
Redlich-Kwong equation of state.
Pressure is held constant at 100 atmospheres.

T (K)	v(m^3/kmol) ideal	v(m^3/kmol) Redlich_Kwong	error percent
350	0.28720	0.17852	60.88
400	0.32823	0.25870	26.88
450	0.36926	0.31985	15.45
500	0.41028	0.37404	9.69
550	0.45131	0.42455	6.30
600	0.49234	0.47279	4.14
650	0.53337	0.51950	2.67
700	0.57440	0.56514	1.64

Example 5.2

In this example, we use MATLAB's `fzero` function to find the roots of the governing equation for the voltage, $v(t)$, of a parallel RLC circuit. The governing equation for $v(t)$ was derived in Project P2.7. For the case when $\left(\dfrac{1}{2RC}\right)^2 < \dfrac{1}{LC}$, the system is underdamped and the governing equation for $v(t)$ is

$$v(t) = \exp\left(-\frac{1}{2RC}t\right)\left\{A\cos\left(\sqrt{\frac{1}{LC}-\left(\frac{1}{2RC}\right)^2}\ t\right) + B\sin\left(\sqrt{\frac{1}{LC}-\left(\frac{1}{2RC}\right)^2}\ t\right)\right\}$$

(5.8)

We wish to determine the first 3 zero crossings of $v(t)$ on the t axis of Equation 5.8. Assume the following parameters:

$$R = 100\ \Omega,\quad L = 10^{-3}\ \text{H},\quad C = 10^{-6}\ \text{F},\quad A = 6.0\ \text{V},\quad B = -9.0\ \text{V}$$

$$0 \le t \le 0.5\ \text{ms}$$

Program

```
% Example_5_2.m
% This script determines the first 3 zero crossings of the
% underdamped voltage of a parallel RLC circuit. The
% underdamped voltage is:
% v(t) = exp(-t/(2RC))*(A*cos(sqrt(1/LC-(1/2RC)^2)t)
%        + B*sin(sqrt(1/LC-(1/2RC)^2)t))
% The circuit component values and initial conditions are:
% R = 100 ohms; L = 1.0e-3 henry; C = 1.0e-6 farads;
% A = 6.0 volt, B = -9.0 volt
% Solve over the interval 0 <= t <= 0.5e-3 sec
clear; clc;
global R L C A B;
R = 100; L = 1.0e-3; C = 1.0e-6; A = 6.0; B = -9;
fprintf('Example 5.2: Find the first 3 zero \n');
fprintf('voltage crossings of the underdamped \n');
fprintf('RLC circuit \n');
tmin = 0.0; tmax = 0.5e-3;
% split up timespan into 50 intervals:
N = 50;
dt = (tmax-tmin)/N;
```

```
% First, calculate t and v(t) at each timestep
for n = 1:N+1
    t(n) = tmin+(n-1)*dt;
    v(n) = func_RLC(t(n));
end
plot(t,v), xlabel('t'), ylabel('v'),
title('v vs t for a RLC circuit'), grid;
% Next, use the search method to find the first 3 time
% intervals where the sign of v(t) changes. When found, use
% the fzero function to determine the root.
nr = 0;
for n = 1:N
    sign = v(n)*v(n+1);
    if sign <= 0.0
        nr = nr+1;
        tr(1) = t(n);
        tr(2) = t(n+1);
        troot(nr) = fzero('func_RLC',tr);
        if nr >= 3
            break;
        end
    end
end
if nr > 0
    fprintf('root #       troot      v(troot) \n');
    fprintf('----------------------------------------------\n');
    for n = 1:nr
        % Check solution
        v(n) = func_RLC(troot(n));
        fprintf('%3i    %10.4e  %10.4e \n',n,troot(n),v(n));
    end
else
    fprintf('\n\n No roots lie within %g <= t <= %g s\n',...
            tmin,tmax);
end
```

```
% func_RLC.m
% This function works with Example_5_2.m
function v = func_RLC(t)
global R L C A B;
arg1 = 1/(2*R*C);
arg3 = sqrt(1/(L*C)-1/(2*R*C)^2);
v = exp(-arg1*t)*(A*cos(arg3*t)+B*sin(arg3*t));
```

Program results

```
Example 5.2: Find the first 3 zero
voltage crossings of the underdamped
RLC circuit.
```

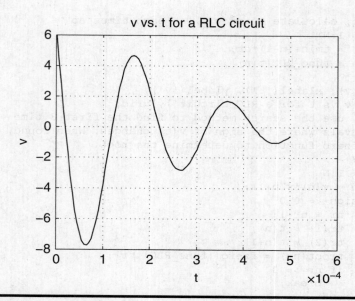

Figure 5.4 **Voltage *v* vs. time *t*.**

```
root number        troot            v(troot)
-----------------------------------------------
1              1.8831e-05       -3.233470e-15
2              1.1944e-04        9.776059e-16
3              2.2005e-04        3.251258e-15
>>
```

See Figure 5.4.

5.5.2 *roots* Function

MATLAB has a function to obtain the roots of a polynomial. The function is roots. To obtain the usage of the function, in the Command window type in

`>> help roots`

This gives (from MathWorks with permission)

```
ROOTS Find polynomial roots.
    ROOTS(C) computes the roots of the polynomial whose
    coefficients are the elements of the vector C. If C has
    N+1 components, the polynomial is C(1)*X^N + ···+ C(N)*X +
    C(N+1).
```

Thus, to find the roots of the polynomial $Ax^4 + Bx^3 + Cx^2 + Dx + E = 0$, run roots([A B C D E]). The roots function will give both real and imaginary roots of the polynomial.

Some additional useful MATLAB functions are:

poly(V) finds the coefficients of the polynomial whose roots are V.

real(X) gives the real part of X.

imag(X) gives the imaginary part of X.

Example 5.3

In this example, MATLAB's roots function is used to find the zeros of the polynomials $f(x) = x^3 - 5.7x^2 - 35.1x + 85.176$ and $g(x) = x^3 - 9x^2 + 23x - 65$.

```
% Example_5_3.m
% This program determines the roots of a polynomial using
% the built in function 'roots'.
% The first polynomial is: f = x^3-5.7*x^2-35.1*x+85.176.
% The roots of this polynomial are all real.
% The second polynomial is: g = x^3-9*x^2+23*x-65. The roots
% of this polynomial are both real and complex. Complex
% roots must be complex conjugates.
% To obtain more info on complex numbers in MATLAB, run
% "help complex" in the Command Window.
clear; clc;
% Define coefficients of first polynomial (real roots)
C = [1.0 -5.7 -35.1 85.176];
fprintf('The first polynomial coefficients are:\n');
C
fprintf('The roots are: \n');
V = roots(C)
fprintf('Polynomial coefficients determined from ');
fprintf('poly(V) are: \n');
C_recalc = poly(V)
fprintf('------------------------------------------\n');
% Define the coefficients of second polynomial
% whose roots are real and complex)
D = [1.0 -9.0 23.0 -65.0];
fprintf('The second polynomial coefficients are:\n');
D
fprintf('The roots are: \n');
W = roots(D)
fprintf('The real and imaginary parts of the roots are:\n');
re = real(W)
im = imag(W)
fprintf('Polynomial coefficients determined from ');
fprintf('poly(W) are: \n');
W_recalc = poly(W)
```

Program results

```
The first polynomial coefficients are:
C =
          1.0000  -5.7000  -35.1000  85.1760
```

```
The roots are:
V =
              8.6247
             -4.9285
              2.0038
Polynomial coefficients determined from poly(V) are:
C_recalc =
         1.0000  -5.7000  -35.1000  85.1760
-----------------------------------------
The second polynomial coefficients are:
D =
         1 -9 23 -65
The roots are:
W =
          7.0449 + 0.0000i
          0.9775 + 2.8759i
          0.9775 - 2.8759i
The real and imaginary parts of the roots are:
re =
          7.0449
          0.9775
          0.9775
im =
             0
          2.8759
         -2.8759
Polynomial coefficients determined from poly(W) are:
W_recalc =
         1.0000  -9.0000  23.0000  -65.0000
>>
```

Review 5.2

1. What is the name of the MATLAB function for determining the roots of a transcendental equation of the form $f(x) = 0$?
2. In MATLAB's function for determining the roots of a transcendental equation, how does one define the function whose roots are to be determined?
3. If you suspect that there is more than one real root, what method should be used in combination with the MATLAB's function to obtain the roots?
4. For the case described in item 3, what can you say about the second argument in MATLAB's function to obtain the roots?
5. What is the purpose of the global statement?
6. If the function $f(x)$ is a polynomial, what MATLAB function should you use to obtain its roots?

Projects

P5.1 Repeat Example 5.1 but replace the Redlich–Kwong equation with Van der Waals' equation [1]. In addition, repeat Example 5.1 for the three gases listed in Table P5.1. In your program, use a `for` loop to select one of the three gases and use an `if-elseif` ladder to select the proper constants for the gas. Van der Waals' equation of state is

$$p = \frac{\bar{R}T}{\bar{v} - b} - \frac{a}{\bar{v}^2} \tag{P5.1}$$

The constants a and b are tabulated in Table P5.1.

P5.2 The temperature distribution of a thick flat plate, initially at a uniform temperature, T_0, and which is suddenly immersed in a huge bath at a temperature T_∞, is given by (see Figure P5.2a)

$$T(x,t) = T_\infty + 2(T_0 - T_\infty)\sum_{n=1}^{\infty} \frac{\sin(\delta_n)\cos\left(\delta_n \frac{x}{L}\right)e^{-a\delta_n^2 t/L^2}}{\cos(\delta_n)\sin(\delta_n) + \delta_n} \tag{P5.2a}$$

where
  L is 1/2 of the plate thickness
  a is the thermal diffusivity of the plate material
  δ_n are the roots of the equation

$$F(\delta) = \tan\delta - \frac{hL}{k\delta} = 0 \tag{P5.2b}$$

where
  h is the convective heat transfer coefficient for the bath
  k is the thermal conductivity of the plate material

There are an infinite number of roots to Equation P5.2b. This can be seen in Figure P5.2b.

Table P5.1   Van der Waals' Constants [1]

Gas #	Gas	a (N-m⁴/kmol²)	b (m³/kmol)	$\bar{R}$ (N-m/K kmol)
1	Air	1.368×10^5	0.0367	8314
2	Oxygen	1.369×10^5	0.0317	8314
3	Carbon dioxide	3.647×10^5	0.0428	8314

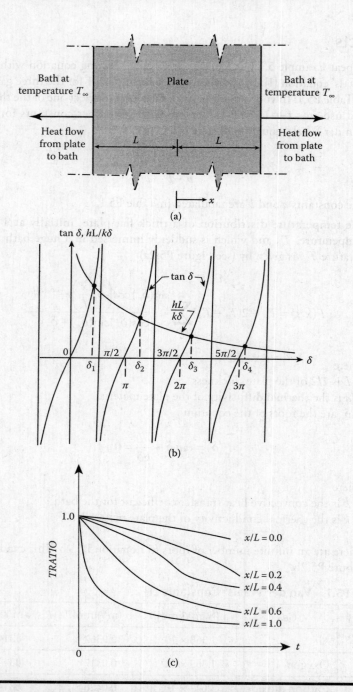

Figure P5.2 (a) Thick plate immersed in a huge bath. (b) Intersection of curves tan δ and $hL/k\delta$. (c) Sample plot of *TRATIO* vs. time t for various values of x/L.

The roots are δ_1, δ_2, δ_3, ..., δ_n. Note that δ_1 lies between 0 and $\pi/2$, δ_2 lies between π and $3\pi/2$, δ_3 lies between 2π and $5\pi/2$, etc. Subtracting T_∞ from Equation P5.2a and dividing by $T_0 - T_\infty$, we obtain the following equation

$$TRATIO = \frac{T\left(\frac{x}{L}, t\right) - T_\infty}{T_0 - T_\infty} = 2 \sum_{n=1}^{\infty} \frac{\sin(\delta_n)\cos\left(\delta_n \frac{x}{L}\right) e^{-a\delta_n^2 t / L^2}}{\cos(\delta_n)\sin(\delta_n) + \delta_n} \qquad \text{(P5.2c)}$$

A plot of *TRATIO* vs. time, for several different values of x/L, should appear as shown in Figure P5.2c.

Finally, the heat transfer ratio, *QRATIO*, from the plate to the bath in time t is given by

$$QRATIO = \frac{Q(t)}{Q_0} = \frac{2hL}{k} \sum_{n=1}^{\infty} \frac{\sin\delta_n \cos\delta_n}{\delta_n^2[\sin\delta_n \cos\delta_n + \delta_n]}\left[1 - e^{-at\delta_n^2/L^2}\right] \qquad \text{(P5.2d)}$$

where

$Q(t)$ is the amount of heat transferred from the plate to the bath in time t
Q_0 is the amount of heat transferred from the plate to the bath in infinite time which equals the change in internal energy in infinite time

1. Write a computer program that will solve for the roots, $\delta_1, \delta_2, \ldots, \delta_{50}$, using MATLAB's `fzero` function. Print out the δ values in 10 rows and 5 columns. Also print out the functional values at the roots, that is, $f(\delta_n)$. Note: Only 50 δ values need to be computed.
2. Solve Equation P5.2c for *TRATIO* for $x/L = 0.0, 0.2, 0.4, 0.6, 0.8, 1.0$ and $t = 0, 10, 20, \ldots, 200$ s. Print out results in table form as shown in Table P5.2. Also use MATLAB to produce a plot similar to Figure P5.2c.

Table P5.2   Temperature Ratio, *TRATIO*

Time (s)	X/L					
	0.0	0.2	0.4	0.6	0.8	1.0
0	1.0	1.0	1.0	1.0	1.0	1.0
10	–	–	–	–	–	–
20	–	–	–	–	–	–
⋮	⋮	⋮	⋮	⋮	⋮	⋮
200	–	–	–	–	–	–

3. Construct a table for *QRATIO* vs. *t* for times 0, 10, 20, 30, ... , 200 s.
4. Use MATLAB to produce a plot of *QRATIO* vs. *t*.
Use the following values for the parameters of the problem:

$$T_0 = 300°C, \quad T_\infty = 30°C, \quad h = 45 \text{ W/m}^2\text{-°C}, \quad k = 10.0 \text{ W/m-°C}$$

$$L = 0.03 \text{ m}, \quad \text{and} \quad a = 0.279 \times 10^{-5} \text{ m}^2/\text{s}$$

P5.3 In this project we consider a semi-infinite slab (such as a thick layer of ice) having a uniform temperature, T_i, that is suddenly subjected to a change in air temperature that is caused by a warm front moving in over the region of interest (see Figure P5.3). The temperature, T, in the slab will be a function of position and time; that is, $T = T(x,t)$. It will also depend on the parameters: h, T_i, T_∞, k and α, where h is the convective heat transfer coefficient, T_i is the initial temperature of the slab and T_∞ is the air temperature, k and α are the thermal conductivity and diffusivity of the slab material respectively. The problem can be solved by Laplace Transforms. The solution is:

$$1 - erf\left(\frac{x}{2\sqrt{\alpha t}}\right) - e^{\left(\frac{hx}{k} + \frac{h^2\alpha t}{k^2}\right)} \times \left[1 - erf\left(\frac{x}{2\sqrt{\alpha t}} + \frac{h\sqrt{\alpha t}}{k}\right)\right] - \frac{T(x,t) - T_i}{T_\infty - T_i} = 0$$

(P5.3)

Given: $T_i = -20°C$, $T_\infty = 10°C$, $k = 2.22$ W/m-C, $\alpha = 12.6 \times 10^{-7}$ m²/s and $h = 100$ W/m²-C.

We wish to determine the time, t, when the temperature in the slab reaches the following temperatures at the following positions.

$$T = [-10, -5, 0]°C \quad \text{and} \quad x = [0, 0.02, 0.04] \text{ m}$$

Use the search method to find an interval in which the root of Equation P5.3 lies.
Then use MATLAB's fzero function to solve for *t* for each condition and print the results with table headings.
Assume that $0 \le t \le 10000$ s with a step size of 10 s.

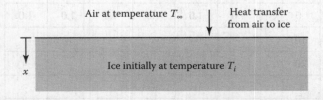

Figure P5.3 Semi-infinite slab subjected to a change in air temperature.

P5.4 A wood circular cylinder, having a specific gravity, S, floats in water as shown in Figure P5.4.

For a floating body, the weight of the floating body equals the weight of fluid displaced; thus,

$$S\gamma_w \pi R^2 L = \gamma_w V \qquad \text{(P5.4a)}$$

where
S is the specific gravity of the wood
γ_w is the specific weight of water
R is the radius of the cylinder
L is the cylinder length
V is the volume of water displaced

We define a cross-sectional differential area dA as shown in Figure P5.4 and calculate the corresponding differential volume dV as follows:

$$dV = L\,dA = 2Lx(y)dy = 2L\sqrt{R^2 - y^2}\,dy$$

$$V = 2L\int_{-R}^{d-R} \sqrt{R^2 - y^2}\,dy = 2L\left\{\frac{1}{2}\left(y\sqrt{R^2 - y^2} + R^2 \sin^{-1}\frac{y}{R}\right)\right\}_{-R}^{d-R} \qquad \text{(P5.4b)}$$

The solution of the integral was obtained from integral tables. Substituting the limits of integration gives

$$V = L\left((d - R)\sqrt{R^2 - (d-R)^2}\right) + R^2 \sin^{-1}\frac{d - R}{R} - R^2 \sin^{-1}(-1)$$

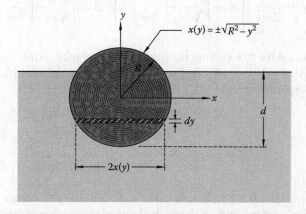

Figure P5.4 Floating wood circular cylinder.

$$V = L\left\{(d - R)\sqrt{R^2 - (d - R)^2} + R^2 \sin^{-1}\frac{d - R}{R} - R^2 \sin^{-1}(-1)\right\} \quad \text{(P5.4c)}$$

Take $\sin^{-1}(-1) = -\pi/2$

Substituting Equation P5.4c into Equation P5.4a and rearranging terms and dividing by R^2 gives

$$f\left(\frac{d}{R}\right) = \left(\frac{d}{R} - 1\right)\sqrt{2\frac{d}{R} - \left(\frac{d}{R}\right)^2} + \sin^{-1}\left(\frac{d}{R} - 1\right) - (S - 0.5)\pi = 0 \quad \text{(P5.4d)}$$

Create a MATLAB program that will use MATLAB's `fzero` function to solve Equation P5.4d for d for the following parameters: $R = 0.5$ m and $0.3 \le S \le 0.5$ in steps of 0.1. Create a table consisting of S and d.

P5.5 Do parts (a) and (b) of Project P2.4 and then, using MATLAB's `fzero` function, determine

a. The times when the velocity of the piston, described in that project, reaches one-half of its maximum positive velocity during the time span $0 \le t \le 0.01$ s. Print these values to the screen.

b. The times when the acceleration of the piston, described in that project, reaches one-half of its maximum positive acceleration during the time span $0 \le t \le 0.01$ s.

P5.6 Determine the first root of the displacement, $y(t)$, of the underdamped vibration problem described in Project P2.5. The governing equation for $y(t)$ is

$$y = \exp\left(-\frac{c}{2m}t\right)\left\{A\cos\left(\sqrt{\frac{k}{m} - \left(\frac{c}{2m}\right)^2}\, t\right) + B\sin\left(\sqrt{\frac{k}{m} - \left(\frac{c}{2m}\right)^2}\, t\right)\right\} \quad \text{(P5.6)}$$

Use MATLAB's `fzero` function to find the first root. Print this value to the screen. Also plot y vs. t for $0 \le t \le 20$ s. Use a step size of 0.1 s. Assume the following parameters:

$$m = 25 \text{ kg}, \quad k = 200 \text{ N/m}, \quad c = 5 \text{ N-s/m}$$

$$A = 0.2 \text{ m} \quad \text{and} \quad B = \frac{c}{2m} \times \frac{A}{\sqrt{k/m - (c/2m)^2}}$$

P5.7 In the time span $0 \le t \le 20$ s, determine the number of roots and their values of the displacement, $y(t)$, of the underdamped vibration problem described in Project P5.6.

a. Use the search method to find a small interval in which each root lies.
b. In each found interval, use MATLAB's `fzero` function to find the root value.
c. Print out to the screen the root number and the root value.

P5.8 Determine the first root of the voltage, $v(t)$, of the underdamped parallel RLC circuit described in Project P2.7. The governing equation for $v(t)$ is

$$v(t) = \exp\left(-\frac{1}{2RC}t\right)\left\{A\cos\left(\sqrt{\frac{1}{LC}-\left(\frac{1}{2RC}\right)^2}\,t\right)+B\sin\left(\sqrt{\frac{1}{LC}-\left(\frac{1}{2RC}\right)^2}\,t\right)\right\}$$

(P5.8)

Use MATLAB's `fzero` function to find the first root. Print this value to the screen. Also plot v vs. t. Assume the following parameters:

$$R = 100\ \Omega,\quad L = 10^{-3}\ \text{H},\quad C = 10^{-6}\ \text{F},\quad A = 6.0\ \text{V},\quad B = -9.0\ \text{V}$$

$$0 \le t \le 0.5\ \text{ms}$$

P5.9 In the time span $0 \le t \le 0.5$ ms, determine the number of roots and their values of the voltage, $v(t)$, of the underdamped parallel RLC circuit described in Project P2.7. The governing equation for $v(t)$ is

$$v(t) = \exp\left(-\frac{1}{2RC}t\right)\left\{A\cos\left(\sqrt{\frac{1}{LC}-\left(\frac{1}{2RC}\right)^2}\,t\right)+B\sin\left(\sqrt{\frac{1}{LC}-\left(\frac{1}{2RC}\right)^2}\,t\right)\right\}$$

(P5.9)

a. Use the search method to find a small interval in which each root lies.
b. In each found interval, use MATLAB's `fzero` function to find the root value.
c. Print out to the screen the root number and the corresponding root value. Assume the following parameters:

$$R = 100\ \Omega,\quad L = 10^{-3}\ \text{H},\quad C = 10^{-6}\ \text{F},\quad A = 6.0\ \text{V},\quad B = -9.0\ \text{V}$$

$$0 \le t \le 0.5\ \text{ms}$$

P5.10 In Project P2.7, instead of solving for $v(t)$, we could have solved for the inductor current, $i_L(t)$. For the underdamped case, the solution for $i_L(t)$ would take the same form as $v(t)$ and is given in Equation P5.10a:

$$i_L = \exp\left(-\frac{1}{2RC}t\right)\left\{A\cos\left(\sqrt{\frac{1}{LC}-\left(\frac{1}{2RC}\right)^2}\,t\right)+B\sin\left(\sqrt{\frac{1}{LC}-\left(\frac{1}{2RC}\right)^2}\,t\right)\right\}$$

(P5.10a)

For this project, we wish to consider the *settling time* (described as follows) of the inductor current, i_L, in this parallel RLC circuit. The resistance R_{crit} that makes the circuit critically damped is given by

$$R_{crit} = \sqrt{\frac{L}{4C}} \qquad \text{(P5.10b)}$$

The underdamped case occurs when $R > R_{crit}$.

The coefficients, A and B, are determined by the initial conditions and the component values. The solutions for A and B are

$$A = i_L(0)$$

$$B = \frac{1}{\sqrt{\dfrac{1}{LC} - \left(\dfrac{1}{2RC}\right)^2}} \times \left(\frac{v(0)}{L} + \frac{i_L(0)}{2RC}\right) \qquad \text{(P5.10c)}$$

The settling time of a waveform is the time t_{settle} that it takes the waveform i_L to reach within a specified percentage of some specified value and stay within that range (see Figure P5.10). For this problem, we will take the settling time for $i_L(t)$ to be within ±10% of the initial value, $i_L(0)$, and stay within that

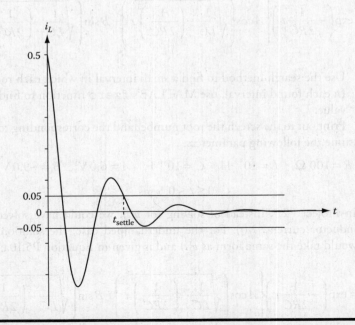

Figure P5.10   Settling time for current i_L.

range. Write a MATLAB program that determines the settling time for the following circuit parameters:

$$R = 4R_{crit}, \quad L = 1.0 \text{ mH}, \quad C = 1.0 \text{ μF}, \quad i_L(0) = 0.5 \text{ A},$$

$$v(0) = -6 \text{ V}, \quad 0 \le t \le 1000 \text{ μs}$$

1. Create a plot of $i_L(t)$ vs. t. On the same plot, create the lines $0.1 i_L(0)$ and $-0.1 i_L(0)$.
2. Use the search method and MATLAB's `fzero` function to determine the settling time of $i_L(t)$. Hint: Use $|i_L(t)|$ and the $0.1 i_L(0)$ line to find the maximum root, thus determining the settling time.

P5.11 The current–voltage relationship of a semiconductor PN diode can be written as follows:

$$i_D = I_S \left(e^{\frac{q}{kT} v_D} - 1 \right) \tag{P5.11a}$$

where

i_D and v_D are the diode current and voltage as defined in Figure P5.11
I_S is a constant (with units of amperes) that is determined by the semiconductor doping concentrations and the device geometry
$q = 1.6 \times 10^{-19}$ C is the unit electric charge
$k = 1.38 \times 10^{-23}$ J/K is the Boltzmann constant
T is the absolute temperature (in K)

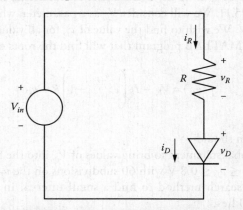

Figure P5.11   Semiconductor PN diode.

We can determine the current and voltage through the diode for this circuit using Kirchhoff's voltage law. We know that the sum of the voltage drops around the circuit must sum to zero:

$$V_{in} - v_R - v_D = 0 \qquad \text{(P5.11b)}$$

Applying Ohm's law for the current through the resistor, $v_R = i_R R$, and observing that the resistor current equals the diode current $i_R = i_D$, we can rewrite Equation P5.11b as

$$V_{in} - I_S \left(e^{\frac{q}{kT} v_D} - 1 \right) R - v_D = 0 \qquad \text{(P5.11c)}$$

Let

$$f(v_D) = V_{in} - I_S \left(e^{\frac{q}{kT} v_D} - 1 \right) R - v_D \qquad \text{(P5.11d)}$$

1. Create a MATLAB function for $f(v_D)$ and plot for the interval $0 \leq v_D \leq 0.8$ V for 10 mV steps (80 subdivisions on the v_D domain).
2. Use the search method to obtain a small interval within which the root of Equation P5.11d lies.
3. Use MATLAB's `fzero` function to obtain a more accurate value for the root. Use the following parameters:

$$T = 300 \text{ K}, \quad I_S = 10^{-14} \text{ A}, \quad V_{in} = 5 \text{ V}, \quad R = 1000 \ \Omega$$

4. Print the root value to the screen.

P5.12 We wish to determine the DC transfer characteristic for the diode circuit of Figure P5.11. We will consider V_{in} as a parameter, where $5 \leq V_{in} \leq 12$ in steps of 1 V. We wish to find the value of v_D for all values of V_{in}.

Write a MATLAB program that will find the roots of $f(v_D) = 0$, where

$$f(v_D) = V_{in} - I_S \left(e^{\frac{q}{kT} v_D} - 1 \right) R - v_D$$

Your program should
1. Use a global statement to bring values of V_{in} into the function $f(v_D)$.
2. Take $0.2 \leq v_D \leq 0.8$ V with 60 subdivisions on the v_D domain.
3. Use the search method to find a small interval in which the root of $f(v_D) = 0$ lies.

4. Use the MATLAB's `fzero` function to obtain a more accurate value for the root.
5. Construct a table consisting of all values of V_{in} and the corresponding roots of $f(v_D) = 0$.
6. If you did Project P5.11, confirm that your program returns the same result as in Project P5.11 when $V_{in} = 5$ V.

Reference

1. Moran, M.J. and Shapiro, H.N., *Fundamentals of Thermodynamics*, John Wiley & Sons, Hoboken, NJ, 2004.

Chapter 6

Numerical Integration

6.1 Introduction

In this chapter, we cover both the trapezoidal rule and Simpson's rule for approximating the value of definite integrals. We then demonstrate the usage of MATLAB®'s `quad` and `dblquad` functions for evaluating definite integrals. Finally, examples demonstrating the usage of these four methods are given.

6.2 Numerical Integration with the Trapezoidal Rule

We wish to evaluate the integral I where

$$I = \int_a^b f(x)dx = \text{the area under the curve } f(x) \qquad (6.1)$$

by the trapezoidal rule (see Figure 6.1).

The steps for calculating the integral, I, numerically by the trapezoidal rule are as follows:

- Subdivide the x-axis from $x = a$ to $x = b$ into N subdivisions, giving

$$\Delta x = \frac{b - a}{N} \qquad (6.2)$$

- The area, A_j, under the curve in jth interval can be approximated as the area of the trapezoid bounded by $(x_j, f_j, f_{j+1}, x_{j+1})$, which is

$$A_j = \frac{1}{2}(f_j + f_{j+1})\Delta x \qquad (6.3)$$

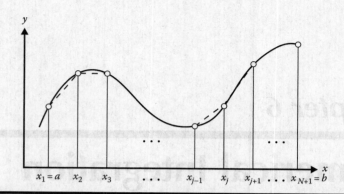

Figure 6.1 Subdivision on *x* domain for trapezoidal rule.

■ Repeat this process for all interval areas, giving

$$A_1 = \frac{1}{2}(f_1 + f_2)\Delta x, \quad A_2 = \frac{1}{2}(f_2 + f_3)\Delta x,$$

$$A_3 = \frac{1}{2}(f_3 + f_4)\Delta x, \cdots, \quad A_N = \frac{1}{2}(f_N + f_{N+1})\Delta x \tag{6.4}$$

■ We now sum all the interval areas $(A_1, A_2, \ldots, A_N)$ to obtain the trapezoidal rule for evaluating the integral, I, which is

$$I = \int_{x_1}^{x_{N+1}} f(x)dx \approx \left(\frac{1}{2}f_1 + f_2 + f_3 + \cdots + f_N + \frac{1}{2}f_{N+1}\right)\Delta x \tag{6.5}$$

Example 6.1

Solve the definite integral, I, by the trapezoidal rule:

$$I = \int_0^{10} (x^3 + 3.2x^2 - 3.4x + 20.2)dx$$

```
% Example_6_1.m
% This program evaluates the integral by the trapezoidal rule
% The integrand is: x^3+3.2*x^2-3.4*x+20.2
% The limits of integration are from 0-10.
clear; clc;
a = 0; b = 10;
N = 100; dx = (b-a)/N;
% Compute values of x and f at each point:
```

```
% An arithmetic expression of vector x produces a vector f.
% Need to use element by element multiplication.
x = a:dx:b;
f = x.^3 + 3.2*x.^2 - 3.4*x + 20.2;
% Calculate the integral as per Equation 6.5
I = dx*(sum(f)-0.5*f(1)-0.5*f(N+1));
% Display results
fprintf('Integrand: x^3 + 3.2*x^2 - 3.4*x + 20.2 \n');
fprintf('Integration limits:%.1f to %.1f \n',a,b);
fprintf('Trapezoidal solution, I = %10.4f \n', I);
% Compare with analytical solution, which is:
% I2 = 1/4*x^4+3.2/3*x^3-2.4/2*x^2+20.2*x with
% limits from 0 to 10.
I2 = 0.25*10^4+3.2/3*10^3-3.4/2*10^2+20.2*10;
fprintf('Analytical solution, I2 =%10.4f \n',I2);
```

Program results

```
Integrand: x^3 + 3.2*x^2 - 3.4*x + 20.2
Integration limits: 0.0 to 10.0
Trapesoidal solution, I = 3598.9700
Analytical solution, I2 = 3598.6667
>>
```

We see that the integration for this integral by the trapezoidal rule gives the correct answer up to four significant figures.

6.3 Numerical Integration and Simpson's Rule

We can also evaluate an integral of a single variable by Simpson's rule, which is a more accurate method than the method using the trapezoidal rule. In Simpson's rule, three points on the curve $f(x)$ are connected by second-degree polynomials (parabolas) and we then sum the areas under the parabolas to obtain the approximate area under the curve (see Figure 6.2). As in the trapezoidal rule case, the x domain is subdivided into N intervals, but in this case, N must be an even number. We proceed by first expanding $f(x)$ in a Taylor series about x_i using three terms; that is,

$$f(x) = a(x - x_i)^2 + b(x - x_i) + c \tag{6.6}$$

Then, the area under two adjacent strips $A_{2\text{strips}}$ is computed by integrating $f(x)$ between x_{i-1} and x_{i+1}:

$$A_{2\text{strips}} = \int_{x_{i-1}}^{x_{i+1}} f(x)dx = \int_{x_i - \Delta x}^{x_i + \Delta x} [a(x - x_i)^2 + b(x - x_i) + c]dx \tag{6.7}$$

Let $\xi = x - x_i$.

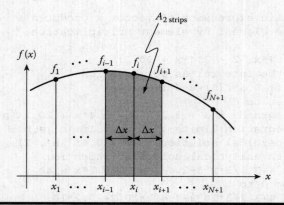

Figure 6.2   Area under 2 adjacent strips.

Then,

$$d\xi = dx$$

When $x = x_i - \Delta x$, $\xi = -\Delta x$, and when $x = x_i + \Delta x$, $\xi = \Delta x$.

Making these substitutions into Equation 6.7 gives

$$A_{2\,\text{strips}} = \int_{x_{i-1}}^{x_{i+1}} f(x)dx$$

$$= \int_{-\Delta x}^{\Delta x} [a\xi^2 + b\xi + c]d\xi$$

$$= \frac{a\xi^3}{3} + \frac{b\xi^2}{2} + c\xi \Big|_{-\Delta x}^{\Delta x}$$

$$= \frac{a}{3}(\Delta x)^3 - \frac{a}{3}(-\Delta x)^3 + \frac{b}{2}(\Delta x)^2 - \frac{b}{2}(-\Delta x)^2 + c(\Delta x) - c(-\Delta x)$$

Thus,

$$A_{2\,\text{strips}} = \frac{2a}{3}(\Delta x)^3 + 2c\Delta x$$

Now, define $f_i = f(x_i)$. Solving Equation 6.6 for f_i, f_{i+1}, and f_{i-1} gives

$$f(x_i) = f_i = c$$

$$f(x_{i+1}) = f_{i+1} = a(\Delta x)^2 + b\Delta x + c$$

$$f(x_{i-1}) = f_{i-1} = a(-\Delta x)^2 + b(-\Delta x) + c$$

Adding the above last two equations gives $f_{i+1} + f_{i-1} = 2a\Delta x^2 + 2c$.

Solving for a gives

$$a = \frac{1}{2\Delta x^2}[f_{i+1} + f_{i-1} - 2f_i]$$

Then,

$$A_{2\,\text{strips}} = \frac{2}{3} \times \frac{1}{2\Delta x^2}[f_{i+1} + f_{i-1} - 2f_i](\Delta x^3) + 2f_i\Delta x = \frac{\Delta x}{3}[f_{i+1} + f_{i-1} - 2f_i + 6f_i]$$

or

$$A_{2\,\text{strips}} = \frac{\Delta x}{3}[f_{i-1} + 4f_i + f_{i+1}] \tag{6.8}$$

To obtain an approximation for the integral I, we need to sum all the 2-strip areas under the curve from $x = A$ to $x = B$ (see Figure 6.3); that is,

$$A_1 = \frac{\Delta x}{3}[f_1 + 4f_2 + f_3]$$

$$A_2 = \frac{\Delta x}{3}[f_3 + 4f_4 + f_5]$$

$$A_3 = \frac{\Delta x}{3}[f_5 + 4f_6 + f_7]$$

$$\vdots$$

$$A_{N/2} = \frac{\Delta x}{3}[f_{N-1} + 4f_N + f_{N+1}]$$

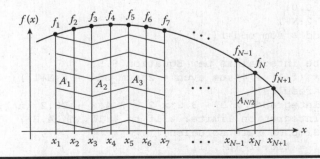

Figure 6.3   Summing all the 2 strip areas in Simpson's rule.

Thus,

$$I = \int_{x_1}^{x_{N+1}} f(x)dx = \frac{\Delta x}{3}[f_1 + 4f_2 + 2f_3 + 4f_4 + 2f_5 + \cdots + 4f_N + f_{N+1}] \qquad (6.9)$$

This is Simpson's rule for integration.

Example 6.2

Solve by Simpson's rule:

$$I = \int_0^{10} (x^3 + 3.2x^2 - 3.4x + 20.2)dx$$

```
% Example_6_2.m
% This program calculates an integral in Example 6.1 by
% Simpson's Rule
% The integrand is: x^3+3.2*x^2-3.4*x+20.2
% The limits of integration are from 0-10.
clear; clc;
A = 0; B = 10;
N = 100; dx = (B-A)/N;
% Compute values of x and f at each point:
% An arithmetic expression of vector x produces a vector f.
% Need to use element by element multiplication.
x = A:dx:B;
f = x.^3 + 3.2*x.^2 - 3.4*x + 20.2;
% Use two separate loops to sum up the even and odd terms
% of Simpson's Rule. Also, exclude endpoints in the loop.
sum_even = 0.0;
for i = 2:2:N
    sum_even = sum_even+f(i);
end
sum_odd = 0.0;
for i = 3:2:N-1
    sum_odd = sum_odd+f(i);
end
% Calculate integral as per Equation 6.5
I = dx/3 * (f(1) + 4*sum_even + 2*sum_odd + f(N+1));
% Display results
fprintf('Integrand: x^3 + 3.2*x^2 - 3.4*x + 20.2 \n');
fprintf('Integration limits: %.1f to %.1f \n',A,B);
fprintf('Simpson rule solution,');
fprintf('I = %10.4f \n', I);
% Compare with analytical solution, which is:
I2 = 0.25*10^4+3.2/3*10^3 - 3.4/2*10^2+20.2*10;
```

```
fprintf('Analytical solution,');
fprintf('I2 = %10.f \n',I2);
```

Program results

```
Integrand: x^3 + 3.2*x^2 - 3.4*x + 20.2
Integration limits: 0.0 to 10.0
Simpson rule solution, I = 3598.6667
Analytical solution, I2 = 3598.6667
>>
```

We see that solving the integral of Example 6.1 by Simpson's rule gives a slightly better answer than solving the same integral by the trapezoidal rule.

Review 6.1

1. What is the formula for evaluating the integral, $I = \int_A^B f(x)dx$, by the trapezoidal rule?

2. What is the formula for evaluating the integral, $I = \int_A^B f(x)dx$, by Simpson's rule?

Exercise

E6.1 Evaluate the following definite integrals by Simpson's rule:

a. $I = \int_0^3 \dfrac{dx}{5e^{3x} + 2e^{-3x}}$

b. $I = \int_{-\pi/2}^{\pi/2} \dfrac{\sin x\, dx}{\sqrt{1 - 4\sin^2 x}}$

c. $I = \int_0^\pi (\sinh x - \cos x)dx$

6.4 Improper Integrals

An integral is improper if the integrand approaches infinity at some point within the limits of integration, including the end points. In many cases, the integration will still result in a finite solution.

Example 6.3

$$I = \int_0^1 \frac{\log(1 + x)}{x}\,dx \tag{6.10}$$

The integral in Equation 6.10 is improper since both the numerator and denominator are zero at the lower limit ($x = 0$). The exact value of I can be obtained by residue theory in complex variables, and in this case the integral, I, evaluates to $\pi^2/12 = 0.822467$.

Let

$$I = \int_0^1 \frac{\log(1+x)}{x} dx = I_1 + I_2$$

where

$$I_1 = \int_\varepsilon^1 \frac{\log(1+x)}{x} dx$$

and

$$I_2 = \int_0^\varepsilon \frac{\log(1+x)}{x} dx$$

and $\varepsilon \ll 1$. To evaluate I_2, expand $\log(1 + x)$ in a Taylor series about $x = 0$, giving

$$\log(1+x) = x - \frac{1}{2}x^2 + \frac{1}{3}x^3 - \frac{1}{4}x^4 + \frac{1}{5}x^5 - \cdots + \cdots$$

then

$$\frac{\log(1+x)}{x} = 1 - \frac{1}{2}x + \frac{1}{3}x^2 - \frac{1}{4}x^3 + \frac{1}{5}x^4 - \cdots + \cdots$$

$$I_2 = \int_0^\varepsilon \left[1 - \frac{1}{2}x + \frac{1}{3}x^2 - \frac{1}{4}x^3 + \frac{1}{5}x^4 - \frac{1}{6}x^5 + \frac{1}{7}x^6 - \cdots + \cdots \right] dx$$

$$I_2 = \varepsilon - \frac{1}{4}\varepsilon^2 + \frac{1}{9}\varepsilon^3 - \frac{1}{16}\varepsilon^4 + \frac{1}{25}\varepsilon^5 - \frac{1}{36}\varepsilon^6 + \frac{1}{49}\varepsilon^7 - \cdots + \cdots \quad (6.11)$$

Evaluate I_1 by Simpson's rule and evaluate I_2 by Equation 6.11. The following program illustrates the method.

```
% Example_6_3.m
% This program evaluates the improper integral: log(1+x)/x
% The limits of integration are from 0 to 1.
% The integrand is undefined (0/0)at x = 0. Thus, the
% integral is broken up into 2 parts: I1 and I2.
% I1 is evaluated from epsilon to 1.
```

```
% I2 is expanded in a Taylor Series and evaluated from 0 to
% epsilon.
clear; clc;
epsilon = 0.0001;
A = 0; B = 1;
N = 100; dx = (B-epsilon)/N;
% Evaluate I1. First, evaluate log(1+x)/x for each value of
% x over the interval [epsilon,1].
x = epsilon:dx:B;
f = log(1+x)./x;
% Next, calculate the even and odd terms for Simpson's Rule.
sumeven = 0.0;
for i = 2:2:N
    sumeven = sumeven+f(i);
end
sumodd = 0.0;
for i = 3:2:N-1
    sumodd = sumodd+f(i);
end
% As per Equation 6.5, I1 is the weighted sum of the even
% and odd terms, plus the end terms.
I1 = dx/3 * (f(1) + 4*sumeven + 2*sumodd + f(N+1));
fprintf('Integrand = log(1+x)/x n');
fprintf('I1 limits of integration are from %.4f to %.4f\n',...
        epsilon,B);
fprintf('I2 limits of integration are from %.4f to %.4f\n',...
        A,epsilon);
fprintf('I1 =%16.6f \n',I1);
% Calculate I2 via first 4 terms of Taylor series expansion.
I2 = epsilon - 1/4*epsilon^2 + 1/9*epsilon^3...
     -1/16*epsilon^4;
fprintf('I2 =%16.6f \n',I2);
I = I1+I2;
fprintf('I = I1+I2 =%10.6f \n',I);
I_exact = 0.822467;
fprintf('I_exact =%10.6f \n',I_exact);
```

Program results

```
Integrand = log(1+x)/x
I1 limits of integration are from 0.0001 to 1.0000
I2 limits of integration are from 0.0000 to 0.0001
I1 =        0.822367
I2 =        0.000100
I = I1+I2 = 0.822467
I_exact =   0.822467
```

As we see the method works quite well.

6.5 MATLAB®'s quad Function

The MATLAB function for evaluating integrals is the function quad. A description of the function can be obtained by typing **help quad** in the Command window (from MathWorks, with permission):

```
Q = QUAD(FUN,A,B) tries to approximate the integral of scalar-
valued function FUN from A to B to within an error of 1.e-6
using recursive adaptive Simpson quadrature. FUN is a function
handle. The function Y = FUN(X) should accept a vector argument
X and return a vector result Y, the integrand evaluated at each
element of X.

Q = QUAD(FUN,A,B,TOL) uses an absolute error tolerance of TOL
instead of the default, which is 1.e-6. Larger values of TOL
result in fewer function evaluations and faster computation,
but less accurate results. The QUAD function in MATLAB 5.3 used
a less reliable algorithm and a default tolerance of 1.e-3.
```

Thus, the quad function takes as arguments a function handle FUN that defines the integrand in a *.m* file and the limits of integration. Alternatively, if the integrand is not very large and can be expressed in a single line, then you can define the integrand within your script with an anonymous function (see Examples 6.4 and 6.5). If the integrand involves very small numbers or very large numbers, you might wish to change the default tolerance by adding a third argument to quad (as shown in the above second usage description). The quad function is able to evaluate certain improper integrals (see Exercises E6.2d, E6.2e, and E6.2f). It does this by selecting limits of integration that are very close to the singular points, but not on them, thus removing the singularity.

Example 6.4

We will now repeat Example 6.2, but this time we will use MATLAB's quad function to do the integration. The integral I in Example 6.2 is

$$I = \int_0^{10} (x^3 + 3.2x^2 - 3.4x + 20.2)dx$$

The program follows:

```
% Example_6_4.m
% This program evaluates the integral of the function 'f1'
% between A and B by MATLAB's quad function. Since the
% function 'f1' is just a single line, we can use the
% anonymous form of the function.
clear; clc;
f1 = @(x).(x.^3+3.2*x.^2-3.4*x+20.2);
```

```
A = 0.0; B = 10.0;
I = quad(f1,A,B);
fprintf('Integration of f1 over [%.0f,%.0f] ',A,B);
fprintf('by MATLAB"s quad function:\n');
fprintf('f1 = x^3 + 3.2*x^2 - 3.4*x + 20.2 \n');
fprintf('integral = %10.4f \n', I);
```

Program results

```
Integration of f1 over [0,10] by MATLAB's quad function:
f1 = x^3 + 3.2*x^2 - 3.4*x + 20.2
integral = 3598.6667
>>
```

We see that the results are the same as those obtained in Example 6.2.

Review 6.2

1. What is the name of MATLAB's function for integration of a single variable?
2. In MATLAB's function for integration, how does one define the function to be integrated?
3. If the integrand contains nonlinear terms, how must they be treated?
4. Will MATLAB's quad function treat improper integrals?

Example 6.5

$$\text{Evaluate} \int_0^1 \frac{t}{t^3 + t + 1}\, dt$$

```
% Example_6_5.m
% This program evaluates the integral of Example 6.5 by
% MATLAB's quad function. Since the integrand can be
% expressed in a single line, we can use the anonymous form
% of the function.
clear; clc;
A = 0.0; B = 1.0;
f2 = @(t) t./(t.^3 +t + 1.0);
I2 = quad(f2,A,B);
fprintf('Integration of f2 over [%.0f,%.0f] ',A,B);
fprintf('by MATLAB"s quad function:\n');
fprintf('f2 = t/(t^3 + t + 1) \n');
fprintf('integral = %f \n',I2);
```

Program results

```
Integration of f2 over [0,1] by MATLAB's quad function:
f2 = t/(t^3 + t + 1)
integral = 0.260068
>>
```

Let us repeat Example 6.3 in which we evaluated the improper integral described by Equation 6.10, but this time we will evaluate the integral using MATLAB's quad function with limits from 0 to 1. Recall that the function $\log(1+x)/x$ is undefined at $x = 0$. In addition, we will use the inline method for specifying the function.

Example 6.6

```
% Example_6_6.m
% This program evaluates the improper integral log(1+x)/x
% with limits from 0 to 1 using MATLAB's quad function.
clear; clc;
A = 0; B = 1;
fprintf('This program uses the quad function to evaluate \n');
fprintf('the integral of log(1+x)/x from %2.0f to %2.0f \n,...
      A,B);
I = quad(inline('log(1+x)./x'), A, B);
fprintf('I =%10.6f \n', I);
```

Program results

```
This program uses the quad function to evaluate
the integral of log(1+x)/x from 0 to 1.
I = 0.822467
>>
```

This answer is the same as shown in Example 6.3.

Exercises

E6.2 Use MATLAB's quad function to evaluate the following integrals. Note that integrals d, e, and f are improper integrals:

a. $I = \int_0^3 \dfrac{dx}{5e^{3x} + 2e^{-3x}}$

b. $I = \int_{-\pi/2}^{\pi/2} \dfrac{\sin x\,dx}{\sqrt{1 - 4\sin^2 x}}$

c. $I = \int_0^{\pi} (\sinh x - \cos x)dx$

d. $I = \int_0^1 \dfrac{3e^x\,dx}{\sqrt{1 - x^2}}$

e. $I = \int_0^1 \frac{\log(1+x)dx}{(1-x)}$

f. $I = \int_0^1 \frac{\log(1+x)dx}{(1-x^2)}$

E6.3 This exercise is from thermodynamics. The entropy change of an ideal gas from state (T_1, p_1) to state (T_2, p_2) is given by

$$s(T_2, p_2) - s(T_1, p_1) = \int_{T_1}^{T_2} c_p(T)\frac{dT}{T} - R\ln\frac{p_2}{p_1} \qquad (6.12)$$

where
  s is the entropy (kJ/kg-K)
  c_p is the specific heat at constant pressure (kJ/kg-K)
  p is the pressure (kPa)
  T is the absolute temperature (K)
  R is the gas constant (kJ/kg-K)

The specific heat, $c_p(T)$, can be approximated by a fourth-degree polynomial [1], that is,

$$c_p(T) = R(a_1 + a_2T + a_3T^2 + a_4T^3 + a_5T^4) \qquad (6.13)$$

$$R = \frac{\bar{R}}{\mathcal{M}} \qquad (6.14)$$

where
  $\bar{R}$ = Universal gas constant = 8.314 kJ/kmol-K
  $\mathcal{M}$ is the molal mass (kg/kmol)

For carbon dioxide [1],

$$a_1 = 2.401, \quad a_2 = 8.735 \times 10^{-3}, \quad a_3 = -6.607 \times 10^{-6},$$

$$a_4 = 2.002 \times 10^{-9}, \quad a_5 = 0.0$$

$$\mathcal{M} = 44.01 \text{ kg/kmol}$$

Use MATLAB's quad function to calculate the change in entropy, $s(T_2, p_2) - s(T_1, p_1)$, for $(T_1, p_1) = (400 \text{ K}, 1.0 \text{ atm})$ and $(T_2, p_2) = (900 \text{ K}, 10.0 \text{ atm})$. Print the results to the screen.

Note: 1 atm = 1.0132×10^5 N/m².

E6.4 Project P5.3 involved a heat transfer problem in which the surface of a thick ice slab was subjected to a sudden change in temperature. Equation P5.3 provided the means of determining the temperature distribution in the slab. The following equation is a rearrangement of Equation P5.3.

$$T_\infty - T(x,t) = (T_\infty - T_i)\left(erf\left(\frac{x}{2\sqrt{\alpha t}}\right) + e^{\left(\frac{hx}{k} + \frac{h^2 \alpha t}{k^2}\right)} \times \left[1 - erf\left(\frac{x}{2\sqrt{\alpha t}} + \frac{h\sqrt{\alpha t}}{k}\right)\right]\right)$$

(6.15)

where
  k = the thermal conductivity of the slab material
  h = the convective heat transfer coefficient
  α = the thermal diffusivity of the slab material
  T_∞ = the air temperature (10°C)
  T_i = the initial slab temperature, assumed to be a constant at temperature –20°C

In this exercise we wish to determine the amount of heat per unit surface area, q, transferred to the slab per unit surface area in a specified time period from 0 to t_f. Heat transfer to the slab occurs at the boundary. The governing equation for q is

$$q = \int_0^{t_f} h(T_\infty - T(0,t))\, dt$$

(6.16)

Setting x = 0 in Equation P5.3 and taking

$$\lim_{x=0,t\to 0} erf\left(\frac{x}{2\sqrt{\alpha t}}\right) = 0, \quad \text{and} \quad \lim_{x=0,t\to 0}\left(\frac{x}{2\sqrt{\alpha t}}\right) = 0$$

Note: $erf(0)$ = 0.

After some manipulation, Equation 6.15 reduces to

$$q = h(T_\infty - T_i)\int_0^{t_f}\left(e^{\left(\frac{h^2 \alpha t}{k^2}\right)} \times \left[1 - erf\left(\frac{h\sqrt{\alpha t}}{k}\right)\right]\right) dt$$

(6.17)

Assume: $T_i = -20°C$, $T_\infty = 10°C$, $k = 2.2$ W/m-C, $\alpha = 12.6 \times 10^{-7}$ m²/s, and $h = 100$ W/m²-C, $t_f = 792$ s.

Develop a MATLAB program using Simpson's rule to evaluate q. Print the result to the screen.

E6.5 We wish to determine the x component of the electric field at position (x_o, y_o, z_o) due to a line of point charges extending along the z-axis from $z = -0.01$ m to $z = +0.01$ m (see Project P6.3 for the derivation of the governing equations for the electric field). Assume that $dQ = \lambda dz_p$, with $\lambda = 2 \times 10^{-9}$ C/m. Here dQ is the strength of the point charge distribution. The x component of the electric field at position (x_o, y_o, z_o) is given by

$$E_x(x_o, y_o, z_o) = \int_{-0.01}^{0.01} \frac{\lambda dz_p}{4\pi\varepsilon_o} \frac{x_o - x_p}{((x_o - x_p)^2 + (y_o - y_p)^2 + (z_o - z_p)^2)^{3/2}} \quad (6.18)$$

where (x_p, y_p, z_p) = the coordinates of the point charge. Use MATLAB's quad function to determine $E_x(0.005, 0, 0)$. The units of E_x are V/m. Take $\varepsilon_o = 8.85 \times 10^{-12}$ F/m.

6.6 MATLAB®'s `dblquad` Function

The MATLAB function for numerically evaluating a double integral is `dblquad`. A description of the function can be obtained by typing **help dblquad** in the Command window (from MathWorks, with permission):

```
Q = DBLQUAD(FUN,XMIN,XMAX,YMIN,YMAX) evaluates the double
integral of FUN(X,Y) over the rectangle XMIN < = X < = XMAX,
YMIN < = Y < = YMAX. FUN is a function handle. The function
Z = FUN(X,Y) should accept a vector X and a scalar Y and
return a vector Z of values of the integrand.

Non-square regions can be handled by setting the integrand to
zero outside of the region of interest. For example, the volume
of a hemisphere of radius R can be determined by the dblquad
function by setting -R <= x <= R and 0 <= y <= R and setting
z = 0 for points (x,y) that lie outside the circle of radius
R around the origin.
```

The usage of `dblquad` is similar to `quad`, except

- FUN must have two arguments, say X and Y and accept X as a vector and Y as a scalar.
- There must be two sets of integration limits (one for the X and one for the Y).

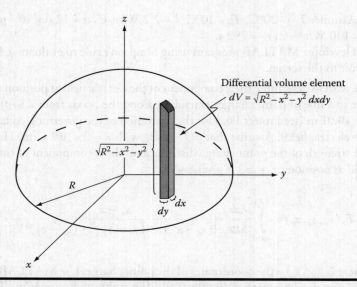

Differential volume element
$dV = \sqrt{R^2 - x^2 - y^2}\, dxdy$

Figure 6.4 Infinitesimal volume inside a hemisphere.

Example 6.7

Calculate the volume of a hemisphere of radius, R, by MATLAB's `dblquad` function.

To find the volume, we define a differential volume element, dV, as follows: $dV = \sqrt{R^2 - x^2 - y^2}\, dxdy$ (as shown in Figure 6.4) and double integrate over the intervals $x = [-R,R]$ and $y = [-R,R]$.

```
% Example_6_7.m
% This program calculates the volume of a hemisphere (with
% R = 1) using MATLAB's dblquad function. The solution is
% compared with the known exact solution for the volume of a
% hemisphere.
clear; clc;
global R;
R = 1;
V = dblquad('hemisphere_V',-R,R,-R,R);
V_exact = 2/3*pi*R^3;
% print results
fprintf('Volume V of a hemisphere of radius %6.4f m \n',R);
fprintf('V by DBLQUAD = %6.4f m^3 \n',V);
fprintf('V exact = %.4f m^3 \n',V_exact);
```

```
% hemisphere_V.m
% This function defines the integrand z(x,y) for determining
% the volume of a hemisphere and is used in MATLAB's dblquad
% function.
```

```
% From analytical geometry, the equation of a sphere is
% X^2 + Y^2 + Z^2 = R^2. Thus, dV = Z*dX*dY, where Z =
% sqrt(R^2-X^2-Y^2).
% Note that the hemisphere is only defined for (X,Y) points
% which lie within a circle of radius R around the origin,
% the region that is the projection of the hemisphere onto
% the X-Y plane. An 'if' statement is used to set Z = 0 for
% (X,Y) points outside the circle of radius R.
% Note: X is a vector and Y is a scalar.
function Z = hemisphere_V(X,Y)
% global R;
for i = 1:length(X)
    if X(i)^2 + Y^2 <= R^2
        Z(i) = sqrt(R^2 - X(i)^2 - Y^2);
    else
        Z(i) = 0;
    end
end
```

Program results

```
Volume V of a hemisphere of radius 1.0000 m
V by DBLQUAD = 2.0944 m^3
V exact      = 2.0944 m^3
>>
```

Exercises

E6.6 The object shown in Figure 6.5 is enclosed by two curves, one of which is a straight line and the other is a parabola.

   a. Use MATLAB to create a plot of Figure 6.5.

   b. Use MATLAB's dblquad function to estimate the area enclosed by the object shown in Figure 6.5.

   c. Suppose the object material is steel with a mass density, $\rho = 8.0$ kg/m³, and the thickness, z, of the region varies with x and is given by $z = x + 3$ cm. Use MATLAB's dblquad function to determine the mass, m, of the object $\left(m = \oiiint_V \rho \, dV = \iint_A \rho z \, dx \, dy \right)$. Assume that the (x,y) coordinates in the figure are in meters.

E6.7 The position of the center of mass (x_c, y_c) of the object described in Exercise E6.6 is given by

$$mx_c = \iint_A x \, \rho z(x) \, dx \, dy \quad my_c = \iint_A y \, \rho z(x) \, dx \, dy \tag{6.19}$$

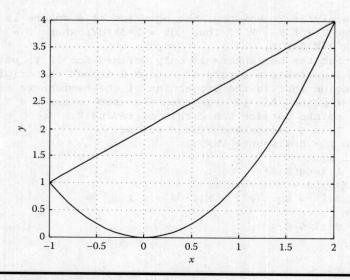

Figure 6.5   Object enclosed by two curves.

where

    ρ is the mass density of the material

    m is the mass of the object

    z is the thickness of the object, which is a function of x

Create a MATLAB program that will calculate (x_c, y_c) using MATLAB's dbl-quad function. Print the results to the screen.

E6.8 Using the infinitesimal volume shown in Figure 6.4 and MATLAB's dblquad function, determine the centroid position, z_c, of the hemisphere described in Example 6.7. By symmetry, we can assume that $x_c = 0$ and $y_c = 0$. Note that z_c for the infinitesimal volume is at the center position, that is,

$$z_c dV = \frac{1}{2}\sqrt{R^2 - x^2 - y^2} \times \sqrt{R^2 - x^2 - y^2}\, dxdy$$

$$(6.20)$$

$$z_c V = \int\limits_{-R}^{R} \int\limits_{-R}^{R} \frac{1}{2}(R^2 - x^2 - y^2)\, dxdy$$

where V is the volume of the hemisphere $= (2/3)\pi R^3$.

We can also determine the centroid position analytically by taking an infinitesimal volume shown in Figure 6.6; then

$$dV = \pi r^2 dz$$

$$(6.21)$$

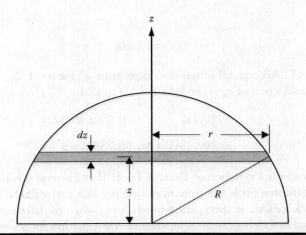

Figure 6.6 Infinitesimal region used to determine the centroid position z_c analytically.

where

$$r = \sqrt{(R^2 - z^2)} \qquad (6.22)$$

Thus,

$$z_c = \frac{\pi}{V} \int_0^R (R^2 z - z^3)\,dz = \frac{\pi}{(2/3)\pi R^3} \times \left[R^2 \frac{z^2}{2} - \frac{z^4}{4} \right]_0^R = \frac{3}{8}R \qquad (6.23)$$

Compare z_c obtained by the use of MATLAB's `dblquad` function with the exact solution. Take $R = 1$.

Projects

P6.1 The solution for the displacement, $Y(x, t)$, from the horizontal of a vibrating string (see Section 13.2) is given by

$$Y(x,t) = \sum_{n=1}^{\infty} a_n \sin\frac{n\pi x}{L} \cos\frac{n\pi ct}{L} \qquad (P6.1a)$$

where

$$a_n = \frac{2}{L} \int_0^L f(x) \sin\frac{n\pi x}{L}\,dx \qquad (P6.1b)$$

and

$$f(x) = Y(x,0) \tag{P6.1c}$$

Use MATLAB's quad function to determine a_n, for $n = 1, 2, \ldots, 10$. Create a table and a plot of a_n vs. n. Take $L = 1.0$ m and

$$f(x) = \begin{cases} 0.4x, & 0 \le x \le 0.75L \\ 0.12 - 0.12x, & 0.75L \le x \le L \end{cases} \tag{P6.1d}$$

P6.2 This project is a variation of Exercise E6.4. That Exercise involved determining by Simpson's rule the amount of heat per unit surface area, q, transferred to a thick ice slab, in time period from 0 to t_f, when the surface was suddenly subjected to a change in air temperature. The equation for q was described in E6.4, which is:

$$q = h(T_\infty - T_i) \int_0^{t_f} \left(e^{\left(\frac{h^2 \alpha t}{k^2} \right)} \times \left[1 - erf\left(\frac{h\sqrt{\alpha t}}{k} \right) \right] \right) dt \tag{P6.2a}$$

where
   k = the thermal conductivity of the slab material
   h = the convective heat transfer coefficient
   α = the thermal diffusivity of the slab material
   T_∞ = the air temperature (10°C)
   T_i = the initial slab temperature, assumed to be a constant at temperature –20°C

Using MATLAB's quad function, develop a MATLAB program that will determine q. Take $t_f = 792$ s.

Assume: $T_i = -20°C$, $T_\infty = 10°C$, $k = 2.2$ W/m-C, $\alpha = 12.6 \times 10^{-7}$ m²/s and $h = 100$ W/m²-C.

P6.3 It is experimentally observed that when two charged bodies are placed in the vicinity of each other, a force will act on the two bodies. If both bodies have like charges, the force will be repulsive; otherwise, the force will be attractive. The force relationship between two charges is similar to Newton's gravitational law. The force relationship between two point charges is

$$\vec{F} = k \frac{Qq}{|\vec{r}|^2} \hat{e}_r$$

Here we take Q to be the first point charge, q as the second point charge, r is the distance between the point charges, k is a constant and $\hat{e}_r$ is a unit vector along the line connecting the two charges. If both charges are positive, then the force on charge q will point away from charge Q. The electric field, $\vec{\mathbf{E}}$, caused by point charge, Q, is defined as

$$\vec{\mathbf{E}} = \frac{\vec{\mathbf{F}}}{q} = \frac{Q}{4\pi\varepsilon_o |\vec{\mathbf{r}}|^2} \hat{\mathbf{e}}_r \qquad \text{(P6.3a)}$$

where $k = 1/(4\pi\varepsilon_o)$ with ε_o is the permittivity of material containing electric field.

In free space, $\varepsilon_o = 8.85 \times 10^{-12}$ F/m. The units of $\vec{\mathbf{E}}$ are newton/coulomb (N/C) or volt/m (V/m). Adding more source charges will alter the electric field distribution. Therefore, $\vec{\mathbf{E}}$ is defined with respect to a particular configuration of source charges.

Electric field problems often involve integration. We begin by assuming that a point charge of Q coulombs is located at the coordinates (x_p, y_p, z_p) in free space. We denote the location of Q by the vector $\vec{\mathbf{r}}_p$ (see Figure P6.3a).

The electric field $\vec{\mathbf{E}}$ at the observation point $\vec{\mathbf{r}}_o$ (corresponding to the coordinates (x_o, y_o, z_o)) due to the charge at $\vec{\mathbf{r}}_p$ is defined by Equation P6.3a. From Figure P6.3a, we can observe the vector sum $\vec{\mathbf{r}}_o = \vec{\mathbf{r}}_p + \vec{\mathbf{r}}$, and thus, $\vec{\mathbf{r}} = \vec{\mathbf{r}}_o - \vec{\mathbf{r}}_p$. Then, Equation P6.3a becomes

$$\vec{\mathbf{E}} = \frac{Q}{4\pi\varepsilon_o |\vec{\mathbf{r}}_o - \vec{\mathbf{r}}_p|^2} \frac{\vec{\mathbf{r}}_o - \vec{\mathbf{r}}_p}{|\vec{\mathbf{r}}_o - \vec{\mathbf{r}}_p|} \qquad \text{(P6.3b)}$$

where we have applied the definition of the unit vector $\hat{\mathbf{e}}_r = (\vec{\mathbf{r}}_o - \vec{\mathbf{r}}_p)/|\vec{\mathbf{r}}_o - \vec{\mathbf{r}}_p|$.

We now break down $\vec{\mathbf{r}}_o$ and $\vec{\mathbf{r}}_p$ into their x, y, and z components so that we can express them in Cartesian coordinates as $\vec{\mathbf{r}}_o = x_o\hat{\mathbf{e}}_x + y_o\hat{\mathbf{e}}_y + z_o\hat{\mathbf{e}}_z$ and $\vec{\mathbf{r}}_p = x_p\hat{\mathbf{e}}_x + y_p\hat{\mathbf{e}}_y + z_p\hat{\mathbf{e}}_z$, where $\hat{\mathbf{e}}_x$, $\hat{\mathbf{e}}_y$, and $\hat{\mathbf{e}}_z$ are unit vectors in the x, y, and z directions, respectively. Then, $|\vec{\mathbf{r}}_o - \vec{\mathbf{r}}_p| = \sqrt{(x_o - x_p)^2 + (y_o - y_p)^2 + (z_o - z_p)^2}$ and Equation P6.3b becomes

$$\vec{\mathbf{E}} = \frac{Q}{4\pi\varepsilon_o} \frac{(x_o - x_p)\hat{\mathbf{e}}_x + (y_o - y_p)\hat{\mathbf{e}}_y + (z_o - z_p)\hat{\mathbf{e}}_z}{((x_o - x_p)^2 + (y_o - y_p)^2 + (z_o - z_p)^2)^{3/2}} \qquad \text{(P6.3c)}$$

Thus, Equation P6.3c gives the electric field at the location (x_o, y_o, z_o) due to a point charge of magnitude Q located at (x_p, y_p, z_p). We now wish to calculate the electric field due to a line of point charges extending along the z-axis from $z = -0.01$ m to $z = +0.01$ m (see Figure P6.3b).

We shall assume that the charges are evenly spaced along the line with a *linear charge density* represented by the symbol λ with units coulomb/meter. We can rewrite Equation P6.3c in differential form to obtain an expression

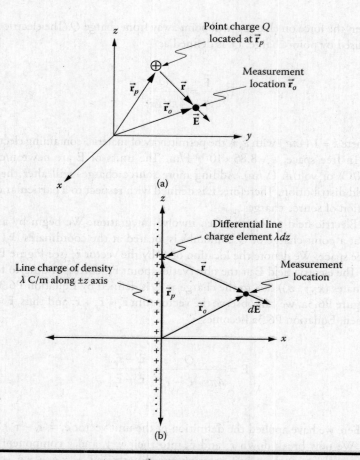

Figure P6.3 **(a) The electric field $\vec{E}$ at observation point $\vec{r}_o$ due to a point charge located at $\vec{r}_p$. (b) The differential electric field $d\vec{E}_r$ at observation point $\vec{r}_o$ due to a differential line charge element λdz located along the z-axis.**

for $d\vec{E}$ in terms of the differential quantity $dQ = \lambda dz_p$, where dz_p represents an infinitesimal segment of the line charge in the z direction:

$$d\vec{E} = \frac{\lambda dz_p}{4\pi\varepsilon_o} \frac{(x_o - x_p)\hat{\mathbf{e}}_x + (y_o - y_p)\hat{\mathbf{e}}_y + (z_o - z_p)\hat{\mathbf{e}}_z}{((x_o - x_p)^2 + (y_o - y_p)^2 + (z_o - z_p)^2)^{3/2}} \qquad \text{(P6.3d)}$$

Separating the individual directional components of Equation P6.3d and integrating -0.01 to 0.01 give

$$E_x = \int\limits_{-0.01}^{0.01} \frac{\lambda dz_p}{4\pi\varepsilon_o} \frac{x_o - x_p}{((x_o - x_p)^2 + (y_o - y_p)^2 + (z_o - z_p)^2)^{3/2}} \qquad \text{(P6.3e)}$$

Table P6.3 Table Format for Presenting E_x Values

E_x Values						
x_o	y_o					
	−0.05	−0.03	−0.01	0.01	0.03	0.05
−0.05	—	—	—	—	—	—
−0.04	—	—	—	—	—	—
−0.03	—	—	—	—	—	—
⋮	⋮	⋮	⋮	⋮	⋮	⋮
0.03	—	—	—	—	—	—
0.04	—	—	—	—	—	—
0.05	—	—	—	—	—	—

$$E_y = \int_{-0.01}^{0.01} \frac{\lambda dz_p}{4\pi\varepsilon_o} \frac{y_o - y_p}{((x_o - x_p)^2 + (y_o - y_p)^2 + (z_o - z_p)^2)^{3/2}} \qquad \text{(P6.3f)}$$

$$E_z = \int_{-0.01}^{0.01} \frac{\lambda dz_p}{4\pi\varepsilon_o} \frac{z_o - z_p}{((x_o - x_p)^2 + (y_o - y_p)^2 + (z_o - z_p)^2)^{3/2}} \qquad \text{(P6.3g)}$$

Calculate and tabulate the electric field components, E_x and E_y, in the (x, y) plane for the intervals $-50 \leq x_o \leq 50$ mm and $-50 \leq y_o \leq 50$ mm with a step size of 10 mm. Omit the point $(x_o, y_o) = (0,0)$. Print E_x and E_y to three decimal places in separate tables using a table format as shown in Table P6.3. Take $\lambda = 2 \times 10^{-9}$ C/m.

P6.4 Modify Project P6.3 to calculate the electric field components, E_x and E_z, at points in the (x, z) plane.

Calculate and tabulate E_x and E_z for $-0.05 \leq x_o \leq 0.05$ m and $-0.05 \leq z_o \leq 0.05$ m with a step size of 0.01 m. Omit the points $(x_o, z_o) = (0, z_o)$. Print the result in the table format shown in Project P6.3.

P6.5 A positive *surface charge density* of magnitude $\sigma = 4 \times 10^{-12}$ C/m² lies in the (x, y) plane and extends $-0.01 \leq x \leq 0.01$ and $-0.01 \leq y \leq 0.01$ as shown in Figure P6.5a. Using a method similar to what was described in Project P6.3, we define a differential charge element $dQ = \sigma dydx$ and then find the resulting electric field by integrating over the sheet of charge for $x_p = [-0.01, 0.01]$, $y_p = [-0.01, 0.01]$, and $z_p = 0$:

$$E_x = \int_{-0.01}^{0.01} \int_{-0.01}^{0.01} \frac{\sigma dy_p dx_p}{4\pi\varepsilon_o} \frac{x_o - x_p}{((x_o - x_p)^2 + (y_o - y_p)^2 + (z_o - z_p)^2)^{3/2}} \quad \text{(P6.5a)}$$

$$E_y = \int_{-0.01}^{0.01} \int_{-0.01}^{0.01} \frac{\sigma dy_p dx_p}{4\pi\varepsilon_o} \frac{y_o - y_p}{((x_o - x_p)^2 + (y_o - y_p)^2 + (z_o - z_p)^2)^{3/2}} \quad \text{(P6.5b)}$$

$$E_z = \int_{-0.01}^{0.01} \int_{-0.01}^{0.01} \frac{\sigma dy_p dx_p}{4\pi\varepsilon_o} \frac{z_o - z_p}{((x_o - x_p)^2 + (y_o - y_p)^2 + (z_o - z_p)^2)^{3/2}} \quad \text{(P6.5c)}$$

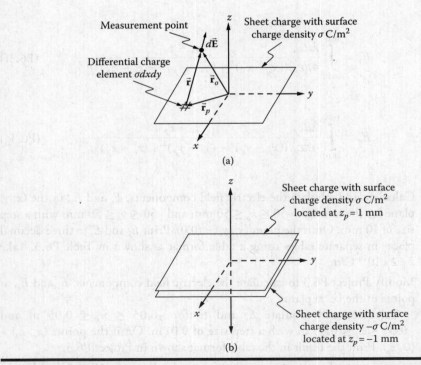

(a)

(b)

Figure P6.5 (a) The differential electric field $d\vec{E}$ at observation point $\vec{r}_o$ due to a differential sheet charge element $\sigma dxdy$ located on the $x-y$ plane. (b) Two parallel and oppositely charged sheet charges located on the planes $z_p = +1$ mm and $z_p = -1$ mm.

1. Write a MATLAB program using the `dblquad` function to calculate E_x, E_y, and E_z at the following (x_o, y_o, z_o) points:

$$x_o, y_o, z_o = [0, 0, 2 \times 10^{-3}] \quad \text{(above the plate)}$$

$$x_o, y_o, z_o = [30 \times 10^{-3}, -5 \times 10^{-3}, -12 \times 10^{-3}] \quad \text{(below the plate)}$$

2. Now assume that there are two sheet charges in the planes parallel to the x–y plane: the first sheet located at $z_p = +1$ mm with surface charge density $\sigma = 4 \times 10^{-12}$ C/m^2 and the second sheet located at $z_p = -1$ mm and oppositely charged with surface charge density $\sigma = -4 \times 10^{-12}$ C/m^2 (see Figure P6.5b). Find the electric field via superposition by calculating the electric field separately for each sheet and then adding them together. Find E_x, E_y, and E_z for these three points:

$$x_o, y_o, z_o = [0, 0, 2 \times 10^{-3}] \quad \text{(above the plates)}$$

$$x_o, y_o, z_o = [10 \times 10^{-3}, -0.5 \times 10^{-3}, 0] \quad \text{(between the plates)}$$

$$x_o, y_o, z_o = [30 \times 10^{-3}, -5 \times 10^{-3}, -12 \times 10^{-3}] \quad \text{(below the plates)}$$

Print out the results. Do these results make sense?

P6.6 The Biot–Savart law relates electrical current to magnetic field (see Figure P6.6a) and is defined for current flow through wires as

$$d\vec{B} = \frac{\mu_o}{4\pi} \frac{I d\vec{\ell} \times \hat{e}_r}{|\vec{r}|^2} \tag{P6.6a}$$

where
  $d\vec{B}$ is the differential magnetic field (a vector)
  μ_o is the permeability of free space ($4\pi \times 10^{-7}$ H/m)
  I is the current (in amperes)
  $d\vec{\ell}$ is a differential wire length with direction corresponding to the current flow
  $\vec{r}$ is the vector from the current element to the measurement point
  $\hat{e}_r$ is the unit vector in the $\vec{r}$ direction
  × represents the vector cross product

In Figure P6.6a, we also define $\vec{r}_p$ as the location of the current element and $\vec{r}_o$ as the point of measurement such that $\vec{r} = \vec{r}_o - \vec{r}_p$.

Figure P6.6b shows a circular wire with radius R centered at the origin and carrying a current I. Using the Biot–Savart law from Equation P6.6a, we can

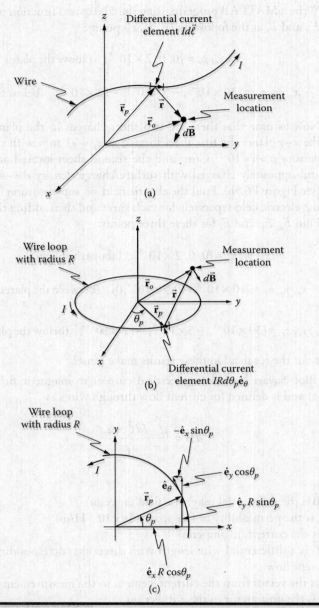

Figure P6.6 (a) The differential magnetic field element $d\vec{B}$ at observation point $\vec{r}_o$ due to a differential wire element carrying current $Id\vec{\ell}$. (b) The differential magnetic field element $d\vec{B}$ at observation point $\vec{r}_o$ due to a wire loop located in the x–y plane. (c) $\vec{r}_p$ and $\hat{e}_{\theta_p}$ may be decomposed into their x and y components in order to solve for the magnetic field with MATLAB®'s quad function.

write the differential magnetic field in terms of a differential current element $Id\vec{\ell} = IRd\theta_p\hat{\mathbf{e}}_\theta$, where $Rd\theta_p$ is the differential length (in cylindrical coordinates) of the current element at location $\vec{\mathbf{r}}_p$ and $\hat{\mathbf{e}}_\theta$ is the unit vector in the θ_p direction. In this type of problem with cylindrical symmetry, the simplest approach is to convert the cylindrical unit vectors into Cartesian unit vectors but still perform the integration using the cylindrical variables. Figure P6.6c shows how $\vec{\mathbf{r}}_p$ and $\hat{\mathbf{e}}_\theta$ can be expressed in terms of their x and y components using the unit vectors $\hat{\mathbf{e}}_x$ and $\hat{\mathbf{e}}_y$:

$$\vec{\mathbf{r}}_p = \hat{\mathbf{e}}_x R\cos\theta_p + \hat{\mathbf{e}}_y R\sin\theta_p \tag{P6.6b}$$

$$\hat{\mathbf{e}}_\theta = -\hat{\mathbf{e}}_x \sin\theta_p + \hat{\mathbf{e}}_y \cos\theta_p \tag{P6.6c}$$

We can now rewrite Equation P6.6a as

$$d\vec{\mathbf{B}} = \frac{\mu_o IR}{4\pi} d\theta_p$$

$$\times \frac{(-\hat{\mathbf{e}}_x \sin\theta_p + \hat{\mathbf{e}}_y \cos\theta_p) \times (\hat{\mathbf{e}}_x (x_o - R\cos\theta_p) + \hat{\mathbf{e}}_y (y_o - R\sin\theta_p) + \hat{\mathbf{e}}_z(z_o - z_p))}{((x_o - R\cos\theta_p)^2 + (y_o - R\sin\theta_p)^2 + (z_o - z_p)^2)^{3/2}}$$

$$\tag{P6.6d}$$

1. Calculate the vector cross product in Equation P6.6d and split the result into three separate equations for the x, y, and z components of the differential magnetic field. Use the resulting expressions for dB_x, dB_y, and dB_z to write three MATLAB functions, which take one argument (θ_p) and which can be used with MATLAB's quad function to determine B_x, B_y, and B_z at any arbitrary position.
2. Create a MATLAB program that will determine B_x, B_y, and B_z at the following positions:
   a. $(x_o, z_o) = (-0.008, -0.002)$.
   b. $(x_o, z_o) = (-0.008, 0.002)$.
   c. $(x_o, z_o) = (0.008, -0.002)$.
   d. $(x_o, z_o) = (-0.002, -0.002)$.
   e. $(x_o, z_o) = (0.002, 0.002)$.
   Print the results to the screen, specifying the (x_o, z_o) position and the values for B_x, B_y, and B_z.
   Assume $R = 2.5$ mm and $I = 1$ ma.

P6.7 A *solenoid* is a coil of wire, which is used to create a magnetic field in order to activate an actuator. Typical uses include electrical relays, electrically controlled water valves, and automotive starter gears. Figure P6.7a shows a solenoid with radius R, length D, turn count N, and current I.

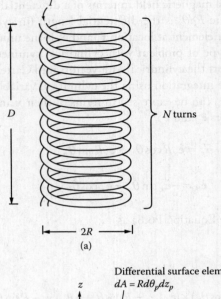

(a)

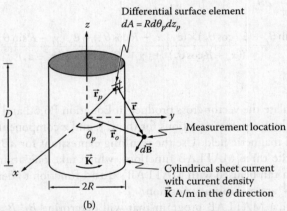

(b)

Figure P6.7 (a) A solenoid consisting of *N* turns, radius *R*, length *D*, and current *I*. (b) The solenoid can be modeled as a sheet current of magnitude *K* A/m.

We will model the solenoid in MATLAB as a cylindrical sheet current with density $\vec{\mathbf{K}} = \dfrac{NI}{D}\hat{\mathbf{e}}_\theta$ A/m with the current density directed in the θ direction (see Figure P6.7b). The Biot–Savart law for a surface current is

$$d\vec{\mathbf{B}} = \frac{\mu_o}{4\pi}\frac{\vec{\mathbf{K}}dA \times \hat{\mathbf{e}}_r}{|\vec{\mathbf{r}}|^2} \tag{P6.7a}$$

where

    dA is a 2-D differential surface element

    $\vec{r}$ is the vector from the differential surface to the measurement point

    $\hat{e}_r$ is the unit vector in the $\vec{r}$ direction

    × represents the vector cross product

    μ_o is the permeability of free space ($4\pi \times 10^{-7}$ H/m)

From Figure P6.7b, we see that in cylindrical coordinates, $dA = Rd\theta_p dz_p$, where $\vec{r}_p$ is the location of the differential surface (with cylindrical components r_p, θ_p, and z_p) and $\hat{e}_\theta$ is the unit vector in the θ_p direction. We also define the observation point $\vec{r}_o = x_o \hat{e}_x + y_o \hat{e}_y + z_o \hat{e}_z$. From the geometry of the problem, we can write $\vec{K}dA$ and $\vec{r}$ in terms of Cartesian unit vectors:

$$\vec{K}dA = \frac{NI}{D} Rd\theta_p dz_p \hat{e}_\theta = \frac{NI}{D} Rd\theta_p dz_p (-\hat{e}_x \sin\theta_p + \hat{e}_y \cos\theta_p) \quad \text{(P6.7b)}$$

$$\vec{r} = \vec{r}_o - \vec{r}_p = \hat{e}_x(x_o - R\cos\theta_p) + \hat{e}_y(y_o - R\sin\theta_p) + \hat{e}_z(z_o - z_p) \quad \text{(P6.7c)}$$

Substituting back into Equation P6.7a gives

$$d\vec{B} = \frac{\mu_o NIR}{4\pi D} d\theta_p dz_p$$

$$\times \frac{\left(-\hat{e}_x \sin\theta_p + \hat{e}_y \cos\theta_p\right) \times \left(\hat{e}_x(x_o - R\cos\theta_p) + \hat{e}_y(y_o - R\sin\theta_p) + \hat{e}_z(z_o - z_p)\right)}{((x_o - R\cos\theta_p)^2 + (y_o - R\sin\theta_p)^2 + (z_o - z_p)^2)^{3/2}}$$

$$\text{(P6.7d)}$$

1. Calculate the vector cross product in Equation P6.7d and split the result into three separate equations for the x, y, and z components of the differential magnetic field. Use the resulting expressions for dB_x, dB_y, and dB_z to write three MATLAB functions, which take two arguments (θ_p and z_p) and compute the differential B field in the x, y, and z directions, respectively.

2. Assume that the solenoid is centered at the origin with radius $R = 2.5$ mm and $D = 15$ mm.

   Use MATLAB's `dblquad` function to calculate B_x, B_y, and B_z at the following positions (be sure to use the correct integration limits). Assume $N = 500$ and $I = 1$ ma:

   a. $(x_o, y_o, z_o) = (8, 8, 40)$ mm.
   b. $(x_o, y_o, z_o) = (4, 4, 0)$ mm.
   c. $(x_o, y_o, z_o) = (0, 0, 20)$ mm.
   d. $(x_o, y_o, z_o) = (-8, -8, 40)$ mm.

Chapter 7

Numerical Integration of Ordinary Differential Equations

7.1 Introduction

In this chapter, we examine several methods for solving ordinary differential equations (ODEs). ODEs can be broken up into two categories:

1. *Initial value problems* in which we know the necessary initial conditions, for example, launching a rocket with a known initial velocity or the circuit node voltage at time $t = 0$.
2. *Boundary value problems* in which we know the conditions at specific coordinates in the problem geometry, for example, the deflection of a beam with known end conditions or the electric potential at both ends of a conductor. This type of problem is covered in Chapter 8.

For initial value problems, we will examine several numerical integration methods including the Euler method, the modified Euler method, the Runge–Kutta method, and MATLAB®'s built-in `ode45` function.

7.2 Initial Value Problem

In an initial value problem, the values of the dependent variable and the necessary derivatives are known at the point at which the integration begins. We will begin with a first-order differential equation of the general form such that the derivative is a known function of x and y and the initial condition, that is, $y(0) = Y_I$, is known:

$$\frac{dy}{dx} = f(x, y)$$

$$y(0) = Y_I$$

(7.1)

There are several techniques for solving this type of problem, including Euler's method, which is simple but not used very often; the modified Euler method; and the Runge–Kutta method. Each technique has pros and cons with respect to simplicity, accuracy, and computational efficiency.

7.3 Euler Algorithm

The general approach to solving differential equations numerically is to subdivide the x domain into N subdivisions giving $x_1, x_2, x_3, \ldots, x_{N+1}$ and then "march" in the x direction over the interval while calculating $y_2, y_3, y_3, \ldots, y_{N+1}$. Note: y_1 is specified and is equal to the initial condition Y_I (see Figure 7.1).

A Taylor series expansion about an arbitrary point x_i gives

$$y(x) = y(x_i) + \frac{y'(x_i)}{1!}(x - x_i) + \frac{y''(x_i)}{2!}(x - x_i)^2 + \cdots$$

(7.2)

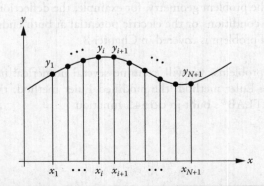

Figure 7.1 Subdivision of x domain $(x_1, x_2, \ldots, x_{N+1})$ with corresponding y values.

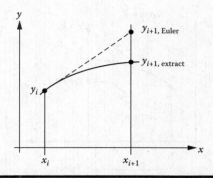

Figure 7.2 Estimate of y_{i+1} by the Euler method.

If we use only the first two terms of the series, we can approximate the value for $y(x_{i+1})$ (which we denote as y_{i+1}) as

$$y_{i+1} \approx y(x_i) + \frac{y'(x_i)}{1}(x_{i+1} - x_i)$$

$$\approx y_i + y_i'h \tag{7.3}$$

where $h = x_{i+1} - x_i$, which is defined as the *step size*, and y_i' is the slope of the curve of $y(x)$ at x_i. Substituting Equation 7.1 into 7.3, we obtain

$$y_{i+1} = y_i + hf(x_i, y_i) \tag{7.4}$$

As can be seen for the configuration shown in Figure 7.2, the prediction of y_{i+1} by the Euler method overshoots the true value of y_{i+1}.

Thus, starting with $i = 1$, we obtain

$$y_2 = y_1 + hf(x_1, y_1) = Y_1 + hf(x_1, Y_1)$$

Similarly, y_3 can be determined by setting $i = 2$ into Equation 7.4 giving

$$y_3 = y_2 + hf(x_2, y_2)$$

The process is continued obtaining $y_4, y_5, \ldots, y_{N+1}$. The Euler method is an example of an *explicit* method because each new value for y_{i+1} depends solely on previous values of x_i and y_i.

7.4 Modified Euler Method with Predictor–Corrector Algorithm

The modified Euler method is a more accurate way for calculating the value of y_{i+1} than the simple linear approximation of the Euler method in the previous section. Again we consider a differential equation of standard form $y' = f(x, y)$ with initial condition

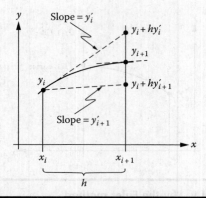

Figure 7.3   Estimates of y_{i+1} by using $y'(x_{i+1})$ and the Euler method.

$y(0) = Y_I$, and again we subdivide the x domain into N subdivisions and march in the x direction. We saw that in the Euler method, taking $y_{i+1} = y_i + hy'_i = y_i + hf(x_i, y_i)$ overshoots the true value of y_{i+1} (as shown in Figure 7.2). Now, suppose we were able to determine $y'(x_{i+1})$, which would be the slope of the curve at x_{i+1}. If we were to predict y_{i+1} by using $y'(x_{i+1})$ in Equation 7.3, that is,

$$y_{i+1} = y_i + hy'_{i+1} = y_i + hf(x_{i+1} y_{i+1})$$

then for the configuration shown in Figure 7.3, we would undershoot the true value of y_{i+1}.

Here we have constructed a tangent to the curve at y_{i+1} and drawn a parallel line passing through point (x_i, y_i) to obtain the predicted value for y_{i+1}. Since using y'_i in Equation 7.3 overshoots the true value of y_{i+1} and using y'_{i+1} in Equation 7.3 undershoots the true value of y_{i+1}, we see that a better estimate for y_{i+1} would be obtained by using an average of the two derivatives in Equation 7.3, that is,

$$y_{i+1} = y_i + h\left(\frac{y'_i + y'_{i+1}}{2} \right) \tag{7.5}$$

Unfortunately, Equation 7.5 is no longer explicit because we do *not* know the value of y'_{i+1}. The use of Equation 7.5 in solving the differential Equation 7.1 is an example of an *implicit* method. However, we can approximate a value for y'_{i+1} by using the *predictor–corrector* method. To apply this method, we rewrite Equation 7.5 as follows:

$$y^C_{i+1} = y_i + h\left(\frac{y'_i + \left(y'_{i+1}\right)^P}{2} \right) \tag{7.6}$$

where
the P superscript indicates the *predicted* value
the C superscript indicates the *corrected* value

Equation 7.6 is called the *corrector equation* and can be used to iteratively estimate the value for y_{i+1}. Substituting Equation 7.1 into 7.6 gives

$$\left(y'_{i+1}\right)^{P} = f(x_{i+1}, y^{C}_{i+1}) \tag{7.7}$$

The predictor–corrector technique proceeds as follows:

1. Use the Euler method to determine a first predicted value for $\left(y'_{i+1}\right)^{P}$, that is,

$$\left(y'_{i+1}\right)^{P_1} = f(x_{i+1}, y_i + y'_i h) \tag{7.8}$$

2. Calculate the first corrected value $y^{C_1}_{i+1}$ by using Equation 7.8 in Equation 7.6.
3. Use $y^{C_1}_{i+1}$ to obtain a new predicted value $\left(y'_{i+1}\right)^{P_2}$, that is, in Equation 7.7.
4. Calculate a new corrected value $y^{C_2}_{i+1}$ by using $\left(y'_{i+1}\right)^{P_2}$ in Equation 7.6.
5. Repeat steps 3 and 4 until $\left| y^{C_{n+1}}_{i+1} - y^{C_n}_{i+1} \right| < \varepsilon$, where ε is an error tolerance that depends on the desired accuracy and is typically a fraction of a percent of the last corrected value, for example, $\varepsilon = y_i \times 10^{-4}$.

Example 7.1

In this example, we find the exact solution of a first-order differential equation and then compare it with the corresponding solutions found by the Euler and modified Euler methods. We will use the problem described in Exercise E2.5 as our example. In that exercise, we determined the velocity, V, of a spherical ball bearing dropped in a viscous fluid. The governing equation describing the velocity of the ball bearing as it moves through the fluid is

$$\frac{dV}{dt} = \frac{1}{m}(W - B - D) \tag{7.9}$$

where
   m is the mass of the ball bearing
   W is the weight of the ball bearing
   D is the drag
   B is the buoyancy
   t is the time

The drag, D, is governed by Stokes' law, which is

$$D = 6\pi R \mu V \tag{7.10}$$

where
   R is the radius of the sphere
   μ is the viscosity of the fluid

The weight $W = mg = \rho \mathbb{V} g$, where $\mathbb{V}$ is the volume of the sphere and ρ is the mass density of the ball bearing material. The buoyancy, B, equals the weight of the fluid displaced. Assuming that the material of the ball bearing is steel and that the fluid is oil, Equation 7.9 reduces to

$$\frac{dV}{dt} = g - \frac{\rho_{oil}\, g}{\rho_{steel}} - \frac{6\pi R \mu V}{\rho_{steel}\, \mathbb{V}} \tag{7.11}$$

$$\mathbb{V}_{sphere} = \frac{4}{3}\pi R^3 \tag{7.12}$$

The closed-form solution was given in Exercise E2.5, which is

$$V = V_T (1 - e^{-(6\pi R \mu g/W)t}) \tag{7.13}$$

where

$$V_T = \text{the terminal velocity} = \frac{(\rho_{steel} - \rho_{oil})g\mathbb{V}}{6\pi R \mu} \tag{7.14}$$

The following program demonstrates the use of the Euler and modified Euler methods:

```
% Example_7_1.m
% This program demonstrates the use of the Euler and modified
% Euler methods in solving a first order differential equation.
% The problem is to determine the velocity of a ball bearing
% as it moves through a viscous liquid.
% An exact solution is available.
% The units are: rho in kg/m^3, mu in N-s/m^2, g in m/s^2, t
% in s, velocity in m/s, volume in m^3.
clear; clc;
t = 0:0.01:1;
rho_steel = 7910; rho_oil = 888; g = 9.81; R = 0.01; mu = 3.85;
v(1) = 0.0;
dt = 0.01;
vol = 4/3*pi*R^3;
m = rho_steel*vol;
w = m*g;
arg = 6*pi*R*mu*g/w;
```

```
vt = (rho_steel-rho_oil)*g*vol/(6*pi*R*mu);
f = @ (v) (g-rho_oil/rho_steel*g-6*pi*R*mu/m*v);
% Euler and exact methods
for i = 1:length(t)-1
    v(i+1) = v(i)+dt*f(v(i));
    vexact(i+1) = vt*(1-exp(-arg*t(i+1)));
end
% Modified Euler method (iterative scheme)
vmod(1) = 0.0;
for i = 1:length(t)-1
    vp(i) = f(vmod(i));
    vp1(i+1) = f(vmod(i)+f(vmod(i))*dt);
    vc1(i+1) = vmod(i)+(vp(i)+vp1(i+1))*dt/2;
    test = vmod(i)*10e-5;
    for j = 1:50
        vp2(i+1) = f(vc1(i+1));
        vc2(i+1) = vmod(i)+(vp(i)+vp2(i+1))*dt/2;
        if abs(vc2(i+1)-vc1(i+1)) < test
            break;
        else
            vp1(i+1) = vp2(i+1);
            vc1(i+1) = vc2(i+1);
        end
    end
    vmod(i+1) = vmod(i)+(vp(i)+vp2(i+1))*dt/2;
end
fo = fopen('output.txt','w');
fprintf(fo,'Comparing results from Euler and modified \n');
fprintf(fo,'Euler methods with exact solution \n\n');
fprintf(fo,'  t(s)      v(m/s)      v(m/s)          v(m/s) \n');
fprintf(fo,'            Euler   Modified Euler   Exact \n');
fprintf(fo,'----------------------------------------------- \n');
for i = 1:5:length(t)
    fprintf(fo,'%5.2f %10.4f %10.4f %10.4f \n',...
        t(i),v(i),vmod(i),vexact(i));
end
for i = 1:2
    subplot(1,2,i),
    if i == 1
        plot(t,vexact,t,v,'x'), xlabel('t'), ylabel('v'), grid,
        title('velocity vs time'), legend('v-exact','v-Euler');
    end
    if i == 2
        plot(t,vexact,t,vmod,'x'), xlabel('t'), ylabel('v'), grid,
        title('velocity vs time'), legend('v-exact','v-modied');
    end
end
```

Program results

Comparing results from Euler and modified Euler methods with
exact solution

t(s)	v(m/s) Euler	v(m/s) Modified Euler	v(m/s) Exact
0.00	0.0000	0.0000	0.0000
0.05	0.2821	0.2652	0.2646
0.10	0.3640	0.3535	0.3531
0.15	0.3879	0.3829	0.3827
0.20	0.3948	0.3927	0.3926
0.25	0.3968	0.3960	0.3959
0.30	0.3974	0.3971	0.3971
0.35	0.3975	0.3974	0.3974
0.40	0.3976	0.3975	0.3975
0.45	0.3976	0.3976	0.3976
0.50	0.3976	0.3976	0.3976
⋮	⋮	⋮	⋮

See Figure 7.4.

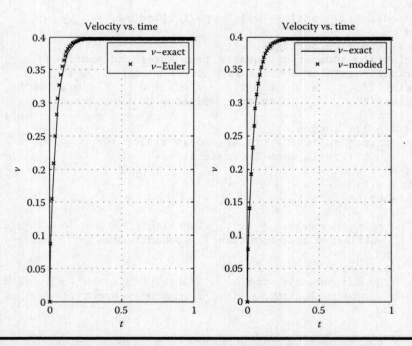

Figure 7.4 **Velocity of ball bearing dropped in a vat of oil determined by Euler and modified Euler methods.**

We see that terminal velocity is reached at approximately 0.45 s. We also see that the modified Euler method produced an answer that was closer to the exact solution than the Euler method, especially in the early stages in the process.

Exercises

E7.1 A simple *RC* circuit is "driven" by a voltage source V_D as shown in Figure 7.5. Writing Kirchhoff's voltage law around the loop gives

$$V_D - v_R - v_C = 0 \tag{7.15}$$

Applying Ohm's law ($v_R = iR$) and the constituent relation for capacitor current $\left(i = C \dfrac{dv_C}{dt} \right)$, we can rewrite Equation 7.15 as

$$V_D - RC \frac{dv_C}{dt} - v_C = 0$$

Rearranging gives

$$\frac{dv_C}{dt} = \frac{1}{\tau}(V_D - v_C) = f(t, v_C) \tag{7.16}$$

where $\tau = RC$ is the *time constant* for the *RC* circuit. Equation 7.16 is a first-order differential equation of v_C with respect to time. We will assume that at $t = 0$, the initial capacitor voltage, v_C, is zero and that the driving voltage is a ramp function, that is, $V_D(t) = V_o t$:

a. Show that the exact solution to Equation 7.16 is

$$v_C(t) = V_o \left(t - \tau(1 - e^{-t/\tau}) \right) \tag{7.17}$$

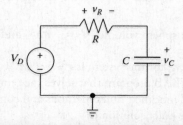

Figure 7.5   *RC* circuit.

    b.   Solve Equation 7.16 by both the Euler and modified Euler methods and compare the results to the exact solution. Take $R = 1\ k\Omega$, $C = 1\ \mu F$, and $V_o = 10\ V$. Choose $t = [0, 5\tau]$ and $\Delta t = \tau/50$.

In practice, the simplest method for choosing the step size is to start with some fraction of the fastest known parameter in the system. For example, when solving an *RC* circuit, start with a step size of one-hundredth of the time constant or $\Delta t = RC/100$. Then, double the step size and rerun the program. If the two solutions are identical, then the step size is sufficiently small, and if the program runs slowly, then you might consider doubling the original step size and comparing results. If the results are sufficiently close, you can use that new step size to obtain a faster execution time without losing accuracy.

E7.2 A small aluminum sphere, initially at temperature T_o, is suddenly immersed in a cool large bath whose temperature is essentially constant at T_∞. For a small good heat conductor body, we may take the temperature of the body to be uniform. This method of analysis is called the lumped parameter method. The governing equation for the temperature, *T*, of the sphere is

$$mc\frac{dT}{dt} = \bar{h}A_s(T_\infty - T) \tag{7.18}$$

$$T(0) = T_o \tag{7.19}$$

where
    $\bar{h}$ is the convective heat transfer coefficient of the fluid
    A_s is the surface area of the sphere
    m is the mass of the sphere
    c is the specific heat of the sphere

Equation 7.18 is a statement that the rate of increase in the internal energy of the body is equal to the rate that heat is carried to the body by convection. If the right-hand side of Equation 7.18 is negative, then the internal energy of the body will be decreasing and thus the temperature of the body will also be decreasing.

Take $\bar{h} = 890\ W/m^2\text{-}°C$, $c = 896\ J/kg\text{-}°C$, sphere density, $\rho_{AL} = 2707\ kg/m^3$, $T_o = 150°C$, $T_\infty = 20°C$, $t = (0, 3\ s)$ in steps of $0.01s$, and sphere radius, $R = 0.2\ m$.

Also note that the sphere volume, $\mathbb{V}_s = \frac{4}{3}\pi R^3$, and the sphere surface area, $A_s = 4\pi R^2$.

The time constant, τ, for the system is $\tau = mc/\bar{h}A_s$.

Create a MATLAB program that solves the temperature of the sphere by both the Euler and modified Euler methods and compare the results with the exact solution. The exact solution is

$$T(t) = (T_o - T_\infty)e^{-t/\tau} + T_\infty$$

7.5 Fourth-Order Runge–Kutta Method

The fourth-order Runge–Kutta method uses a weighted average of derivative estimates within the interval of interest in order to calculate a value for y_{i+1}. We again start with the first-order differential equation of Equation 7.1 with known initial condition, that is,

$$\frac{dy}{dx} = f(x, y)$$

$$y(0) = Y_I \tag{7.20}$$

In the modified Euler method, we used $y_{i+1} = y_i + \frac{h}{2}(y_i' + y_{i+1}')$. In the Runge–Kutta method, instead we use

$$y_{i+1} = y_i + \frac{h}{6}(k_1 + 2k_2 + 2k_3 + k_4) \tag{7.21}$$

where

$$
\begin{aligned}
k_1 &= f(x_i, y_i) && \text{(value of } y' \text{ at } x_i) \\
k_2 &= f\left(x_i + \frac{h}{2}, y_i + \frac{h}{2}k_1\right) && \left(\text{estimate of } y' \text{ at } x_i + \frac{h}{2}\right) \\
k_3 &= f\left(x_i + \frac{h}{2}, y_i + \frac{h}{2}k_2\right) && \left(\text{a second estimate of } y' \text{ at } x_i + \frac{h}{2}\right) \\
k_4 &= f(x_i + h, y_i + hk_3) && \text{(estimate of } y' \text{ at } x_{i+1})
\end{aligned}
$$

The Runge–Kutta method is an explicit algorithm and thus is simple to compute with MATLAB.

Example 7.2

This example is a variation of Example 7.1. In Example 7.1, we determined the velocity, V, of a spherical ball bearing dropped in a viscous fluid by the Euler and the modified Euler methods. In this example, we will solve the problem by the Runge–Kutta method. In Example 7.1, we reduced the governing differential equation to (see Equations 7.11 through 7.14)

$$\frac{dV}{dt} = g - \frac{\rho_{oil}\, g}{\rho_{steel}} - \frac{6\pi R\mu V}{\rho_{steel}\, \mathbb{V}}$$

$$\mathbb{V}_{sphere} = \frac{4}{3}\pi R^3$$

The closed-form solution was given in Exercise E2.5, which is

$$V = V_T(1 - e^{-\frac{6\pi R\mu g}{W}t})$$

where

$$V_T = \text{the terminal velocity} = \frac{(\rho_{steel} - \rho_{oil})g\mathbb{V}}{6\pi R\mu}$$

The program follows:

```
% Example_7_2.m
% This program demonstrates the use of the Runge-Kutta method
% in solving a first order differential equation.
% The problem is to determint the velocity of a ball bearing
% as it is dropped in a viscous liquid.
% An exact solution is available.
% Rho is in kg/m^3, g is in m/s^2
% R is in m and mu is in N-S/m^2
clear; clc;
t = 0:0.01:1;
rho_steel = 7910; rho_oil = 888; g = 9.81; R = 0.01; mu = 3.85;
v(1) = 0.0;
dt = 0.01;
vol = 4/3*pi*R^3;
m = rho_steel*vol;
w = m*g;
arg = 6*pi*R*mu*g/w;
vt = (rho_steel-rho_oil)*g*vol/(6*pi*R*mu);
f = @ (v) (g-rho_oil/rho_steel*g-6*pi*R*mu/m*v);
for i = 1:length(t)-1
    v_exact(i+1) = vt*(1-exp(-arg*t(i+1)));
    k1 = f(v(i));
    k2 = f(v(i)+dt/2*k1);
    k3 = f(v(i)+dt/2*k2);
    k4 = f(v(i)+dt*k3);
    v(i+1) = v(i)+dt/6*(k1+2*k2+2*k3+k4);
end
fprintf(' t(s)      v(m/s)      v_exact(m/s) \n');
fprintf('         Runge-Kutta                \n');
fprintf('-------------------------------------\n');
for i = 1:5:length(t)
    fprintf('%5.2f  %10.4f  %10.4f \n', t(i),v(i),v_exact(i));
end
plot(t,v_exact,t,v,'x'), xlabel('t)'),ylabel('v,v-exact'),
grid, title('v(m/s) vs. t(s)'),
legend('v-exact','v-Runge-Kutta');
```

Program results

t(s)	v(m/s) Runge-Kutta	v_exact(m/s)
0.00	0.0000	0.0000
0.05	0.2646	0.2646
0.10	0.3531	0.3531
0.15	0.3827	0.3827
0.20	0.3926	0.3926
0.25	0.3959	0.3959
0.30	0.3971	0.3971
0.35	0.3974	0.3974
0.40	0.3975	0.3975
0.45	0.3976	0.3976
0.50	0.3976	0.3976
0.55	0.3976	0.3976
0.60	0.3976	0.3976
⋮	⋮	⋮

See Figure 7.6.

We see that there is good agreement between the exact solution and the solution obtained by the Runge–Kutta method.

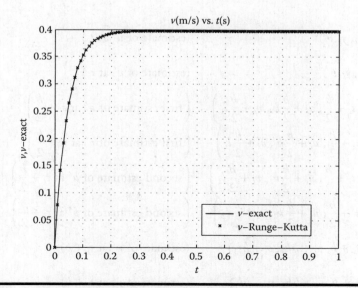

Figure 7.6 Velocity of ball bearing dropped in a vat of oil determined by the Runge–Kutta method.

Exercises

E7.3 Solve the *RC* circuit of Exercise E7.1 by using the Runge–Kutta method. Take $R = 1\ \text{k}\Omega$, $C = 1\ \mu\text{F}$, $V_o = 10\ \text{V}$, $v_c(0) = 0$, and $t = [0, 5\tau]$ with a $\Delta t = \tau/50$.

E7.4 Repeat Exercise E7.2, but this time solve the problem using the Runge–Kutta method and compare the result with the exact solution.

7.6 System of Two First-Order Differential Equations

Consider the following two first-order ODEs:

$$\frac{du}{dt} = f(t, u, v); \quad u(0) = u_0$$

$$\frac{dv}{dt} = g(t, u, v); \quad v(0) = v_0$$

(7.22)

To solve a system of two first-order differential equations by the Runge–Kutta method, take

$$u_{i+1} = u_i + \frac{h}{6}(k_1 + 2k_2 + 2k_3 + k_4)$$

(7.23)

$$v_{i+1} = v_i + \frac{h}{6}(l_1 + 2l_2 + 2l_3 + l_4)$$

where

$k_1 = f(t_i, y_i, v_i)$ (estimate of u' at t_i)

$l_1 = g(t_i, y_i, v_i)$ (estimate of v' at t_i)

$k_2 = f\left(t_i + \dfrac{h}{2}, u_i + \dfrac{h}{2}k_1, v_i + \dfrac{h}{2}l_1\right)$ $\left(\text{first estimate of } u' \text{ at } t_i + \dfrac{h}{2}\right)$

$l_2 = g\left(t_i + \dfrac{h}{2}, u_i + \dfrac{h}{2}k_1, v_i + \dfrac{h}{2}l_1\right)$ $\left(\text{first estimate of } v' \text{ at } t_i + \dfrac{h}{2}\right)$

$k_3 = f\left(t_i + \dfrac{h}{2}, u_i + \dfrac{h}{2}k_2, v_i + \dfrac{h}{2}l_2\right)$ $\left(\text{second estimate of } u' \text{ at } t_i + \dfrac{h}{2}\right)$

$l_3 = g\left(t_i + \dfrac{h}{2}, u_i + \dfrac{h}{2}k_2, v_i + \dfrac{h}{2}l_2\right)$ $\left(\text{second estimate of } v' \text{ at } t_i + \dfrac{h}{2}\right)$

$k_4 = f(t_i + h, u_i + hk_3, v_i + hl_3)$ (estimate of u' at $t_i + h$)

$l_4 = g(x_i + h, u_i + hk_3, v_i + hl_3)$ (estimate of v' at $t_i + h$)

(7.24)

and $h = \Delta t = t_{i+1} - t_i$.

Example 7.3

In this example, we examine the temperature of a small object dropped into a fluid contained within a vertical circular cylinder of radius R. We will assume that the body is a solid aluminum sphere of radius r and the fluid depth is L. We will neglect any heat transfer to the container walls. The governing equations for this problem are

$$\left(mc\frac{dT}{dt}\right)_B = \bar{h}A_s(T_F - T_B)$$

$$\left(mc\frac{dT}{dt}\right)_F = \bar{h}A_s(T_B - T_F)$$

where
    m is mass
    c is specific heat
    $\bar{h}$ is the convective heat transfer coefficient of the fluid
    A_s is the surface area of the sphere
    T is temperature
    The subscripts B and F correspond to the body and fluid respectively

The preceding equations state that the heat lost by the body equals the heat gained by the fluid.

We need to rewrite the equations to put them in the form required by the Runge–Kutta method:

$$\frac{dT_B}{dt} = \frac{\bar{h}A_s}{(mc)_B}(T_f - T_B) = F(T_B, T_f) \tag{7.25}$$

$$\frac{dT_F}{dt} = -\frac{\bar{h}A_s}{(mc)_F}(T_F - T_B) = g(T_B, T_F) \tag{7.26}$$

We will use the following parameters,

$$\rho_B = 2707\,\text{kg/m}^3, \quad \rho_F = 880\,\text{kg/m}^3, \quad c_B = 0.896\,\text{kJ/kg-°C}, \quad c_F = 2.05\,\text{kJ/kg-°C}$$

$$T_B(0) = 150°\text{C}, \quad T_F(0) = 20°\text{C}, \quad r = 0.2, \quad R = 0.5\,\text{m}, \quad L = 0.5\,\text{m}, \quad \bar{h} = 890\,\text{W/m}^2\text{-}°\text{C}$$

$$t = [0, 1]\,\text{s}, \quad \text{and} \quad dt = 0.005\,\text{s}$$

The program follows:

```
% Example_7_3a.m
% method for solving a system of first order differential
% equations.
% The problem is to determine the temperature of a sphere
% that is suddenly immerse in a bath whose temperature
% varies as heat from the sphere enters the fluid.
% The units are: rho in kg/m^3, c in kJ/kg-C, t in s,
% volume in m^3, h in W/m^2-C, R,r and L in m, As in m^2
clear; clc;
rho_al = 2707; c_al = 0.896e3; r = 0.2; Talo = 150.0;
rho_f = 880; c_f = 2.05e3; R = 0.50; L = 0.5; Tfo = 20;
T_al(1) = Talo; T_f(1) = Tfo;
vol_al = 4/3*pi*r^3;
vol_f = pi*R^2*L-vol_al;
As = 4*pi*r^2;
m_al = rho_al*vol_al;
m_f = rho_f*vol_f;
h = 890.0;
tau_al = m_al/h*c_al/As;
tau_f = m_f/h*c_f/As;
fprintf('tau_al=%10.5f tau_f=%10.5f \n',tau_al,tau_f);
F = @ (T_al,T_f) ((T_f-T_al)/tau_al);
G = @ (T_al,T_f) (-(T_f-T_al)/tau_f);
t = 0:1000
dt = 1
% Runge_Kutta method
for i = 1:length(t)-1
    k1 = F(T_al(i),T_f(i));
    L1 = G(T_al(i),T_f(i));
    k2 = F(T_al(i)+dt/2*k1,T_f(i)+dt/2*L1);
    L2 = G(T_al(i)+dt/2*k1,T_f(i)+dt/2*L1);
    k3 = F(T_al(i)+dt/2*k2,T_f(i)+dt/2*L2);
    L3 = G(T_al(i)+dt/2*k2,T_f(i)+dt/2*L2);
    k4 = F(T_al(i)+dt*k3,T_f(i)+dt*L3);
    L4 = G(T_al(i)+dt*k3,T_f(i)+dt*L3);
    T_al(i+1) = T_al(i)+dt/6*(k1+2*k2+2*k3+k4);
    T_f(i+1) = T_f(i)+dt/6*(L1+2*L2+2*L3+L4);
end
fo = fopen('output.txt','w');
fprintf(fo,'Determining the temperature of the Aluminum \n');
```

```
fprintf(fo,'sphere and the temperature of the fluid \n\n');
fprintf(fo,' t(s)      T(C)        T(C)       \n');
fprintf(fo,'          sphere      fluid      \n');
fprintf(fo,'-----------------------------------------------\n');
for i = 1:10:length(t)
    fprintf(fo,'%17.2f %10.2f %10.2f \n',...
            t(i),T_al(i),T_f(i));
end
plot(t,T_al,t,T_f,'--'), xlabel('t'), ylabel('T-al,T-f'), grid,
title('T-al(C) and T-f(C) vs. t(s)'), legend('al','f');
```

Program results

```
Determining the temperature of the Aluminum
sphere and the temperature of the fluid
```

t(s)	T(C) sphere	T(C) fluid
0.00	150.00	20.00
10.00	143.06	20.87
20.00	136.54	21.69
30.00	130.41	22.46
40.00	124.65	23.18
50.00	119.23	23.86
60.00	114.14	24.50
70.00	109.36	25.10
80.00	104.86	25.66
90.00	100.63	26.19
100.00	96.66	26.69
:	:	:
900.00	34.93	34.43
910.00	34.90	34.44
920.00	34.88	34.44
930.00	34.85	34.44
940.00	34.83	34.45
950.00	34.81	34.45
960.00	34.79	34.45
970.00	34.77	34.45
980.00	34.76	34.46
990.00	34.74	34.46
1000.00	34.72	34.46

See Figure 7.7.

We see that in 1000 s, the sphere and the fluid are nearly in thermal equilibrium.

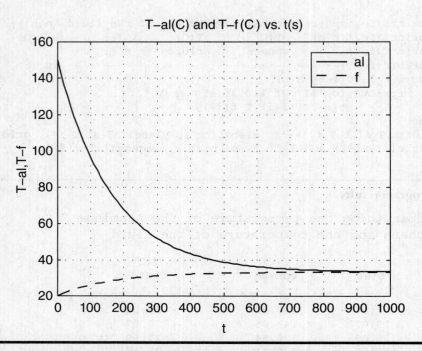

Figure 7.7 Temperatures of aluminum sphere and fluid vs. time solved by the Runge–Kutta method.

Exercise

E7.5 Solve the following system of first-order differential equations by the Runge–Kutta method over the interval $0 \leq t \leq 10$:

$$\frac{dr}{dt} = 2re^{-0.1t} - 2ry = f(t, r, y)$$

$$\frac{dy}{dt} = -y + ry = g(r, y)$$

(7.27)

The initial conditions are $r(0) = 1.0$ and $y(0) = 3.0$.

7.7 Single Second-Order Equation

For a single second-order ODE, the method of solution is to reduce the equation to a system of two first-order equations. Given the following second-order differential equation with initial conditions,

$$u'' = \frac{d^2u}{dt^2} = f(t, u, u')$$

$$u(0) = u_0 \quad \text{and} \quad u'(0) = u_0'$$

(7.28)

Let $u' = v$, then $u'' = \dfrac{dv}{dt} = v' = f(t,u,v)$. Also, $u' = g(t,u,v) = v$, giving

$$u' = v$$

$$v' = f(t,u,v) \tag{7.29}$$

Thus, we have converted Equation 7.28 into two first-order differential equations of the same form as Equation 7.19, and thus the same solution techniques can be applied; that is,

$$k_1 = f(t_i, u_i, v_i) \qquad\qquad (v' \text{ at } t_i)$$

$$l_1 = u'(t_i) = v_i \qquad\qquad (u' \text{ at } t_i)$$

$$k_2 = f\left(t_i + \frac{h}{2}, u_i + \frac{h}{2}l_1, v_i + \frac{h}{2}k_1\right) \qquad \left(\text{first estimate of } v' \text{ at } t_i + \frac{h}{2}\right)$$

$$l_2 = u'\left(t_i + \frac{h}{2}\right) = v_i + \frac{h}{2}k_1 \qquad \left(\text{first estimate of } u' \text{ at } t_i + \frac{h}{2}\right)$$

$$k_3 = f\left(t_i + \frac{h}{2}, u_i + \frac{h}{2}l_2, v_i + \frac{h}{2}k_2\right) \qquad \left(\text{second estimate of } v' \text{ at } t_i + \frac{h}{2}\right)$$

$$l_3 = u'\left(t_i + \frac{h}{2}\right) = v_i + \frac{h}{2}k_2 \qquad \left(\text{second estimate of } u' \text{ at } t_i + \frac{h}{2}\right)$$

$$k_4 = f(t_i + h, u_i + hl_3, v_i + hk_3) \qquad (\text{estimate of } v' \text{ at } t_i + h)$$

$$l_4 = u'(t_i + h) = v_i + hk_3 \qquad (\text{estimate of } u' \text{ at } t_i + h)$$

The values of u and v at the next time step are given by

$$u_{i+1} = u_i + \frac{h}{6}(l_1 + 2l_2 + 2l_3 + l_4)$$

$$v_{i+1} = v_i + \frac{h}{6}(k_1 + 2k_2 + 2k_3 + k_4) \tag{7.30}$$

Example 7.4

We will illustrate this method by applying it to the mass–spring–dashpot system described in Project P2.5. Equation 7.31 gives the governing equation for the displacement, y, of the mass from the equilibrium position, that is,

$$y'' + \frac{c}{m} y' + \frac{k}{m} y = 0 \tag{7.31}$$

Let $y' = v$,
then

$$y'' = \frac{dv}{dt} = -\frac{c}{m} v - \frac{k}{m} y = f(y, v, t) \tag{7.32}$$

and

$$\frac{dy}{dt} = v = g(y, v, t) \tag{7.33}$$

For the underdamped case with no forcing function, the exact solution is

$$y_{exact} = \exp\left(-\frac{c}{2m} t\right) \left\{ A \exp\left(\sqrt{\left(\frac{c}{2m}\right)^2 - \frac{k}{m}} \, t \right) + B \exp\left(-\sqrt{\left(\frac{c}{2m}\right)^2 - \frac{k}{m}} \, t \right) \right\} \tag{7.34}$$

We will use the following parameters for the system:

$m = 25$ kg, $k = 200$ N/m, $c = 5$ N-s/m, and initial conditions $y(0) = 5$ m and $y'(0) = 0$ m/s. Also

$$A = 5\,\text{m} \quad \text{and} \quad B = \frac{c}{2m} \times \frac{A}{\sqrt{\dfrac{k}{m} - \left(\dfrac{c}{2m}\right)^2}} \tag{7.35}$$

We want to create a MATLAB program that will solve Equation 7.31 by the Runge–Kutta method and plot y vs. t and y_{exact} vs. t on the same graph.

The program follows:

```
% Example_7_4.m
% This program solves the motion of a mass-spring-dashpot system.
% The governing equation is a Second Order Ordinary
% Differential (ODE).
% The second order DEQ is reduced to 2 first order ODEs.
% Equation (y vs. t) is solved by the Runge Kutta method.
% m = 25 kg, k = 200 N/m, c = 5 N-s/m, v = dy/dt
% y(1) = 5, v(1) = 0
clear; clc;
m = 25; k = 200; c = 5;
arg1 = sqrt(k/m-(c/2/m)^2);
arg2 = c/2/m;
A = 5;
B = arg2*A/arg1;
y(1) = 5; v(1) = 0; y_exact(1) = 5;
t = 0:0.05:20;
dt = 0.05;
f = @(y,v) (-c/m*v-k/m*y);%dvdt
g = @(v) (v);            %dy/dt
for i = 1:length(t)-1
    k1 = f(y(i),v(i)); %dv/dt
    L1 = g(v(i));        %dy/dt
    k2 = f(y(i)+dt/2*L1,v(i)+dt/2*k1);
    L2 = g(v(i)+dt/2*k1);
    k3 = f(y(i)+dt/2*L2,v(i)+dt/2*k2);
    L3 = g(v(i)+dt/2*k2);
    k4 = f(y(i)+dt*L3,v(i)+dt*k3);
    L4 = g(v(i)+dt*k3);
    y(i+1) = y(i)+dt/6*(L1+2*L2+2*L3+L4);
    v(i+1) = v(i)+dt/6*(k1+2*k2+2*k3+k4);
    y_exact(i+1) = exp(-arg2*t(i))*(A*cos(arg1*t(i))...
                  +B*sin(arg1*t(i)));
end
fo = fopen('output','w');
fprintf(fo,'    t(s)          y(m)         v(m/s)          \n');
fprintf(fo,'--------------------------------------------- \n');
for i = 1:10:length(t)
    fprintf(fo,'%6.2f %10.4f %10.4f \n',t(i),y(i),v(i));
end
plot(t,y,t,y_exact,'x'), xlabel('t'), ylabel('y'), grid,
title('y & y_exact vs. t'), legend('y-Runge-Kutta','y-exact');
figure;
plot(t,v), xlabel('t'), ylabel('v'), title('v vs. t'), grid;
```

Program results

See Figure 7.8a and b.

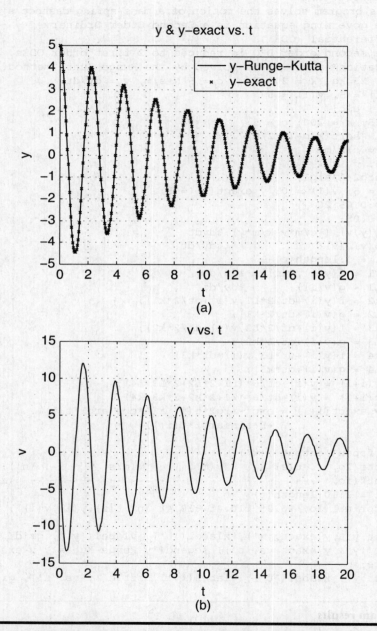

Figure 7.8   Solution of mass–spring–dashpot system by the Runge–Kutta method (a) *y* vs. *t* and (b) *v* vs. *t*.

Example 7.5

Another example illustrating the method of reducing a single second-order ODE to two first-order ODEs and then solving the system by the Runge–Kutta method is the problem involving the series RLC circuit of Project P2.12. The governing equation for the capacitor voltage, v_C (with no driving voltage), is

$$\frac{d^2 v_C}{dt^2} + \frac{R}{L}\frac{dv_C}{dt} + \frac{1}{LC}v_C = 0 \tag{7.36}$$

We wish to solve for v_C for the component values $R = 10\ \Omega$, $L = 1$ mH, and $C = 2\,\mu\text{F}$ and initial conditions $v_C(0) = 10$ and $\dfrac{dv_C}{dt}(0) = \dfrac{6}{\sqrt{LC}}$.

We first reduce this single second-order differential equation into two first-order equations.

Let $\dfrac{dv_C}{dt} = u$. Then, the two coupled equations are

$$\frac{dv_C}{dt} = u$$

$$\frac{du}{dt} = \frac{d^2 v_C}{dt^2} = -\frac{1}{LC}v_C - \frac{R}{L}v_2 \tag{7.37}$$

Assuming an underdamped system, the exact solution has the form

$$v_C = \exp\left(-\frac{R}{2L}t\right)\left\{\alpha\cos\left(\sqrt{\frac{1}{LC}-\left(\frac{R}{2L}\right)^2}\,t\right) + \beta\sin\left(\sqrt{\frac{1}{LC}-\left(\frac{R}{2L}\right)^2}\,t\right)\right\} \tag{7.38}$$

For the given initial conditions, $\alpha = 0$ and $\beta = \dfrac{\dfrac{dv_C}{dt}(0)}{\sqrt{\dfrac{1}{LC}-\left(\dfrac{R}{2L}\right)^2}}$.

The program follows:

```
% Example_7_5.m
% This program solves a single second order ordinary
% differential equation by the Runge-Kutta Method.
```

```
% The differential equation describes the capacitor voltage
% in a series RLC circuit.
% Let dvc/dt = u, then du/dt = -vc/LC -R/L*u = f(vc,u)
clear; clc;
% Component values:
R = 10; L = 1.0e-3; C = 2.0e-6;
% Initial conditions:
vc(1) = 0; u(1) = 6*sqrt(1/(L*C));
% Define time interval and step size. Since the natural
% frequency sqrt(1/LC) is 3559 Hz, choose a time interval to
% capture several oscillations. 3559 Hz has period of 281
% usec, so choose 1000 usec for interval.
tmax = 1000e-6; steps = 100; h = tmax/steps;
arg1 = sqrt(1/(L*C) - (R/(2*L))^2);
arg2 = R/2/L;
vc_exact(1) = 0;
t = 0:h:tmax;
% Define u' = f(t,x,y):
u_prime = @(vc,u) -(1/(L*C))*vc - (R/L)*u;
vc_prime = u;
% Do Runge-Kutta algorithm:
for i = 1:length(t)-1
    K1 = u_prime(vc(i), u(i));
    L1 = u(i);
    K2 = u_prime(vc(i)+h/2*L1, u(i)+h/2*K1);
    L2 = u(i)+h/2*K1;
    K3 = u_prime(vc(i)+h/2*L2, u(i)+h/2*K2);
    L3 = u(i)+h/2*K2;
    K4 = u_prime(vc(i)+h*L3, u(i)+h*K3);
    L4 = u(i)+h*K3;
    vc(i+1) = vc(i)+h/6*(L1+2*L2+2*L3+L4);
    u(i+1) = u(i)+h/6*(K1+2*K2+2*K3+K4);
    % Calculate exact solution
    vc_exact(i+1) = exp(-arg2*t(i+1))*u(1)/...
    arg1*sin(arg1*t(i+1));
end
% Print results to screen.
fprintf('Runge-Kutta solution for a single second order ODE \n');
fprintf('      t(s)         vc(v)          vc(v) \n');
fprintf('                  R-K Sol''n      Exact Sol''n \n');
fprintf('-----------------------------------------------\n');
for i = 1:10:steps+1 % just print every tenth step
    fprintf('%10.3e %8.4f %10.4f \n',t(i),vc(i), vc_exact(i));
end
% Plot results:
plot(t,vc,t,vc_exact,'x'), xlabel('t (seconds)'), ylabel('volts'),
title('Capacitor voltage of series RLC circuit');
legend('Runge-Kutta','exact');
```

Program results

```
Runge-Kutta solution for a single second order ODE

    t(s)            vc(v)              vc(v)
                  R-K Sol'n          Exact Sol'n
-----------------------------------------------------
0.000e+00         0.0000             0.0000
1.000e-04         3.0634             3.0632
2.000e-04        -2.1247            -2.1246
3.000e-04         0.3466             0.3467
4.000e-04         0.5413             0.5411
5.000e-04        -0.5030            -0.5029
6.000e-04         0.1497             0.1497
7.000e-04         0.0812             0.0812
8.000e-04        -0.1114            -0.1114
9.000e-04         0.0474             0.0474
1.000e-03         0.0081             0.0081
>>
```
See Figure 7.9.

A comparison of the Runge–Kutta solution with the exact solution is shown in Figure 7.9.

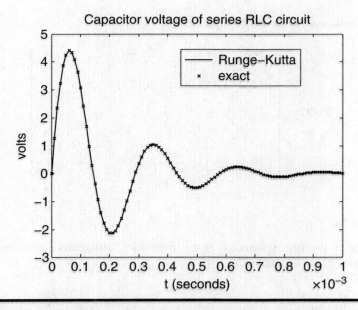

Figure 7.9   Solution of the voltage across a capacitor in a series circuit by the Runge–Kutta method (v_c vs. t).

7.8 MATLAB's ODE Function

MATLAB has several built-in ode functions that solve a system of first order ordinary differential equations, including `ode23` and `od45`. In this chapter, we will demonstrate the `ode45` function, which is based on the fourth- and fifth-order Runge-Kutta methods. A description of the `ode45` function follows. The syntax for the function is:

$$[\text{TOUT},\text{YOUT}] = \text{ODE45}(\text{ODEFUN},\text{TSPAN},\text{Y0})$$

Thus, `ode45` takes as arguments: a handle to a function describing the differential equations (ODEFUN), a vector describing a time interval (TSPAN), and vector describing the initial conditions (Y0). We will assume that the independent variable is t and the dependent variables are: $y_1, y_2, ..., y_n$. The function ODEFUN must take two input arguments: a time t and variables $y_1, y_2, ..., y_n$ which describes the system of n differential equations. The system of n differential equations must be in standard form, that is: $y'_j = f(t, y_1, y_2, ..., y_n), j = 1, 2, 3, ..., n$. To obtain solutions at specific times T0,T1,...,TFINAL (all increasing or all decreasing), use TSPAN = [T0 T1 .... TFINAL]. The function `ode45` will return two vectors: a list of time-points

$$\text{TOUT} = [\text{T0 T1 . ... TFINAL}]$$

and the solution vector YOUT equal to

$$
\text{YOUT} =
\begin{bmatrix}
y_1(\text{T0}) & y_2(\text{T0}) & ...... & y_n(\text{T0}) \\
y_1(\text{T1}) & y_2(\text{T1}) & ...... & y_n(\text{T1}) \\
. & & . & . \\
. & & . & . \\
y_1(\text{TFINAL}) & y_2(\text{TFINAL}) & ... & y_n(\text{TFINAL})
\end{bmatrix}
$$

Example 7.6

Solve the following system of three first-order differential equations using MATLAB's `ode45` function:

$$y'_1 = y_2 y_3 t$$

$$y'_2 = -y_1 y_3 \tag{7.39}$$

$$y'_3 = -0.51 y_1 y_2$$

The initial conditions are $y_1(0) = 0$, $y_2(0) = 1.0$, and $y_3(t) = 1.0$.

```
% Example_7_6.m
% This program solves a system of 3 ordinary differential
% equations by using MATLAB's ode45 function.
% y1' = y2*y3*t, y2' = -y1*y3, y3' = -0.51*y1*y2
% y1(0) = 0, y2(0) = 1.0, y3(0) = 1.0
clear; clc;
initial = [0.0 1.0 1.0];
tspan = 0.0:0.1:10.0;
[t,Y] = ode45(@dydt3,tspan,initial);
y1 = Y(:,1);
y2 = Y(:,2);
y3 = Y(:,3);
fid = fopen('output.txt','w');
fprintf(fid,'    t              y1           y2          y3        \n');
fprintf(fid,'-------------------------------------------------\n');
for i = 1:2:101
    fprintf(fid,'%7.2f %10.4f %10.4f %10.4f \n',...
            t(i),y1(i),y2(i),y3(i))
end
fclose(fid);
plot(t,y1,t,y2,'-.',t,y3,'--'), xlabel('t'),
ylabel('y1,y2,y3'),title('(y1, y2, y3) vs. t'), grid,
text(5.2,-0.8,'y1'), text(7.7,-0.25,'y2'),
text(4.2,0.85,'y3');
```

```
% dydt3.m
% This function works with example_7_6.m
% y1' = y2*y3*t, y2' = -y1*y3, y3' = -0.51*y1*y2
% y1 = Y(1), y2 = Y(2), y3 = Y(3).
function Yprime = dydt3(t,Y)
Yprime = zeros(3,1);
Yprime(1) = Y(2)*Y(3)*t;
Yprime(2) = -Y(1)*Y(3);
Yprime(3) = -0.51*Y(1)*Y(2);
```

Program results

The calculated solutions for y_1, y_2, and y_3 are shown in Figure 7.10.

Exercises

E7.6 Repeat Example 7.5, but this time solve the series RLC circuit for the capacitor voltage, v_C, by using MATLAB's ode45 function. The governing differential equation (see Project P2.11) for the capacitor voltage is

$$\frac{d^2 v_C}{dt^2} + \frac{R}{L}\frac{dv_C}{dt} + \frac{1}{LC}v_C = 0 \qquad (7.40)$$

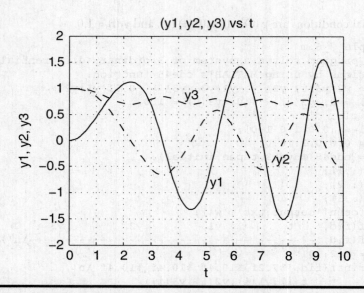

Figure 7.10   Solution to Equations 7.39 using MATLAB®'s ode45 function.

Take $R = 10\ \Omega$, $L = 1$ mH, and $C = 2\ \mu$F and initial conditions $v_C(0) = 10$ and $\dfrac{dv_C}{dt}(0) = \dfrac{6}{\sqrt{LC}}$.

The exact general solution is

$$v_{C,exact} = \exp\left(-\frac{R}{2L}t\right)\left\{\alpha\cos\left(\sqrt{\frac{1}{LC}-\left(\frac{R}{2L}\right)^2}\,t\right)+\beta\sin\left(\sqrt{\frac{1}{LC}-\left(\frac{R}{2L}\right)^2}\,t\right)\right\} \quad (7.41)$$

For the given initial conditions, $\alpha = 0$ and $\beta = \dfrac{\dfrac{dv_C}{dt}(0)}{\sqrt{\dfrac{1}{LC}-\left(\dfrac{R}{2L}\right)^2}}$.

Plot the solution obtained by MATLAB's `ode45` function and the exact solution on the same graph for $0 \le t \le 4\ \mu$s.

E7.7 The governing equation for the inductor current, i_L, in the parallel RLC circuit of Figure P2.7 is

$$\frac{d^2 i_L}{dt^2}+\frac{1}{RC}\frac{di_L}{dt}+\frac{1}{LC}i_L = 0 \quad (7.42)$$

Solve the parallel RLC circuit of Exercise E2.8 by using MATLAB's ode45 function. Assume $R = 50\ \Omega$, $L = 1\ \mu H$, $C = 10\ nF$, $v(0) = 3.3\ V$, and $i_L(0) = 0\ A$. The exact solution is given by

$$i_{L,exact} = \exp\left(-\frac{1}{2RC}t\right)\left\{\alpha\cos\left(\sqrt{\frac{1}{LC}-\left(\frac{1}{2RC}\right)^2}\,t\right)+\beta\sin\left(\sqrt{\frac{1}{LC}-\left(\frac{1}{2RC}\right)^2}\,t\right)\right\}$$

(7.43)

with $\alpha = 0$ and $\beta = -\dfrac{3.3}{L\sqrt{\dfrac{1}{LC}-\left(\dfrac{1}{2RC}\right)^2}}$

Plot i_L obtained by MATLAB's ode45 function and the exact solution on the same graph for $0 \le t \le 5\ \mu s$.

Projects

P7.1 This project is a modification of Exercise E2.4 in which a boy on a snowboard (considered as one unit), initially at rest, slides down a smooth hill, which makes an angle ϑ with the horizontal (see Figure 2.12). The boy's weight, W, is 650 N, and the friction coefficient, μ, between the snowboard and the snow is 0.05. In this project, the drag force on the boy and his snowboard is included. The x and y components of Newton's second law are

$$F_x = ma_x = m\frac{dV_x}{dt} = m\frac{d^2x}{dt^2}\quad\text{and}\quad F_y = N - W\cos\vartheta = 0$$

$$F_x = W\sin\vartheta - f - D, \quad .a_y = V_y = 0$$

where

F_x is the unbalanced force acting on the boy–snowboard unit in the x direction

F_y is the unbalanced force acting on the boy–snowboard unit in the y direction

V_x, V_y are the x and y velocity components of the boy–snowboard unit

a_x, a_y are the x and y acceleration components of the boy–snowboard unit

m is the mass of the boy–snowboard unit

f is the friction force

D is the drag on the boy–snowboard unit

N is the normal force on the bottom of the snowboard

The friction force $f = \mu N$, $W = mg$, and

$$D = C_d \left(\frac{\rho V_x^2}{2} \right) A$$

where
$\quad C_d$ is the drag coefficient
$\quad \rho$ is the air mass density
$\quad A$ is the frontal area of the boy–snowboard unit

Use the following parameters: $A = 0.5 \text{ m}^2$, $C_d = 1.0$, $\rho = 1.225 \text{ kg/m}^3$, $\vartheta = 15°$
$V_x(0) = 0$, and $x(0) = 0$.

a. Use MATLAB's ode45 function to solve for x and V_x at intervals of 0.10 s for $0 \le t \le 60.0$ s.

b. Create plots of x and V_x vs. t on separate graphs.

P7.2 An airplane flying horizontally at 50 m/s and at an altitude of 300 m is to drop a food package weighing 2000 N to a group of people stranded in an inaccessible area resulting from an earthquake. Let $\vec{V}$ be the velocity of the package. A wind velocity, $\vec{V}_w$, of 20 m/s flows horizontally in the opposite airplane direction (see Figure P7.2). A drag force, $\vec{D}$, acts on the package in the direction of the free stream, $\vec{V}_\infty$, as seen from the package (see Figure P7.2). We wish to determine (t, x, y, u, v) as a function of time and when the package hits the ground, where

$\quad (x, y)$ = the position of the package at time t.

$\quad (u, v)$ = the horizontal and vertical components of the package velocity, respectively.

The governing equations are

$$M \frac{d\vec{V}}{dt} = Mg\hat{e}_y + \vec{D} \qquad \text{(P7.2a)}$$

$$\vec{V} = u\hat{e}_x + v\hat{e}_y = \frac{dx}{dt}\hat{e}_x + \frac{dy}{dt}\hat{e}_y \qquad \text{(P7.2b)}$$

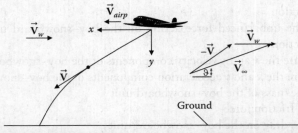

Figure P7.2 Path of a food package dropped from an airplane.

$$\vec{D} = C_d \rho \frac{V_\infty^2}{2} A \hat{e}_\vartheta \qquad (P7.2c)$$

$$V_\infty = \sqrt{(u + V_w)^2 + v^2} \qquad (P7.2d)$$

where

$\hat{e}_x$ and $\hat{e}_y$ are unit vectors in the x and y directions, respectively

$\hat{e}_\vartheta$ is unit vector in the direction of the free-stream velocity as seen from the package

C_d is the drag coefficient

ρ is the air density

A is the frontal area of the package

The equations reduce to

$$\frac{du}{dt} = -\frac{C_d \, \rho V_\infty^2 \, A}{2 \, M} \cos \vartheta \qquad (P7.2e)$$

$$\frac{dv}{dt} = g - \frac{C_d \, \rho V_\infty^2 \, A}{2 \, M} \sin \vartheta \qquad (P7.2f)$$

$$\frac{dx}{dt} = u \qquad (P7.2g)$$

$$\frac{dy}{dt} = v \qquad (P7.2h)$$

$$\cos \vartheta = \frac{u + V_w}{V_\infty}, \quad \text{and} \quad \sin \vartheta = \frac{v}{V_\infty} \qquad (P7.2i)$$

Initial conditions

$$x(0) = 0, \quad y(0) = 0, \quad u(0) = 50 \, \text{m/s}, \quad v(0) = 0$$

Use the following parameters:

$$C_d = 0.8, \quad \rho = 1.225 \text{ kg/m}^3, \quad \text{and} \quad A = 1.0 \text{ m}^2$$

Use MATLAB's ode45 function to solve for (t, x, y, u, v) at intervals of 0.10 s for $0 \le t \le 10.0$ s:

a. Create plots of x and y vs. t both on the same graph.

b. Create plots of u and v vs. t both on the same graph.

c. Create a table containing (t, x, y, u, v) at intervals of 0.10 s. Stop printing table the first time $y > 300$ m.

d. Use MATLAB's function `interp1` to interpolate for the (t, x, u, v) values when the package hits the ground. Print out these values.

P7.3 Figure P7.3a shows a third-order RLC circuit. In order to run a time-domain transient analysis, we transform the circuit into differential equations using the following method:

First, write Kirchhoff's current law for each node in the circuit:

$$\text{At node 1:} \, i_{C1} + i_{C2} - i_L = 0 \qquad \text{(P7.3a)}$$

$$\text{At node 2:} \, i_R + i_{C2} = 0. \qquad \text{(P7.3b)}$$

Second, write Kirchhoff's voltage law for the two loops in the circuit:

$$\text{For loop 1:} \, V_S - v_{C1} - v_L = 0 \qquad \text{(P7.3c)}$$

$$\text{For loop 2:} \, v_L + v_{C2} - v_R = 0. \qquad \text{(P7.3d)}$$

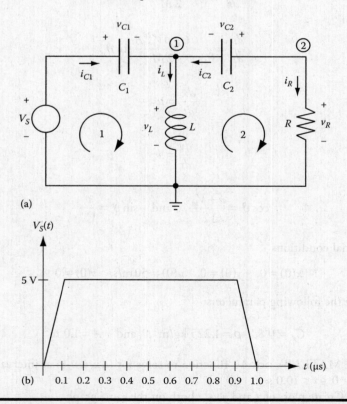

(a)

(b)

Figure P7.3 **A third-order RLCC circuit: (a) circuit configuration and (b) pulse input.**

Third, write the constituent relations for each circuit element:

$$\text{For } C_1, \ i_{C1} = C_1 v'_{C1} \tag{P7.3e}$$

$$\text{For } C_2, \ i_{C2} = C_2 v'_{C2} \tag{P7.3f}$$

$$\text{For } L, \ v_L = Li'_L \tag{P7.3g}$$

$$\text{For } R, \ v_R = i_R R. \tag{P7.3h}$$

We will choose the three voltages and currents with derivative terms (v_{C1}, v_{C2}, and i_L) as the *state variables* for this problem:

a. Rearrange Equations P7.3a through P7.3h into the following standard form:

$$\frac{dv_{C1}}{dt} = f\left(v_{C1}, v_{C2}, i_L, V_S(t)\right) \tag{P7.3i}$$

$$\frac{dv_{C2}}{dt} = g\left(v_{C1}, v_{C2}, i_L, V_S(t)\right) \tag{P7.3j}$$

$$\frac{di_L}{dt} = h\left(v_{C1}, v_{C2}, i_L, V_S(t)\right) \tag{P7.3k}$$

where the functions f, g, and h involve parameters C_1, C_2, L, and R.

b. Solve Equations P7.3i through P7.3k with the Runge–Kutta algorithm for the following circuit parameters: $C_1 = 1$ μF, $C_2 = 0.001$ μF, $R = 100$ kΩ, and $L = 0.01$ mH. Use a time interval of $0 \le t \le 10$ μs and a step size of 0.01 μs. Assume $V_S(t)$ is a 5 V pulse starting at time $t = 0$ with rise time of 0.1 μs, an "on" time of 0.8 μs, and fall time of 0.1 μs (as shown in Figure P7.3b). The initial conditions are $v_{C1}(0) = 0$, $v_{C2}(0) = 0$, and $i_L(0) = 0$.

c. Plot on separate graphs $v_{C1}(t)$, $v_{C2}(t)$, $i_L(t)$, and $V_S(t)$ vs. time.

P7.4 A small rocket with an initial mass of 350 kg, including a mass of 100 kg of fuel, is fired from a rocket launcher (see Figure P7.4). The rocket leaves the launcher at velocity v_o and at an angle of θ_o with the horizontal. Neglect the fuel consumed inside the rocket launcher. The rocket burns fuel at the rate of 10 kg/s and develops a thrust $T = 6000$ N. The thrust acts axially along the rocket and lasts for 10 s. Assume that the drag force also acts axially and is

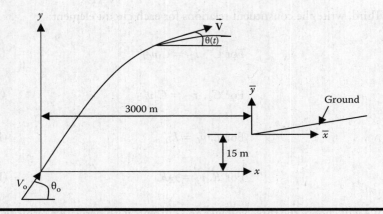

Figure P7.4   Launching of a rocket.

proportional to the square of the rocket velocity. The governing differential equations describing the position and velocity components of the rocket are as follows:

$$\frac{dv_x}{dt} = \frac{T}{m}\cos(\theta) - \frac{Kv^2}{m}\cos(\theta) \qquad \text{(P7.4a)}$$

$$\frac{dv_y}{dt} = \frac{T}{m}\sin(\theta) - \frac{Kv^2}{m}\sin(\theta) - g \qquad \text{(P7.4b)}$$

$$\frac{dx}{dt} = v_x \qquad \text{(P7.4c)}$$

$$\frac{dy}{dt} = v_y \qquad \text{(P7.4d)}$$

where
$v^2 = v_x^2 + v_y^2$
θ is the angle the velocity vector makes with the horizontal
m is the mass of the rocket (varies with time)
v_x, v_y are the x and y components of the rocket's velocity relative to the ground
K is the drag coefficient
g is the gravitational constant
(x, y) are the position of the rocket relative to the ground
t is the time of rocket flight
$\cos\theta = v_x/v$ and $\sin\theta = v_y/v$

Substituting for $\cos\theta$ and $\sin\theta$ in Equations P7.3a and P7.3b, they become

$$\frac{dv_x}{dt} = \frac{v_x T}{m\sqrt{v_x^2 + v_y^2}} - \frac{v_x K \sqrt{v_x^2 + v_y^2}}{m} \qquad \text{(P7.4e)}$$

$$\frac{dv_y}{dt} = \frac{v_y T}{m\sqrt{v_x^2 + v_y^2}} - \frac{v_y K \sqrt{v_x^2 + v_y^2}}{m} - g \qquad \text{(P7.4f)}$$

The target lies on ground that has a slope of 5%. The ground elevation relative to the origin of the coordinate system of the rocket is given by

$$y_g = 15 + 0.05(x - 3000) \qquad \text{(P7.4g)}$$

Using Equations P7.4e, P7.4f, P7.4c, and P7.4d, write a computer program in MATLAB using the fourth-order Runge–Kutta method described in Sections 7.5 through 7.8 that will solve for x, y, v_x and v_y for $0 \leq t \leq 60$ s. Use Equation P7.4g to solve for y_g.

Use a fixed time step of 0.01 s. Take $x(0) = 0$, $y(0) = 0$, $v_x(0) = v_o\cos\theta_o$, $v_y(0) = v_o\sin\theta_o$, $v_o = 150$ m/s, $\theta_o = 60°$, $K = 0.045$ N-s^2/m^2, and $g = 9.81$ m/s^2:

a. Print out a table for x, y, y_g, v_x, v_y at every 1.0 s. Run the program for $0 \leq t \leq 60$ s.

b. Use MATLAB to plot x, y, and y_g vs. t on the same graph and v_x and v_y vs. t on the same graph.

c. Assume a linear trajectory between the closest two data points where the rocket hits the ground. The intersection of the two straight lines gives the (x, y) position of where the rocket hits the ground.

P7.5 Repeat Problem P7.4, but this time use MATLAB's ode45 function to solve the problem. Use a $0 \leq t \leq 60$ s with a step size of 1 s.

P7.6 The *Sallen–Key* circuit (Figure P7.6) is commonly used to implement second-order (or higher) filters. Although there are several ways to model this circuit,

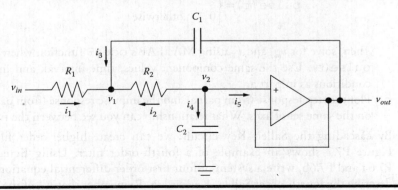

Figure P7.6   Sallen–Key circuit.

we will analyze it here directly in the time domain using differential equations. If we assume that the op amp is ideal, then $v_2 = v_{out}$ and also $i_5 = 0$. Applying Kirchhoff's current law at the nodes labeled v_1 and v_2 and using the constituent relations for resistors ($v_R = i_R R$) and capacitors $\left(i_C = C \dfrac{dv_C}{dt} \right)$, we get

At node v_1: $i_1 + i_2 + i_3 = 0 \Rightarrow \dfrac{v_{in} - v_1}{R_1} + \dfrac{v_2 - v_1}{R_2} + C_1 \dfrac{d(v_{out} - v_1)}{dt} = 0$

At node v_2: $i_2 + i_4 = 0 \Rightarrow \dfrac{v_2 - v_1}{R_2} + C_2 \dfrac{dv_2}{dt} = 0$

Rearranging into a system of two first-order ODEs and setting $v_2 = v_{out}$, we get

$$\frac{dv_{out}}{dt} = \left(\frac{-1}{R_2 C_2} \right) v_{out} + \left(\frac{1}{R_2 C_2} \right) v_1 \tag{P7.6a}$$

$$\frac{dv_1}{dt} = \left(\frac{1}{R_2 C_1} - \frac{1}{R_2 C_2} \right) v_{out} + \left(\frac{1}{R_2 C_2} - \frac{1}{R_1 C_1} - \frac{1}{R_2 C_1} \right) v_1 + \left(\frac{1}{R_1 C_1} \right) v_{in} \tag{P7.6b}$$

a. Solve for v_{out} and v_1 using MATLAB's ode45 function. Assume that the input to the circuit v_{in} is a step voltage that changes from 0 to 1 V at time $t = 0^+$. Assume the following values for the circuit elements: $R_1 = 5000\ \Omega$, $R_2 = 5000\ \Omega$, $C_1 = 2200$ pF, and $C_2 = 1100$ pF. Use a time interval of $t = [0, 100\ \mu s]$ and assume $v_{out}(0) = v_1(0) = 0$.

b. Find the impulse response of the circuit by first creating a MATLAB function pulse(t), which returns the following values:

$$\text{pulse}(t) = \begin{cases} 10^6 & \text{for } 0 < t < 10^{-6} \\ 0 & \text{otherwise} \end{cases}$$

Then, solve for v_{out} and v_1 using MATLAB's ode45 function where $v_i =$ pulse(t). Use the same component values, time interval, and initial conditions as in part a.

c. Plot the step response (from part a) and the impulse response (from part b) on the same set of axes. What relationship can you see between the two?

P7.7 By cascading the Sallen–Key circuit, we can create higher-order filters. Figure P7.7 shows an example of a fourth-order filter. Using Equation P7.6a and P7.6b, write a system of four first-order differential equations to describe the fourth-order Sallen–Key circuit. Remember that v_{out} of the first stage will be equal to v_{in} of the second stage.

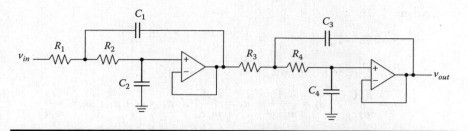

Figure P7.7 Fourth-order filter using cascaded Sallen–Key circuit.

a. Solve for v_{out} using MATLAB's ode45 function. Assume that the input to the circuit v_{in} is a step voltage that changes from 0 to 1 V at time $t = 0^+$. Assume the following values for the circuit elements: $R_1 = R_2 = R_3 = R_4 = 5{,}000\ \Omega$, $C_1 = 18{,}000$ pF, $C_2 = 15{,}000$ pF, $C_3 = 4.3\ \mu$F, and $C_4 = 10$ pF. Use a time interval of $t = [0, 0.005]$ s and assume $v_o(0) = v_1(0) = v_2(0) = v_3(0) = 0$.

b. Plot v_{out} and v_{in} with respect to t on the same set of axes.

P7.8 We wish to examine the time temperature variation of a fluid, T_f, enclosed in a container with a heating element and a thermostat. The walls of the container are pure copper. The fluid is engine oil and its temperature T_f varies with time. The thermostat is set to cut off power from the heating element when T_f reaches 65°C and to resume supplying power when T_f reaches 55°C.

Wall properties:

$$k = 386.0\ \text{W/m-C}, \quad c = 0.3831 \times 10^3\ \text{J/kg-C}, \quad \rho = 8954\ \text{kg/m}^3$$

Engine oil properties:

$$k = 0.137\ \text{W/m-C}, \quad c = 2.219 \times 10^3\ \text{J/kg-C}, \quad \rho = 840\ \text{kg/m}^3$$

The inside size of the container is (0.5 m × 0.5 m × 0.5 m)

The wall thickness is 0.01 m. Thus, the inside surface area $A_{s,i} = 1.5$ m², outside surface area $A_{s,o} = 1.5606$ m², engine oil volume $V_{oil} = 0.125$ m³, and wall volume $V_{wall} = 0.0153$ m³.

The power, Q, of the heating element = 20,000 W.

The inside convective heat transfer coefficient $h_i = 560$ W/m²-C.

The outside convective heat transfer coefficient $h_o = 110$ W/m²-C.

Using a lump parameter analysis (assume that the engine oil is well mixed) and the first law of thermodynamics, the governing equations describing the time temperature variation of both materials are as follows:

$$\frac{d\theta_f}{dt} = -a_1(\theta_f - \theta_w) + a_5 \tag{P7.8a}$$

$$\frac{d\theta_w}{dt} = a_2(\theta_f - \theta_w) - a_3\theta_w = a_2\theta_f - a_4\theta_w \tag{P7.8b}$$

where

$$\theta_f = T_f - T_\infty$$

$$\theta_w = T_w - T_\infty$$

$$a_1 = \frac{h_i A_{s,i}}{m_f c_f}, \quad a_2 = \frac{h_i A_{s,i}}{m_w c_w}, \quad a_3 = \frac{h_o A_{s,o}}{m_w c_w}, \quad a_4 = a_2 + a_3, \quad a_5 = \frac{Q}{m_f c_f}$$

Initial conditions

$$T_f(0) = T_w(0) = 15°C$$

$$T_\infty = 15°C$$

Using MATLAB's `ode45` function, construct a simulation of this system. Run the time for 3600 s with a time step of 3 s. Print out values of T_f and T_w vs. t at every 100 s. Construct plots of T_f and T_w vs. t.

P7.9 We wish to determine the altitude and velocity of a helium-filled spherically shaped balloon as it lifts off from its mooring. We will assume that atmospheric conditions can be described by the U.S. Standard Atmosphere. We will assume that there is no change in the balloon's volume. The governing equation describing the motion of the balloon is

$$M \frac{d^2 z}{dt^2} = \left(B - W - \mathrm{sgn}\left(\frac{dz}{dt}\right) \times D\right) \tag{P7.9a}$$

where

    z is the altitude of the centroid of the balloon
    B is the buoyancy force acting on the balloon
    M is the total mass of the balloon material, ballast, and the gas
    W is the total weight of the balloon material, ballast, and the gas = Mg
    D is the drag on the balloon
    $\mathrm{sgn}(x) = \begin{cases} +1 & \text{if } x \geq 0 \\ -1 & \text{if } x < 0 \end{cases}$

The U.S. Standard Atmosphere as applied to this balloon problem consists of the following governing equations:

$$\frac{dp}{dz} = -\gamma = -\rho g, \quad \frac{dp}{dz} = \frac{dp}{dt}\frac{dt}{dz} = \frac{dp}{dt}\frac{1}{v} = -\rho g$$

or

$$\frac{dp}{dt} = -\frac{p}{RT} g v \tag{P7.9b}$$

$$T = T_i - \lambda z \qquad \text{(P7.9c)}$$

$$\rho = \frac{p}{RT} \qquad \text{(P7.9d)}$$

where
    p is the outside air pressure at an elevation of the centroid of the balloon
    ρ is the outside air density at an elevation of the centroid of the balloon
    g is the gravitational constant that varies with altitude
    R is the gas constant for air
    T is the outside air temperature at an elevation of the centroid of the
        balloon
    v is the vertical velocity of the balloon
    T_i is the temperature at the Earth's surface = 288.15 (K)
    λ is the lapse rate

The second-order ODE, Equation P7.9a, can be reduced to two first-order differential equations, by letting

$$\frac{dz}{dt} = v \qquad \text{(P7.9e)}$$

Then

$$\frac{dv}{dt} = \frac{1}{M}(B - W - \text{sgn}(v) \times D) \qquad \text{(P7.9f)}$$

The three Equations P7.9e, P7.9f, and P7.9b represent three coupled ODEs that can be solved using MATLAB's `ode45` function. The buoyancy force, B, is given by

$$B = \rho g \mathbb{V} \qquad \text{(P7.9g)}$$

and

$$g = g_0 \left(\frac{r_e}{z + r_e} \right) \qquad \text{(P7.9h)}$$

where
    $\mathbb{V}$ is the volume of the balloon
    r_b is the radius of the balloon
    r_e is the radius of the Earth
    g_0 is the gravitational acceleration near the Earth's surface
    g is the gravitational acceleration at an elevation of the centroid of the balloon
    D is the drag force
    v is the velocity of the balloon

For a low Reynolds number, Re, less than 0.1, the drag force is given by Stokes' formula, which is

$$D = 6\pi\mu v r_b \tag{P7.9i}$$

For flow speeds with Re > 0.1, use

$$D = C_d \frac{\rho}{2} v^2 A \tag{P7.9j}$$

where
C_d is the drag coefficient
A is the frontal area of the balloon = πr_b^2

The drag coefficient, C_d, is given by

$$C_d = \frac{24}{\text{Re}} + \frac{6}{1.0 + \sqrt{\text{Re}}} + 0.4 \tag{P7.9k}$$

where

$$\text{Re} = \frac{2\rho v r_b}{\mu} \tag{P7.9l}$$

μ is the fluid viscosity

The fluid viscosity, μ, can be determined by the Sutherland formula, which is

$$\mu = \mu_0 \left(\frac{T}{T_0}\right)^{1.5} \left(\frac{T_0 + S}{T + S}\right) \tag{P7.9m}$$

For air, $S = 110.4$ K, $\mu_0 = 1.71 \times 10^{-5}$ N-s/m², and $T_0 = 273$ K.

Write a computer program, using MATLAB's ode45 function, that will determine the balloon's altitude as a function of time. Create plots of z vs. t, v vs. t, and p vs. t. Use the following values:

$$M = 2200 \text{ kg}, \quad r_b = 7.816 \text{ m}, \quad \mathbb{V} = (4/3)\pi r_b^3 \text{ m}^3, \quad R = 287 \text{ J/(kg-K)}, \quad T_i = 288.15 \text{ K},$$

$$\lambda = 0.0065 \text{ (K/m)}, \quad g_0 = 9.81 \text{ m/s}^2, \quad \text{and} \quad r_e = 6371 \times 10^3 \text{ m}$$

Take $0 \le t \le 1000$ s in steps of 0.1 s and the following initial conditions:

$$z(0) = r_b, \quad v(0) = 0, \quad \text{and} \quad p(0) = 1.0132 \times 10^5$$

P7.10 A small tank with its longitudinal axis in a vertical position is connected to a pressurized air supply system as shown in Figure P7.10. The tank contains two gate valves, one which controls the pressurization of the tank and

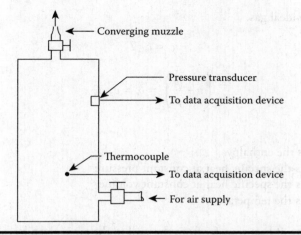

Figure P7.10 Air-pressurized tank.

the other controls the discharge of the tank through a converging nozzle. The tank is instrumented with a copper–constantan thermocouple and a pressure transducer. We wish to predict the temperature and pressure time histories of the air inside the tank as it is being discharged. We shall assume that the air properties inside the tank are uniform. Due to its large heat capacity, as compared to the air inside the tank, we shall assume that the wall temperature remains nearly constant during the discharge phase of the problem. We will also assume that the change in kinetic and potential energy of the air inside the tank is negligible. Applying the energy equation to a control volume enclosing the interior of the tank gives

$$\frac{d}{dt}(m_a u_a) = -\dot{m}_e \left(u + \frac{p}{\rho} + \frac{V^2}{2} \right)_e + \left(\frac{\delta Q}{\delta t} \right)_{w \to a} \tag{P7.10a}$$

where
  m is the mass
  u is the internal energy
  p is the pressure
  ρ is the density
  V is the velocity
  $\dot{m}_e$ is the mass flow rate of air exiting the tank
  $\left(\dfrac{\delta Q}{\delta t} \right)_{w \to a}$ is the rate of heat transfer from wall to air inside the tank
  Subscript a is the air
  Subscript w is the wall
  Subscript e refers to the conditions at the exit

For an ideal gas,

$$u_a = c_{v,a}T_a \qquad (P7.10b)$$

$$\left(u + \frac{p}{\rho}\right)_e = h_e = c_{p,e}T_e \qquad (P7.10c)$$

where
  h is the enthalpy
  c_p is the specific heat at constant pressure
  c_v is the specific heat at constant volume
  T is the temperature

The rate of heat transfer from the wall to the air is given by

$$\left(\frac{\delta Q}{\delta t}\right)_{w \to a} = \bar{h}A_{s,i}(T_w - T_a) \qquad (P7.10d)$$

where
  $\bar{h}$ is the convective transfer coefficient
  $A_{s,i}$ is the interior surface area of the tank

The equation describing the rate of change of mass in the tank is

$$\frac{dm_a}{dt} = -\dot{m}_e \qquad (P7.10e)$$

Substituting Equation P7.10b through P7.10e into Equation P7.10a gives

$$\frac{dT_a}{dt} = \frac{T_a}{m_a}\dot{m}_e + \frac{A_{s,i}\bar{h}_i}{c_{v,a}m_a}(T_w - T_a) - \frac{\dot{m}_e}{m_a}k_aT_e - \frac{\dot{m}_eV_e^2}{2c_{v,a}m_a} \qquad (P7.10f)$$

where k_a is the ratio of specific heats for air = $\dfrac{c_{p,a}}{c_{v,a}}$.

We have assumed that $c_{v,a}$ and $c_{p,a}$ do not vary significantly in the temperature range of the problem. The functional relation for $\dot{m}_e$ in terms of the other variables is obtained from 1-D compressible flow through nozzles. Two possible cases exist, depending on the ratio of p_b/p_a. The pressure p_b is the back pressure, which is the surrounding air pressure.

Case 1:

$$\frac{p_b}{p_a} > 0.528$$

then

$$p_e = p_b$$

$$M_e = \frac{2}{k_a - 1}\left(1 - \left(\frac{p_e}{p_a}\right)^{(k_a-1)/k_a}\right)$$

$$T_e = \frac{T_a}{1 + \dfrac{k_a - 1}{2}M_e^{2}}$$

$$\bar{c}_e = \sqrt{k_a R_a T_e}$$

$$V_e = M_e \bar{c}_e$$

$$\dot{m}_e = \frac{p_e}{R_a T_e}A_e V_e$$

Case 2:

$$\frac{p_b}{p_a} \leq 0.528$$

then

$$p_e = 0.528\,p_a$$

$$M_e = 1.0$$

$$T_e = \frac{T_a}{1 + \dfrac{k_a - 1}{2}}$$

$$V_e = \bar{c}_e = \sqrt{k_a R_a T_e}$$

$$\dot{m}_e = \frac{p_e}{R_a T_e}A_e V_e$$

In the preceding equations, $\bar{c}$ is the speed of sound, R is the gas constant, and M is the Mach number.

From these relations, it can be seen that $\dot{m}_e$ is an implicit function of the variables T_a, m_a, and t. These equations, along with Equations P7.10e and P7.10f, form two coupled differential equations of the form

$$\frac{dT_a}{dt} = f_1(t, T_a, m_a)$$

$$\frac{dm_a}{dt} = f_2(t, T_a, m_a)$$

$$p_a = \left(\frac{mRT}{\mathbb{V}}\right)_a$$

where $\mathbb{V}$ is the volume inside the tank $= \pi D^2 L/4$, L is the tank length and D is the diameter .

This system of equations can be solved using the MATLAB's ode45 function.

However, before this can be done, we need to determine $\bar{h}$. We shall assume that the heat transfer from the wall to the air inside the tank occurs by natural convection. We will also assume that the wall temperature remains nearly constant, since it has a much larger heat capacity than the air. The empirical relation for natural convection for vertical plates and cylinders is

$$\bar{h} = \frac{\bar{k}_a}{L}\left[0.825 + \frac{0.387(\mathrm{Gr}\,\mathrm{Pr})^{1/6}}{\left[1 + (0.492/\mathrm{Pr})^{9/16}\right]^{8/27}}\right]^2 \tag{P7.10g}$$

where

$$\mathrm{Gr} = \text{Grashof number} = \frac{g\beta(T_w - T_a)L^3}{(\mu_a/\rho_a)^2} \tag{P7.10h}$$

and

Pr is the Prandtl number
$\bar{k}_a$ is the thermal conductivity of air
μ_a is the viscosity of air
β is the coefficient of expansion
g is the gravitational constant

The properties of Pr, $\bar{k}_a$, and μ_a are evaluated at the film temperature, T_f, which is

$$T_f = 0.5(T_w + T_a) \quad \text{and} \quad \rho_f = \left(\frac{p}{RT}\right)_f$$

Since β is a fluid property, take $\beta = 1/T_f$.

Table P7.10 Air Properties vs. Temperature

T (K)	μ (N-s/m^2)	k (W/(m-K))	Pr
100	6.9224×10^{-6}	0.009246	0.77
150	1.0283×10^{-5}	0.013735	0.753
200	1.3289×10^{-5}	0.01809	0.739
250	1.4880×10^{-5}	0.02227	0.722
300	1.9830×10^{-5}	0.02624	0.708
350	2.075×10^{-5}	0.03003	0.697

Air property values for these variables are given in Table P7.10. Use MATLAB's interp1 function to interpolate for air properties of μ, $\bar{k}$, and Pr at air temperature T_f.

Using MATLAB'S ode45 function, determine the air temperature, T_a, air pressure, p_a, and mass, m_a, in the tank as a function of time. Take $0 \le t \le 70$ s with a time step of 0.1 s.

Print out a table of these values at every 1 s. Also, create plots of T_a vs. t, p_a vs. t, and m_a vs. t. Use the following values for the program:

$$D = 0.07772 \text{ m}, \quad L = 0.6340 \text{ m}, \quad d_e = 0.001483 \text{ m}, \quad c_{p,a} = 1.006 \times 10^3 \text{ J/(kg-K)},$$

$$c_{v,a} = 0.721 \times 10^3 \text{ J/(kg-K)}, \quad k_a = 1.401, \quad g = 9.807 \text{ m/s}^2, \quad p_b = p_{atm} = 1.013 \times 10^5 \text{ N/m}^2,$$

$$p_{a,i} = 7.04648 \times 10^5 \text{ N/m}^2, \quad T_{a,i} = 294.5 \text{ K}, \quad T_w = 294.5 \text{ K}, \quad \text{and} \quad R_a = 287.2 \text{ J/(kg-K)}$$

where
  D is the inside diameter of the tank
  L is the inside length of the tank
  d_e is the diameter of the nozzle
  $T_{a,i}$ is the initial air temperature inside the tank
  $p_{a,i}$ is the initial air pressure inside the tank

Chapter 8

Boundary Value Problems of Ordinary Differential Equations

8.1 Introduction

When an ordinary differential equation (ODE) involves boundary conditions instead of initial conditions, then a different numerical approach is used to solve the problem. In a boundary value problem, we essentially need to "fit" a solution into the known boundary conditions as opposed to simply integrating from the initial conditions. An example of this type of problem is the deflection of a beam where boundary conditions at both ends of the beam are specified. Another example of this type of problem is to determine the electric field between the plates of a capacitor with a known charge density between the plates and a fixed voltage across the plates. In both cases, a solution is found by numerically solving a second-order, nonhomogeneous ODE using *finite difference methods*.

8.2 Difference Formulas

To numerically solve a boundary value problem involving an ordinary, linear differential equation, we will employ difference formulas obtained by Taylor series expansion. This will enable us to reduce differential equations to a set of algebraic equations. As we saw in Section 3.1, we can expand $y = g(x)$ in a Taylor series expansion about point x_i; that is,

$$y(x) = y(x_i) + y'(x_i)(x - x_i) + \frac{y''(x_i)}{2!}(x - x_i)^2 + \frac{y'''(x_i)}{3!}(x - x_i)^3 + \cdots \quad (8.1)$$

We define the step size $h = x - x_i$, which gives

$$y(x_i + h) = y(x_i) + y'(x_i)h + \frac{y''(x_i)}{2!}h^2 + \frac{y'''(x_i)}{3!}h^3 + \cdots \quad (8.2)$$

Let $y(x_i + h) = y_{i+1}$ and $y(x_i) = y_i$, $y'(x_i) = y_i'$, etc. Then, the Taylor series expansion equation can be written as

$$y_{i+1} = y_i + y_i'h + \frac{y_i''h^2}{2!} + \frac{y_i'''h^3}{3!} + \cdots \quad (8.3)$$

We will now rewrite Equation 8.2 for the point $x = x_i - h$, that is, the equidistant point on the other side of x_i. This gives

$$y(x_i - h) = y_{i-1} = y_i + y_i'(-h) + \frac{y_i''}{2!}(-h)^2 + \frac{y_i'''}{3!}(-h)^3 + \cdots$$

or

$$y_{i-1} = y_i - y_i'h + \frac{y_i''h^2}{2!} - \frac{y_i'''h^3}{3!} + \cdots - \cdots \quad (8.4)$$

Using only two terms on the right side of Equations 8.3 and 8.4 and solving for y_i', we obtain difference formulas for y_i' with an error of order h, that is,

$$y_i' = \frac{y_{i+1} - y_i}{h} \quad \text{(forward difference formula)} \quad (8.5)$$

and

$$y_i' = \frac{y_i - y_{i-1}}{h} \quad \text{(backward difference formula)} \quad (8.6)$$

Now, let us just keep three terms in Equations 8.3 and 8.4 and subtract the latter from the first, giving

$$y_{i+1} - y_{i-1} = 2y_i'h$$

Solving for y_i' gives

$$y_i' = \frac{y_{i+1} - y_{i-1}}{2h}. \quad (8.7)$$

This is the central difference formula for y_i' with an error of order h^2 (the error is order h^2 because three terms in the Taylor series were used to obtain the formula).

Table 8.1   Summary of Finite Difference Formulas for Boundary Value Problems

$y_i' = \dfrac{y_{i+1} - y_i}{h}$	First-order forward difference formula. Used for y' boundary condition at beginning of interval.
$y_i' = \dfrac{y_i - y_{i-1}}{h}$	First-order backward difference formula. Used for y' boundary condition at end of interval.
$y_i' = \dfrac{y_{i+1} - y_{i-1}}{2h}$	First-order central difference formula. Used for first-order differential equation in middle of interval.
$y_i'' = \dfrac{y_{i+1} + y_{i-1} - 2y_i}{h^2}$	Second-order central difference formula. Used for second-order differential equation in middle of interval.

Now, let us again keep just three terms in Equations 8.3 and 8.4 and add the resulting equations; we obtain

$$y_{i+1} + y_{i-1} = 2y_i + y_i'' h^2$$

Solving for y_i'' gives

$$y_i'' = \frac{y_{i+1} + y_{i-1} - 2y_i}{h^2} \tag{8.8}$$

This is the central difference formula for y_i'' with error of order h^2.

The four difference formulas derived above are summarized in Table 8.1. We will now show how they are applied in order to model a boundary value problem that leads to a tri-diagonal system of linear equations. The formulas used in the following examples are tabulated in Table 8.1.

Example 8.1

In this example, we consider the deflection of a beam subjected to both a uniform load, w, applied over the region $0 \le x \le L_1$ and a concentrated load P applied at $x = L_2$. The beam is shown in Figure 8.1a.

The governing equation for the deflection of a beam is (for a derivation of this equation, see Appendix C)

$$\frac{d^2 y}{dx^2} = \frac{M(x)}{EI(x)} \tag{8.9}$$

where
   y is the deflection of beam
   M is the internal bending moment
   E is the modulus of elasticity of beam material
   I is the moment of inertia of an area

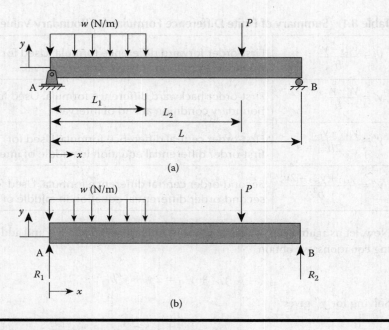

Figure 8.1 (a) Loaded beam. (b) Beam reactions.

To obtain the finite difference form of the governing equation, subdivide the
x-axis into N subdivisions, giving $x_1, x_2, x_3, \ldots, x_{N+1}$.

Let the deflections at these points be $y_1, y_2, y_3, \ldots, y_{N+1}$.

The finite difference form for $\dfrac{d^2 y}{dx^2}$, as discussed in Section 8.2, is

$$\frac{d^2 y}{dt^2}(x_n) = \frac{y_{n+1} + y_{n-1} - 2y_n}{\Delta x^2} \tag{8.10}$$

Thus, the governing differential equation becomes

$$\frac{y_{n+1} + y_{n-1} - 2y_n}{\Delta x^2} = \frac{M_n}{EI_n}$$

or

$$y_{n-1} - 2y_n + y_{n+1} = \frac{M_n \Delta x^2}{EI_n}, \quad \text{for } n = 2, 3, 4, \ldots, N \tag{8.11}$$

The boundary conditions are:

$$y_1 = 0 \tag{8.12}$$

$$y_{N+1} = 0 \tag{8.13}$$

Equations 8.11–8.13 represent a system of linear, algebraic equations which can be solved by using MATLAB's inv function. However, we first need to obtain expressions for the bending moments, M_n, First, solve for the reactions R_1 and R_2 (see Figure 8.1b). Taking the moment about point A will enable us to determine R_2, that is,

$$\sum M_A = 0 = R_2 L - w L_1 \times \frac{L_1}{2} - P L_2$$

Solving for R_2 gives

$$R_2 = \frac{w L_1^2}{2L} + \frac{P L_2}{L} \tag{8.14}$$

Taking the moments about point B will enable us to determine R_1, that is,

$$\sum M_B = 0 = P(L - L_2) + w L_1 \left(L - \frac{L_1}{2} \right) - R_1 L$$

Solving for R_1 gives

$$R_1 = P \left(1 - \frac{L_2}{L} \right) + w L_1 \left(1 - \frac{L_1}{2L} \right) \tag{8.15}$$

The internal bending moments are taken about the neutral axis at section x.

For $0 \leq x \leq L_1$ (see Figures 8.2a and 8.2b), the sum of the moments at section x is:

$$M(x) + wx \frac{x}{2} - R_1 x = 0$$

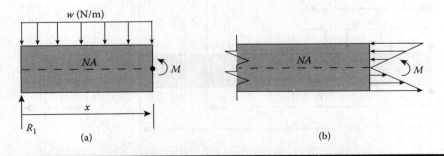

Figure 8.2 (a) **Moment about neutral axis at section** x, $0 \leq x \leq L_1$. (b) **Force distribution at section** x.

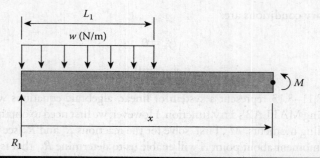

Figure 8.3   Moment about neutral axis at section x, $L_1 < x \le L_2$.

Solving for $M(x)$ and expressing the equation in finite difference form gives

$$M_n = R_1 x_n - \frac{w}{2} x_n^2 \tag{8.16}$$

For $L_1 < x \le L_2$ (see Figure 8.3), the sum of the moments at section x is:

$$M(x) + wL_1\left(x - \frac{L_1}{2}\right) - R_1 x = 0$$

Solving for $M(x)$ and expressing the equation in finite difference form gives

$$M_n = R_1 x_n - wL_1\left(x_n - \frac{L_1}{2}\right) \tag{8.17}$$

For $L_2 < x \le L$ (see Figure 8.4a and b), the sum of the moments at section x is:

$$M(x) + P(x - L_2) + wL_1\left(x - \frac{L_1}{2}\right) - R_1 x = 0$$

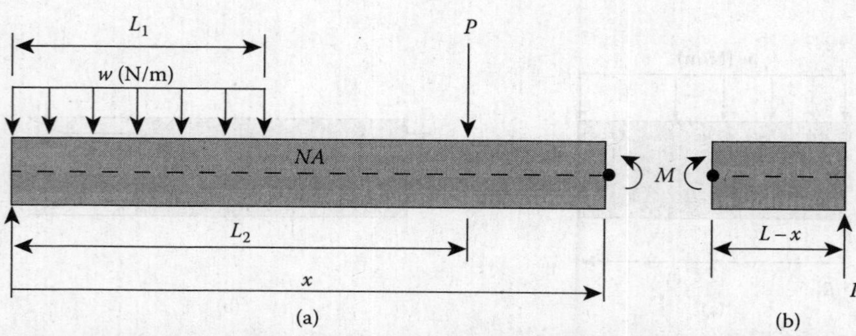

Figure 8.4   (a) Moment about neutral axis at section x, $L_2 \le x \le L$ from the left side and (b) from the right side.

Solving for $M(x)$ and expressing the equation in finite difference form gives

$$M_n = R_1 x_n - P(x_n - L_2) - wL_1\left(x_n - \frac{L_1}{2}\right) \tag{8.18}$$

For this case, it is more convenient to determine the moment from the right side of the beam (see Figure 8.4b), giving

$$-M(x) + R_2(L - x) = 0$$

Solving for $M(x)$ and expressing the equation in finite difference form gives

$$M_n = R_2(L - x_n) \tag{8.19}$$

Equation 8.19 is equivalent to Equation 8.18 for M_n for the region $L_2 < x \leq L$.

As indicated earlier, Equations 8.11–8.13 represent a system of algebraic, linear equations which can be solved by MATLAB's inv function or Gauss Elimination. In matrix notation, the system can be represented as $\mathbf{AY} = \mathbf{C}$, where $\mathbf{A}$ is an $n \times n$ coefficient matrix, $\mathbf{Y}$ is a column vector of the unknown values that we wish to determine and $\mathbf{C}$ is a column vector representing the right hand side of Equations 8.11–8.13. To utilize MATLAB's inv function, we need to write the set of Equations 8.11–8.13 in the form

$$\sum_j a_{n,j}\, y_j = c_n$$

and determine the $a_{n,j}$ and c_n terms. As in the Statics problem of Section 4.4 (see page 137), the first index in the matrix coefficient, $a_{n,j}$ is the equation number and the second index represents the y variable that is multiplied by that coefficient. We now apply this concept to our beam deflection problem.

Eq # Equation

1    $y_1 = 0,$ $a_{1,1} = 1,\ c_1 = 0$

2    $y_1 - 2y_2 + y_3 = \dfrac{M_2 \Delta x^2}{EI},$ $a_{2,1} = 1,\ a_{2,2} = -2,\ a_{2,3} = 1,\ c_2 = \dfrac{M_2 \Delta x^2}{EI}$

3    $y_2 - 2y_3 + y_4 = \dfrac{M_3 \Delta x^2}{EI},$ $a_{3,2} = 1,\ a_{3,3} = -2,\ a_{3,4} = 1,\ c_3 = \dfrac{M_3 \Delta x^2}{EI}$

etc.

We will use the following parameters in Example 8.1: w = 4.0 kN/m, EI = 1.5×10^3 kN-m^2, P = 35 kN, L = 3 m, L_1 = 1 m, L_2 = 2 m and 30 subdivisions on the x domain.

The program follows:

```
% Example_8_1.m
% This program calculates the deflection of a beam which is a
% boundary value problem. Finite difference method is used
% to solve the problem. This method results in a set of
% linear equations which is solved by the inv matrix method.
clear; clc;
w = 40.0; EI = 1.5e3; P = 35; L = 3; L1 = 1; L2 = 2;
N1 = 11; N2 = 21; N = 30; dx = L/N;
R1 = P*(1-L2/L)+w*L1*(1-L1/(2*L));
R2 = w/(2*L)*L1^2 + P*L2/L;
M(1) = 0; M(N+1) = 0; y(1) = 0.0; y(N+1) = 0.0; c(1) = 0; c(N+1) = 0;
x = 0:dx:L;
% determining M(i) values
for i = 2:N1
    M(i) = R1*x(i)-w/2*x(i)^2;
end
for i = N1+1:N2
    M(i) = R1*x(i)-w*L1*(x(i)-L1/2);
end
for i = N2+1:N
    M(i) = R2*(L-x(i));
end
% Establing the coefficient matrix
a = zeros(N+1,N+1); c = zeros(N+1,1);
% Overwrite the coefficients that are not zero.
a(1,1) = 1;
a(N+1,N+1) = 1;
for i = 2:N
    a(i,i-1) = 1;
    a(i,i) = -2;
    a(i,i+1) = 1;
    c(i) = M(i)/(EI)*dx^2;
end
y = inv(a)*c*100;     % Changing y to cm
fprintf('x(m)         y(cm) \n');
fprintf('------------------------\n');
for i = 1:length(x)
    fprintf('%5.2f    %12.4e \n',x(i),y(i));
end
plot(x,y), xlabel('x(m)'), ylabel('y(cm)'),
  title('deflection vs. position'), grid;
```

Program results

```
x (m)              y (m)
----------------------------
0.00        0.0000e+00
0.10       -1.1294e-01
0.20       -2.2490e-01
0.30       -3.3491e-01
0.40       -4.4204e-01
 .              .
 .              .
1.50       -1.1895e+00
1.60       -1.1982e+00
1.70       -1.1939e+00
 .              .
 .              .
>>
```
See Figure 8.5.

From Figure 8.5, we see that the maximum deflection occurs at approximately 1.5 m.

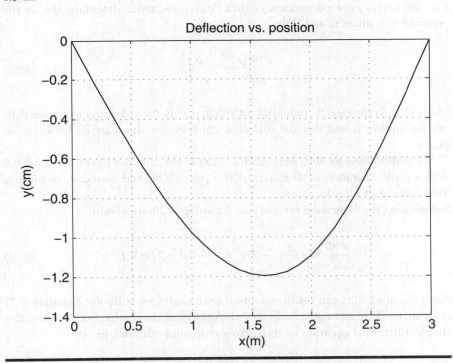

Figure 8.5   Deflection of beam shown in Figure 8.1a.

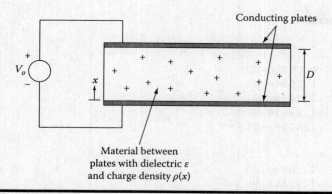

Figure 8.6 Parallel plate capacitor with constant applied voltage V_o.

Example 8.2

Figure 8.6 shows a parallel plate capacitor with constant applied voltage V_o and a fixed charge density ρ between the plates. For cases with planar symmetry such as the parallel plate capacitor where the charge density only changes in the x direction (i.e., there is no y nor z dependency), then Poisson's equation describing the electric potential Φ reduces to an ODE:

$$\frac{d^2\Phi(x)}{dx^2} = -\frac{\rho(x)}{\varepsilon} \tag{8.20}$$

where $\Phi(x)$ is the electric potential (in volts), $\rho(x)$ is the x-dependent charge density (in coul/m³), and ε is the dielectric constant for the material between the plates.

We wish to solve for $\Phi(x)$ between the plates of the capacitor shown in Figure 8.6 with a plate separation of D meters, $\rho(x) = \rho_o(x - D)^2$, and boundary conditions $\Phi(0) = 0$ and $\Phi(D) = V_o$.

Substituting the expression for $\rho(x)$ into Equation 8.20, we obtain

$$\frac{d^2\Phi}{dx^2} = -\frac{\rho_o}{\varepsilon}(x - D)^2 = -\frac{\rho_o}{\varepsilon}(x^2 - 2Dx + D^2) \tag{8.21}$$

Since Equation 8.21 can readily be solved analytically, we will solve Equation 8.21 both analytically and numerically and then compare the results. We can solve this simple differential equation by integrating twice on both sides, giving

$$\Phi(x) = -\frac{\rho_o}{\varepsilon}\left(\frac{1}{12}x^4 - \frac{D}{3}x^3 + \frac{D^2}{2}x^2 + \alpha x + \beta\right) \tag{8.22}$$

where α and β are constants of integration, which are determined by applying the boundary conditions. Applying the first boundary condition $\Phi(0) = 0$ gives

$$0 = -\frac{\rho_o}{\varepsilon}\left(\frac{1}{12}0^4 - \frac{D}{3}0^3 + \frac{D^2}{2}0^2 + \alpha 0 + \beta\right)$$

Thus, $\beta = 0$.

For the second boundary condition, $\Phi(D) = V_o$, we get

$$V_o = -\frac{\rho_o}{\varepsilon}\left(\frac{1}{12}D^4 - \frac{D}{3}D^3 + \frac{D^2}{2}D^2 + \alpha D\right)$$

Solving for α gives

$$\alpha = -\frac{\varepsilon V_o}{D\rho_o} - \frac{D^3}{4}$$

Thus, the exact solution to Equation 8.22 is

$$\Phi(x) = -\frac{\rho_o}{\varepsilon}\left[\frac{1}{12}x^4 - \frac{D}{3}x^3 + \frac{D^2}{2}x^2 - \left(\frac{\varepsilon V_o}{D\rho_o} + \frac{D^3}{4}\right)x\right] \qquad (8.23)$$

To solve Equation 8.21 numerically, we subdivide the region between the plates into N intervals and apply the finite difference formulas of Table 8.1. Dividing the region $0 \leq x \leq D$ into N intervals will result in $N + 1$ values for x and thus $N + 1$ values for Φ. Our goal is to set up the system of equations as a matrix equation, i.e., $A\Phi = \mathbf{C}$.

We begin by applying the second-order central difference formula from Table 8.1 to Equation 8.21, giving

$$\Phi_i'' = \frac{\Phi_{i+1} + \Phi_{i-1} - 2\Phi_i}{h^2} = -\frac{\rho_o}{\varepsilon}(x_i^2 - 2Dx_i + D^2)$$

Rearranging gives

$$\Phi_{i-1} - 2\Phi_i + \Phi_{i+1} = -\frac{\rho_o h^2}{\varepsilon}(x_i^2 - 2Dx_i + D^2) \qquad (8.24)$$

This equation is valid for $i = 2, 3, 4, \ldots, N$.

We now use the boundary conditions to determine the values for the upper-left and lower-right corners of the coefficient matrix. First, we rewrite the boundary conditions as

$$\Phi(0) = 0 \rightarrow \Phi_1 = 0 \qquad (8.25)$$

$$\Phi(D) = V_o \rightarrow \Phi_{N+1} = V_o \qquad (8.26)$$

The system is a set of linear algebraic equations and can be solved by MATLAB's inv function. The following parameters are used in Example 8.2: $N = 40$, $D = 0.4$ mm, $\rho_o = 10^4$ Coul/m^3, $\varepsilon = 1.04 \times 10^{-12}$ and $V_o = 5$ V

The program follows:

```
% Example_8_2.m
% Find the electric potential between a parallel plate
% capacitor with fixed charge density between the plates and
% known boundary conditions.
clear; clc;
N = 40;                 % step count
D =.0004;               % plate separation (meters)
h = D/N;                % step size
rho = 1e4;              % coulomb/m^3
epsilon = 1.04e-12;     % dielectric between the plates
Vo = 5;                 % voltage across plates
% First, calculate all values of x.
x = 0:h:D;
% Determine matrix coefficients, a*phi = c:
a(1,1) = 1; c(1) = 0;
a(N+1,N+1) = 1; c(N+1) = Vo;
for i = 2:N
    a(i,i-1) = 1; a(i,i) = -2; a(i,i+1) = 1;
    c(i) = -rho/epsilon*h^2*(x(i)^2-2*D*x(i)+D^2);
    % Calculate exact solution (Equation 8.23)
    phi_exact(i) = (-rho/epsilon)* (1.0/12*x(i)^4 -...
        (D/3)*x(i)^3 + (D^2/2)*x(i)^2 -...
        (epsilon*Vo/(D*rho)+D^3/4)*x(i));
end
phi_exact(1) = 0; phi_exact(N+1) = Vo;
phi = inv(a)*c';
% Plot numerical and exact solutions for phi:
plot(x,phi,x,phi_exact,'x'), xlabel('x (meters)'),
ylabel('Phi (volts)'), grid, legend('phi','phi-exact');
```

Program results

See Figure 8.7.

In both Examples 8.1 and 8.2, the system of linear equations is "tri-diagonal." This designation refers to the fact that coefficient matrices defined by the systems of linear algebraic Equations 8.11 through 8.13 and 8.24 through 8.26 contain nonzero values for the main diagonal elements and the diagonal elements immediately above and below the main diagonal. Except for these three diagonals, the rest of the elements in the coefficient matrix are all zero. A tri-diagonal system of linear equations can be treated in a more efficient way

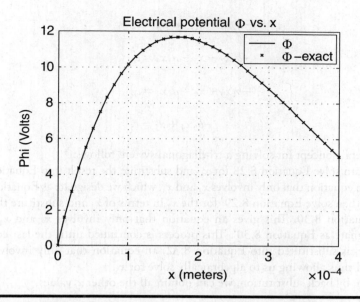

Figure 8.7 Electric potential, Φ vs. x.

than by taking the inverse of the coefficient matrix or by using the Gauss elimination method for solving a system of linear algebraic equations. This becomes important only for very large systems. The treatment of a tri-diagonal system of linear equations follows.

8.3 Solution of a Tri-Diagonal System of Linear Equations

A *tri-diagonal* system of equations has the following form:

$$\begin{bmatrix} 1 & -a_1 & 0 & 0 & 0 \\ -b_2 & 1 & -a_2 & 0 & 0 \\ 0 & -b_3 & 1 & -a_3 & 0 \\ 0 & 0 & -b_4 & 1 & -a_4 \\ 0 & 0 & 0 & -b_5 & 1 \end{bmatrix} \begin{bmatrix} x_1 \\ x_2 \\ x_3 \\ x_4 \\ x_5 \end{bmatrix} = \begin{bmatrix} c_1 \\ c_2 \\ c_3 \\ c_4 \\ c_5 \end{bmatrix} \tag{8.27}$$

where a_i, b_i, and c_i are constant terms and x_i are the unknowns. By multiplying out the matrices in Equation 8.27, the set of equations become

$$x_1 - a_1 x_2 = c_1 \tag{8.28}$$

$$-b_2 x_1 + x_2 - a_2 x_3 = c_2 \qquad (8.29)$$

$$-b_3 x_2 + x_3 - a_3 x_4 = c_3 \qquad (8.30)$$

$$-b_4 x_3 + x_4 - a_4 x_5 = c_4 \qquad (8.31)$$

$$-b_5 x_4 + x_5 = c_5 \qquad (8.32)$$

The general concept for solving a tri-diagonal system follows:

We can solve Equation 8.28 for x_1 and substitute the result into Equation 8.29, giving an equation that only involves x_2 and x_3, which we designate as Equation 8.29′. We can then solve Equation 8.29′ for the x_2 in terms of x_3 and substitute the result into Equation 8.30. This gives an equation that only involves x_3 and x_4, which we designate as Equation 8.30′. This process is continued until the last equation. When x_4 is substituted into Equation 8.32, an equation that only involves x_5, is obtained thus allowing us to algebraically solve for x_5.

Then, by back substitution, we can obtain all the other x_i values.

Method summary for m equations

Arrange the set of equations into the general form:

$$x_i = a_i x_{i+1} + b_i x_{i-1} + c_i \qquad (8.33)$$

Note: For the first and last equations (corresponding to the upper-left and lower-right coefficients in the tri-diagonal matrix), $b_1 = 0$ and $a_m = 0$.

By the substitution procedure outlined earlier, we can obtain a set of equations of the form

$$x_i = d_i + e_i x_{i+1} \qquad (8.34)$$

where d_i and e_i are the coefficients for the "prime" equations. Note that $e_m = 0$.
Then,

$$x_m = d_m$$

$$x_{m-1} = d_{m-1} + e_{m-1} x_m$$

$$\vdots \qquad\qquad\qquad (8.35)$$

$$x_1 = d_1 + e_1 x_2$$

If general expressions for d_i and e_i can be obtained, then we can solve the system for all x_i. We start by rewriting Equation 8.34 to put the $(i-1)$th equation into the form

$$x_{i-1} = d_{i-1} + e_{i-1} x_i \qquad (8.36)$$

Substituting Equation 8.36 into 8.33 gives

$$x_i = a_i x_{i+1} + b_i (d_{i-1} + e_{i-1} x_i) + c_i$$

Solving for x_i gives

$$x_i = \frac{(c_i + b_i d_{i-1})}{(1 - b_i e_{i-1})} + \frac{a_i x_{i+1}}{(1 - b_i e_{i-1})} \tag{8.37}$$

Matching terms between Equations 8.34 and 8.37, we obtain

$$d_i = \frac{c_i + b_i d_{i-1}}{1 - b_i e_{i-1}}$$

$$e_i = \frac{a_i}{1 - b_i e_{i-1}} \tag{8.38}$$

which is valid for $i = 2, 3, \ldots, m$.

The very first equation in the system is already in the form $x_i = d_i + e_i x_{i-1}$, and thus, matching terms between Equations 8.28 and 8.34, we obtain

$$d_1 = c_1$$

$$e_1 = a_1 \tag{8.39}$$

We can now successively apply Equation 8.38 to find $d_2, d_3, \ldots, d_m$ and $e_2, e_3, \ldots, e_m$. Then, $x_m = d_m$, and by back substitution,

$$x_{m-1} = d_{m-1} + e_{m-1} x_m$$

$$x_{m-2} = d_{m-2} + e_{m-2} x_{m-1}$$

$$\vdots$$

$$x_1 = d_1 + e_1 x_2$$

Example 8.3

Solve the following system of equations for all x_i:

$$x_1 \quad +2x_2 \qquad\qquad\qquad = \quad 7 \tag{8.40}$$

$$4x_1 \quad +3x_2 \quad +14x_3 \qquad\qquad = \quad -9 \tag{8.41}$$

$$11x_2 \quad +x_3 \quad -6x_4 \qquad = \quad 3 \tag{8.42}$$

$$-3x_3 \quad +2x_4 \quad -31x_5 \quad = \quad 1 \qquad (8.43)$$

$$4x_4 \quad +x_5 \quad -8x_6 = -19 \qquad (8.44)$$

$$20x_5 \quad +x_6 = \quad 11 \qquad (8.45)$$

First, we need to put the equations in the form of Equation 8.33, which is

$$x_i = a_i x_{i+1} + b_i x_{i-1} + c_i$$

This gives

$$x_1 = -2x_2 + 7$$

$$x_2 = -\frac{14}{3} x_3 - \frac{4}{3} x_1 - \frac{9}{3}$$

$$x_3 = 6x_4 - 11x_2 + 3$$

$$x_4 = -\frac{31}{2} x_5 + \frac{3}{2} x_3 + \frac{1}{2}$$

$$x_5 = 8x_6 - 4x_4 - 19$$

$$x_6 = -20x_5 + 11$$

We see that

$$a = [-2 \quad -14/3 \quad 6 \quad -31/2 \quad 8 \quad 0]$$
$$b = [0 \quad -4/3 \quad -11 \quad 3/2 \quad -4 \quad -20]$$
$$c = [7 \quad -3 \quad 3 \quad 1/2 \quad -19 \quad 11]$$

A MATLAB program to solve Equations 7.40 through 7.45 for all x_i follows.

```
% Example_8_3.m
% Example of solving a tri-diagonal system of equations
clear; clc;
a = [-2 -14/3 6 -31/2 8 0];
b = [0 -4/3 -11 3/2 -4 -20];
c = [7 -3 3 1/2 -19 11];
m = length(a);
% compute d and e coefficients
d(1) = c(1);
e(1) = a(1);
for i = 2:m
    d(i) = (c(i) + b(i)*d(i-1))/(1 - b(i)*e(i-1));
    e(i) = a(i)/(1 - b(i)*e(i-1));
end
% compute x
```

```
x(m) = d(m);
for i = (m-1):-1:1
    x(i) = d(i) + e(i)*x(i+1);
end
% display the solution:
x
```

Program results

```
x =

   37.3362  -15.1681  -8.0600  -29.6515  1.1653  -12.3051
```

Note that this example contains only arithmetic operations and no matrix operations (e.g., solving by Gauss elimination or finding an inverse matrix). Thus, the algorithm will be extremely fast, even for systems with many equations (e.g., order of m = 1000 or more).

Exercises

E8.2 Solve the beam problem of Example 8.1 by the tri-diagonal method described in Section 8.3.

E8.3 Solve the parallel plate capacitor problem of Example 8.2 by the tri-diagonal method described in Section 8.3.

Projects

P8.1 A heat sink is often attached to electronic components in order to dissipate excess heat generated by the device in order to prevent it from overheating (Figure P8.1a). A sketch of one heat sink fin is shown in Figure P8.1b, where the trapezoidal fin has bottom width W_1, top width W_2, height L_1, and depth D and the coordinates x and y are as drawn in the diagram. The temperature distribution in the fin is governed by a 1-D analysis of the heat equation in a solid. The analysis includes an empirical relation that involves the convective heat transfer coefficient, h. This empirical relation is introduced to separate the heat conduction problem within the solid from the heat transfer problem in the surrounding gas. The convective heat transfer coefficient, h, is determined by either experiment or by analytical/numerical methods. The governing equation for the fin is

$$\frac{d}{dx}\left(A\frac{dT}{dx}\right) = \frac{hP}{k}(T - T_\infty) \qquad \text{(P8.1a)}$$

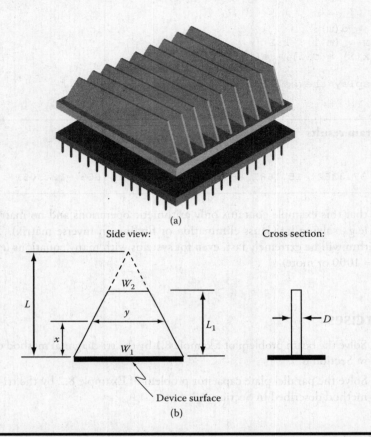

Figure P8.1 **(a) Sketch of a heat sink attached to an electronic component and (b) dimensions of a single fin.**

or

$$\frac{dA}{dx}\frac{dT}{dx} + A\frac{d^2T}{dx^2} = \frac{hP}{k}(T - T_\infty) \qquad \text{(P8.1b)}$$

where

　　　T is the temperature in the fin at position x
　　　T_∞ is the surrounding ambient air temperature
　　　A is the fin cross-sectional area
　　　h is the corrective heat transfer coefficient
　　　k is the thermal conductivity of the fin material
　　　P is the fin cross-sectional perimeter

We begin by writing equations for the area and perimeter in terms of the measurements of the fin:

$$A = yD$$
$$P = 2y + 2D$$

By the geometry of similar triangles, we can determine that $y/W_1 = (L - x)/L$, and thus, the area and perimeter as functions of x are

$$A(x) = \frac{W_1 D}{L}(L - x)$$

$$P(x) = \frac{2W_1}{L}(L - x) + 2D$$

The governing differential equation becomes

$$\frac{W_1 D}{L}(L - x)\frac{d^2 T}{dx^2} - \frac{W_1 D}{L}\frac{dT}{dx} = \frac{h}{k}\left[\frac{2W_1}{L}(L - x) + 2D\right](T - T_\infty) \quad \text{(P8.1c)}$$

The first boundary condition is that at the surface of the device, the temperature T_W is known:

$$T(0) = T_W \tag{P8.1d}$$

To obtain the second boundary condition, we employ the concept of energy conservation, that is, the rate that heat leaves the fin at $(x = L_1)$ per unit surface area by conduction = the rate that heat is carried away by convection per unit surface area.

This statement can be written mathematically as

$$-k\frac{dT}{dx}(L_1) = h\left[T(L_1) - T_\infty\right] \tag{P8.1e}$$

We wish to solve this problem numerically using the finite difference method.

First, subdivide the x-axis into N subdivisions, giving $x_1, x_2, x_3, \ldots, x_{N+1}$. Take the temperature at x_i to be T_i. The finite difference formulas for $d^2 T/dx^2$ and dT/dx (see Table 8.1) are

$$\frac{d^2 T}{dx^2}(x_i) = \frac{T_{i+1} + T_{i-1} - 2T_i}{\Delta x^2}$$

$$\frac{dT}{dx}(x_i) = \frac{T_{i+1} - T_i}{\Delta x}$$

The finite differential form of Equation P8.1c is

$$\frac{W_1 D}{L}(L - x_i)\left(\frac{T_{i+1} + T_{i-1} - 2T_i}{\Delta x^2}\right) - \frac{W_1 D}{L}\left(\frac{T_{i+1} - T_i}{\Delta x}\right)$$

$$= \frac{h}{k}\left[\frac{2W_1}{L}(L - x_i) + 2D\right](T_i - T_\infty)$$

Solving for T_i gives

$$T_i = \left\{\frac{2W_1 D}{L}(L - x_i) - \frac{W_1 D}{L}\Delta x + \frac{h\Delta x^2}{k}\left(\frac{2W_1}{L}(L - x_i) + 2D\right)\right\}^{-1}$$

$$\times \left\{\left[\frac{W_1 D}{L}(L - x_i) - \frac{W_1 D}{L}\Delta x\right]T_{i+1} + \frac{W_1 D}{L}(L - x_i)T_{i-1}\right.$$

$$\left. + \frac{h\Delta x^2}{k}\left[\frac{2W_1}{L}(L - x_i) + 2D\right]T_\infty\right\} \tag{P8.1f}$$

Equation P8.1f is valid for $i = 2, 3, \ldots, N$.

The finite difference form for Equation P8.1e is

$$-k\frac{T_{N+1} - T_N}{\Delta x} = h[T_{N+1} - T_\infty]$$

Solving for T_{N+1} gives

$$T_{N+1} = \frac{1}{1 + (h\Delta x/k)}\left\{T_N + \frac{h\Delta x}{k}T_\infty\right\} \tag{P8.1g}$$

Also,

$$T_1 = T_W \tag{P8.1h}$$

All the heat transfer from the fin to the surrounding air passes through the base of the fin. Thus, the rate of heat loss, Q, through the fin is given by

$$Q = -\frac{kA_1}{\Delta x}(T_2 - T_1)$$

where A_1 is the cross-sectional area at the base of the fin, that is, at $x = 0+$.

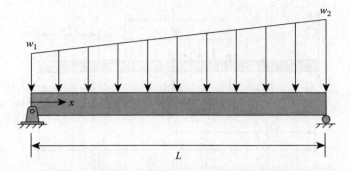

Figure P8.2 **Loaded beam with a linearly distributed load of w_1 N/m at $x = 0$ and w_2 N/m at $x = L$.**

Using the method described in Section 8.3 for a tri-diagonal system of equations, write a MATLAB program that will solve for the temperature distribution in the fin and the rate of heat loss through the fin. Use the following values:

$$T_W = 200°C, \quad T_\infty = 40°C, \quad k = 204\ \text{W/m-K}, \quad h = 60\ \text{W/m}^2\text{-K},$$

$$N = 50, \quad W_1 = 2\ \text{cm}, \quad L = 6\ \text{cm}, \quad L_1 = 4\ \text{cm}, \quad D = 0.2\ \text{cm}$$

The output of your program should include the values of h, k, T_w, T_∞, Q, a table of T vs. x at every 0.1 cm, and a plot of T vs. x.

P8.2 For the beam shown in Figure P8.2, determine the beam deflection, $y(x)$, by the finite difference method utilizing MATLAB's inverse matrix function or MATLAB's Gauss Elimination function. Print the results in a table format. Also determine the approximate maximum deflection. Use the following parameters.

$$w_1 = 10\ \text{kN/m}, \quad w_2 = 20\ \text{kN/m}, \quad EI = 1.5 \times 10^3\ \text{kN-m}^2, \quad L = 5\ \text{m},$$

for $0 \le x \le 5$ m in steps of 0.1 m.

Hint: The load can be considered as the sum of a uniform load and a triangular load. For the triangular load, the resultant force equals $(w_2 - w_1)L/2$ located $2L/3$ from the apex of the triangle.

P8.3 We wish to obtain the reactions, the bending moment, and the deflection of the statically indeterminate beam as shown in Figure P8.3a. The problem can be solved by the method of superposition. First, solve for the deflection $y(x)$ by the finite difference method utilizing the tri-diagonal method to obtain a solution for the statically determinate structure shown in Figure P8.3b. Then, determine the deflection $y(L_1)$ for the structure shown in Figure P8.3b. Next, determine the value of F in the structure shown in

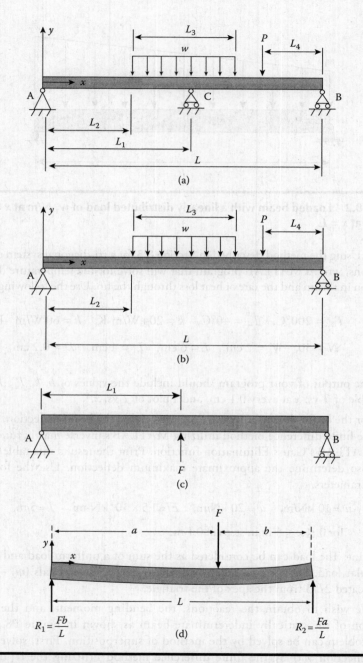

Figure P8.3 **(a) Indeterminate beam, (b) beam with center support removed, (c) beam with a single concentrated point load, and (d) reactions due to a single concentrated point load.**

Figure P8.3c that would cause the deflection at L_1 in that structure to be $-y(L_1)$. You may use the following formula to determine the F value that would give the required deflection at $x = L_1$ (see Figure P8.3d):

$$y(x) = \frac{Fbx}{6LEI}\left[x^2 - (L^2 - b^2)\right] \qquad (x \le a)$$

$$= \frac{Fb}{6LEI}\left[x^3 - (L^2 - b^2)x - \frac{L}{b}(x - a)^3\right] \qquad (x \ge a)$$

Finally, superimpose both solutions to give the true values for the reactions, R_A, R_B, and R_C (at points A, B, and C), the bending moment $M(x)$, and the deflection $y(x)$.

Print out the final reactions, R_A, R_B, and R_C. Print out a table of $M(x)$ and $y(x)$ vs. x at every other node. Use MATLAB to plot $M(x)$ and $y(x)$ vs. x.

Take $w = 40$ kN/m, $EI = 1.5 \times 10^3$ kN-m^2, $P = 35$ kN, $L = 3$ m, $L_1 = 1.3$ m, $L_2 = 0.5$ m, $L_3 = 1.5$ m, and $L_4 = 0.5$ m.

Take the number of subdivisions on the x-axis to be 150.

P8.4 For cantilevered beams (see Figure P8.4), both the deflection and slope at the wall are both zero and thus can be considered an initial value problem. The solution can be obtained analytically by simple integration or by MATLAB's ode45 function. Determine the beam deflection $y(x)$ by MATLAB's ode45 function. Use the following parameters:

$$w = 40 \text{ kN/m}, \quad EI = 1.5 \times 10^3 \text{ kN-m}^2, \quad L = 3.0 \text{ m}, \quad \text{and} \quad \Delta x = 0.1 \text{ m}$$

The exact solution, y_{exact}, is

$$y_{exact}(x) = \frac{1}{EI}\left(\frac{wLx^3}{6} - \frac{wL^2x^2}{4} - \frac{wx^4}{24}\right)$$

Create a table containing (x, y, y_{exact}). Also plot y vs. x and y_{exact} vs. x on the same graph, and make y a solid line and y_{exact} as "x."

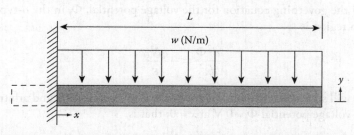

Figure P8.4   Loaded cantilevered beam.

P8.5 A common boundary value problem in electrical engineering is solving for the electric potential in the depletion region of a PN junction. PN junctions are formed by two adjacent regions of oppositely doped semiconductor (typically silicon) and are used to create electronic devices such as bipolar transistors, MOSFETs, and JFETs. Here we will analyze the simplest device, the PN diode.

Figure P8.5a shows the circuit representation of a diode with applied voltage V_A and resulting current i_D. Figure P8.5b shows a physical representation of the diode as adjacent regions of p-type semiconductor (where current is primarily carried by mobile holes with density N_A carriers/cm³) and n-type semiconductor (where current is primarily carried by mobile electrons with density N_D carriers/cm³). In the area immediately surrounding the junction (shown in the figure detail), the mobile charge carriers of the (normally neutral) p-type and n-type regions will recombine, leaving a net fixed charge density layer around the junction called the *depletion region*. The charge density in the depletion region is dependent on the *doping profile*, that is,

$$\rho_P = qN_A \quad \text{for } -x_P < x < 0 \tag{P8.5a}$$

$$\rho_N = -qN_D \quad \text{for } 0 < x < x_N \tag{P8.5b}$$

where ρ_P and ρ_N are the charge densities in the p- and n-type regions, respectively, and q is the unit electric charge (1.6×10^{-19} coulomb). (Note that the charge density on the P side of the depletion region is negative because it is depleted of positive carriers. Similarly, the N side will have a positive charge density.) We will also assume that the depletion region has fixed boundaries as shown in Figure P8.5c. Thus, the depletion region ranges from $x = -x_P$ in the p-type region to $x = x_N$ in the n-type region.

The governing equation for the voltage potential, Φ, in the p-type depletion region is

$$\frac{d^2\Phi}{dx^2} = \frac{qN_A}{\varepsilon} \tag{P8.5c}$$

and the governing equation for the voltage potential, Φ, in the n-type depletion region is

$$\frac{d^2\Phi}{dx^2} = -\frac{qN_D}{\varepsilon} \tag{P8.5d}$$

We will set $x = 0$ at the exact point of the P-to-N transition and arbitrarily set the voltage potential $\Phi = 0$ V at $x = 0$; that is,

$$\Phi(0) = 0 \tag{P8.5e}$$

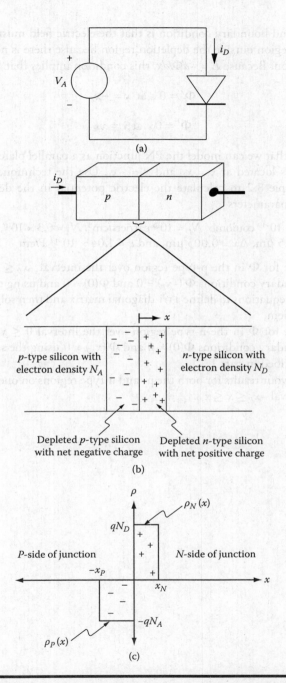

Figure P8.5 **(a) Diode circuit, (b) close-up of the depletion region of a PN junction, and (c) charge densities $P_N(x)$ and $P_p(x)$ at the PN junction.**

The second boundary condition is that the electric field must be zero in the neutral region outside the depletion region because there is no net charge in that region. Because $E = -d\Phi/dx$, this condition implies that

$$\Phi' = 0 \quad \text{at } x = -x_P \tag{P8.5f}$$

$$\Phi' = 0 \quad \text{at } x = x_N \tag{P8.5g}$$

Assume that we can model the PN junction as a parallel plate capacitor with the plates located at $x = x_N$ and $x = -x_P$. Use the technique demonstrated in Example 8.2 to calculate the electric potential in the depletion region. Assume parameters

$q = 1.6 \times 10^{-19}$ coulomb, $N_D = 10^{16}$ carriers/cm^3, $N_A = 0.3 \times 10^{16}$, $x_N = 0.15$ μm, $x_P = 0.5$ μm, $\Delta x = 0.005$ μm, and $\varepsilon = 1.04 \times 10^{-12}$ F/cm

1. Solve for Φ in the p-type region over the interval $-x_P \leq x \leq 0$ with the boundary conditions $\Phi'(-x_P) = 0$ and $\Phi(0) = 0$ and using the finite difference equations to define a tri-diagonal matrix and then solving the matrix problem.
2. Solve for Φ in the n-type region over the interval $0 \leq x \leq x_N$ with the boundary conditions $\Phi(0) = 0$ and $\Phi'(x_N) = 0$ using the same method as described in part 1.
3. Plot your results for both the p- and n-type regions on one graph over the interval $-x_P \leq x \leq x_N$.

Chapter 9

Curve Fitting

9.1 Introduction

There are many occasions in engineering that require experiment to determine the behavior of a particular phenomenon. The experiment may produce a set of data points, which represents a relationship between the variables involved in the phenomenon. We may then wish to express this relationship analytically. A mathematical expression, which describes the data, is called an *approximating function*. There are two approaches to determining an approximating function:

1. The approximating function graphs as a smooth curve. The approximating curve will generally not pass through all the data points. However, we will seek to minimize the resulting error in order to get the best fit curve. A plot of the data on linear, semilog, or log-log coordinates can often suggest an appropriate form for the approximating function.
2. The approximating function passes through all data points (as described in Section 9.5). However, if there is some scatter in the data points, this approximating function may not be satisfactory.

9.2 Method of Least Squares

9.2.1 Best-Fit Straight Line

In the method of least squares, we seek to find a straight line that best fits a given set of n data points (x_1, y_1), (x_2, y_2), ..., (x_n, y_n).

We wish to represent the approximating curve, y_c, as a straight line of the form

$$y_c = c_1 + c_2 x \tag{9.1}$$

where c_1 and c_2 are unknown constants to be determined. Let D be the sum of the square of the errors between the approximating line and the actual points. Then,

$$D = \sum_{i=1}^{n} [y_i - y_c(x_i)]^2 = \sum_{i=1}^{n} [y_i - (c_1 + c_2 x_i)]^2 \tag{9.2}$$

or

$$D = [y_1 - (c_1 + c_2 x_1)]^2 + [y_2 - (c_1 + c_2 x_2)]^2 + \cdots + [y_n - (c_1 + c_2 x_n)]^2 \tag{9.3}$$

To obtain the best-fit straight-line approximating function, minimize D by taking $\frac{\partial D}{\partial c_1} = 0$ and $\frac{\partial D}{\partial c_2} = 0$. Taking the partial derivative of Equation 9.3 with respect to c_1 gives

$$\frac{\partial D}{\partial c_1} = 0 = \sum_{i=1}^{n} 2[y_i - (c_1 + c_2 x_i)][-1]$$

$$0 = \sum_{i=1}^{n} y_i - c_2 \sum_{i=1}^{n} x_i - nc_1$$

or

$$nc_1 + \left(\sum_{i=1}^{n} x_i \right) c_2 = \sum_{i=1}^{n} y_i \tag{9.4}$$

Taking the partial derivative of Equation 9.3 with respect to c_2 gives

$$\frac{\partial D}{\partial c_2} = 0 = \sum_{i=1}^{n} 2[y_i - (c_1 + c_2 x_i)][-x_i]$$

$$0 = \sum_{i=1}^{n} x_i y_i - c_1 \sum_{i=1}^{n} x_i - c_2 \sum_{i=1}^{n} x_i^2$$

or

$$\left(\sum_{i=1}^{n} x_i \right) c_1 + \left(\sum_{i=1}^{n} x_i^2 \right) c_2 = \sum_{i=1}^{n} x_i y_i \tag{9.5}$$

Equations 9.4 and 9.5 describe a system of two algebraic equations in two unknowns, which can be solved by the method of determinants (Cramer's rule):

$$c_1 = \frac{\begin{vmatrix} \sum y_i & \sum x_i \\ \sum x_i y_i & \sum x_i^2 \\ \hline n & \sum x_i \\ \sum x_i & \sum x_i^2 \end{vmatrix}}{} = \frac{\left(\sum y_i\right)\left(\sum x_i^2\right) - \left(\sum x_i\right)\left(\sum x_i y_i\right)}{n\sum x_i^2 - \left(\sum x_i\right)\left(\sum x_i\right)} \tag{9.6}$$

$$c_2 = \frac{\begin{vmatrix} n & \sum y_i \\ \sum x_i & \sum x_i y_i \\ \hline n & \sum x_i \\ \sum x_i & \sum x_i^2 \end{vmatrix}}{} = \frac{n\sum x_i y_i - \left(\sum x_i\right)\left(\sum y_i\right)}{n\sum x_i^2 - \left(\sum x_i\right)\left(\sum x_i\right)} \tag{9.7}$$

9.2.2 Best-Fit mth-Degree Polynomial

We can generalize the earlier approach for an mth-degree polynomial fit. In this case, take the approximating curve, y_c, to be

$$y_c = c_1 + c_2 x + c_3 x^2 + c_4 x^3 + \cdots + c_{m+1} x^m \tag{9.8}$$

where $m \leq n - 1$ and n is the number of data points.

The measured values are (x_i, y_i) for $i = 1, 2, \ldots, n$.

Let $y_{c,i} = y_c(x_i)$ be the approximated value of y_i at the point (x_i, y_i). Then,

$$D = \sum_{i=1}^{n} [y_i - y_{c,i}]^2 = \sum_{i=1}^{n} \left[y_i - \left(c_1 + c_2 x_i + c_3 x_i^2 + \cdots + c_{m+1} x_i^m\right) \right]^2 \tag{9.9}$$

To minimize D, take

$$\frac{\partial D}{\partial c_1} = 0, \quad \frac{\partial D}{\partial c_2} = 0, \ldots, \frac{\partial D}{\partial c_{m+1}} = 0$$

Then,

$$\frac{\partial D}{\partial c_1} = 0 = \sum_{i=1}^{n} 2\Big[y_i - \big(c_1 + c_2 x_i + \cdots + c_{m+1} x_i^m\big)\Big][-1]$$

$$\frac{\partial D}{\partial c_2} = 0 = \sum_{i=1}^{n} 2\Big[y_i - \big(c_1 + c_2 x_i + \cdots + c_{m+1} x_i^m\big)\Big][-x_i]$$

$$\frac{\partial D}{\partial c_3} = 0 = \sum_{i=1}^{n} 2\Big[y_i - \big(c_1 + c_2 x_i + \cdots + c_{m+1} x_i^m\big)\Big]\big[-x_i^2\big]$$

$$\vdots$$

$$\frac{\partial D}{\partial c_{m+1}} = 0 = \sum_{i=1}^{n} 2\Big[y_i - \big(c_1 + c_2 x_i + \cdots + c_{m+1} x_i^m\big)\Big]\big[-x_i^m\big]$$

This set of equations reduces to

$$nc_1 + \Big(\sum x_i\Big)c_2 + \Big(\sum x_i^2\Big)c_3 + \cdots + \Big(\sum x_i^m\Big)c_{m+1} = \sum y_i$$

$$\Big(\sum x_i\Big)c_1 + \Big(\sum x_i^2\Big)c_2 + \Big(\sum x_i^3\Big)c_3 + \cdots + \Big(\sum x_i^{m+1}\Big)c_{m+1} = \sum x_i y_i$$

$$\vdots$$

$$\Big(\sum x_i^m\Big)c_1 + \Big(\sum x_i^{m+1}\Big)c_2 + \cdots + \Big(\sum x_i^{2m}\Big)c_{m+1} = \sum x_i^m y_i$$

(9.10)

Equation 9.10 can be solved by Gauss elimination (as described in Chapter 4).

Alternatively, MATLAB®'s `polyfit` function (discussed in Section 9.4) provides a solution to Equation 9.10, which represents the best-fit polynomial of degree m for the (x_i, y_i) set of data points.

9.3 Curve Fitting with the Exponential Function

Many physical systems can be modeled as exponential functions. If your experimental data appear to fall into this category, it can be fitted with a function of the form

$$y_c = \alpha_1 e^{-\alpha_2 x}$$

(9.11)

where α_1 and α_2 are real constants.

Let us assume that a set of n measured data points $(x_1, y_1), (x_2, y_2), \ldots, (x_n, y_n)$ exists. Then, let $z_i = \ln y_i$ and $z_c = \ln y_c = \ln \alpha_1 - \alpha_2 x$, and also let $c_1 = \ln \alpha_1$ and $c_2 = -\alpha_2$. Then, taking the log of both sides of Equation 9.11 and making the earlier substitutions, we obtain the linear equation

$$z_c = c_1 + c_2 x \qquad (9.12)$$

For the data points $(x_1, y_1), (x_2, y_2), \ldots, (x_n, y_n)$, the new set of data points becomes $(x_1, z_1), (x_2, z_2), \ldots, (x_n, z_n)$.

As we derived in the previous section, the best-fit approximating straight-line curve by the method of least squares gives

$$c_1 = \frac{\left(\sum z_i\right)\left(\sum x_i^2\right) - \left(\sum x_i\right)\left(\sum x_i z_i\right)}{n \sum x_i^2 - \left(\sum x_i\right)^2} \qquad (9.13)$$

and

$$c_2 = \frac{n \sum x_i z_i - \left(\sum x_i\right)\left(\sum z_i\right)}{n \sum x_i^2 - \left(\sum x_i\right)^2} \qquad (9.14)$$

Then, $\alpha_1 = e^{c_1}$ and $\alpha_2 = -c_2$.

The above analysis can be used to determine the damping constant in a mass–spring–dashpot system. This is accomplished by examining the oscilloscope graph of free damped vibration (see Figure 9.1). The governing equation of the envelope is

$$y = y_0 e^{-(c/2m)t} \qquad (9.15)$$

where

c is the damping constant
m is the mass
y is the mass displacement from the equilibrium position
y_0 is the initial position of the mass

Comparing Equation 9.15 with Equation 9.11, we see that

$$\alpha_1 = y_0 \quad \text{and} \quad \alpha_2 = \frac{c}{2m} \text{ with } t \text{ replacing } x$$

Therefore, $c = 2m\alpha_2$ (see Project P2.5).

By measuring n coordinates on the envelope, that is, $(t_1, y_1), (t_2, y_2), \ldots, (t_n, y_n)$, we can determine the best-fit value for α_2 giving our best estimate for the damping factor c.

A similar analysis can be used to determine the damping constant ζ in a parallel RLC circuit (as analyzed in Project P2.9). This can be accomplished by examining

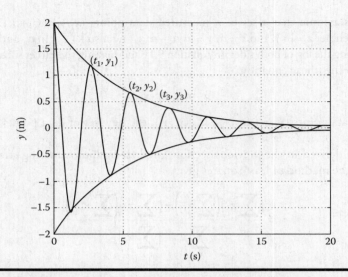

Figure 9.1 Oscilloscope graph of free damped oscillations of a mass–spring–dashpot system for determining the damping factor.

the oscilloscope output of the step response (see Figure 9.2). The governing equation of the envelope of the response is of the form

$$v = v_0 e^{-\zeta \omega_n t} \tag{9.16}$$

where
    v is the capacitor voltage
    v_0 is the initial capacitor voltage (dependent on initial conditions)
    ζ is the damping factor
    ω_n is the natural frequency = $1/\sqrt{LC}$

Comparing Equation 9.16 with Equation 9.11, we see that

$$\alpha_1 = v_0$$

$$\alpha_2 = \zeta \omega_n$$

with t replacing x. Therefore,

$$\zeta = \frac{\alpha_2}{\omega_n} = \alpha_2 \sqrt{LC} \tag{9.17}$$

By measuring n coordinates on the envelope, that is, (t_1, y_1), (t_2, y_2), ..., (t_n, y_n), we can determine the best-fit value for α_2 giving our best estimate for the damping factor ζ.

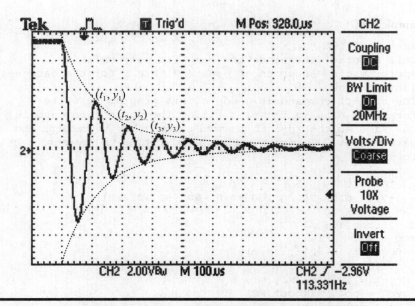

Figure 9.2   Oscilloscope graph of voltage output in a parallel RLC circuit used to determine the damping constant ζ. (Courtesy of Tektronix, Inc.)

9.4 MATLAB®'s Curve Fitting Functions

MATLAB calls curve fitting with a polynomial by the name "polynomial regression." The function `polyfit(x,y,m)` returns a vector of $(m+1)$ coefficients, a_i, that represent the best-fit polynomial of degree m for the (x_i, y_i) set of n data points. The coefficient order corresponds to decreasing powers of x; that is,

$$y_c = a_1 x^m + a_2 x^{m-1} + a_3 x^{m-2} + \cdots + a_m x + a_{m+1} \tag{9.18}$$

To obtain y_c at $(x_1, x_2, \ldots, x_n)$, use the MATLAB function `polyval(a,x)`.
`polyval(a,x)` returns a vector of length n giving $y_{c,i}$ where

$$y_{c,i} = a_1 x_i^m + a_2 x_i^{m-1} + a_3 x_i^{m-2} + \cdots + a_m x_i + a_{m+1} \tag{9.19}$$

MATLAB measures the precision of the fit with a function named MSE, which calculates the mean square error (MSE) and which is defined as follows:

$$\text{MSE} = \frac{1}{n} \sum_{i=1}^{n} (y_i - y_{c,i})^2 \tag{9.20}$$

where n is the number of data points.

Example 9.1

```
% Example_9_1.m
% This program determines the best fit polynomial
% approximating function of orders 2 thru 5 for the data set
% listed below.
% The sprintf command is used in this program to write
% formatted data in the plot title. The sprintf command is
% the same as the fprintf command except that it returns the
% data in a MATLAB string rather than writing to the screen or
% to a file.
clear; clc;
x = -10:2:10;
y = [-980 -620 -70 80 100 90 0 -80 -90 10 220];
x2 = -10:0.5:10;
MSE = zeros(4);
for m = 2:5
    fprintf('m = %i \n',m);
    coef = zeros(m+1);
    coef = polyfit(x,y,m);
    yc2 = polyval(coef,x2);
    yc = polyval(coef,x);
    MSE(m) = sum((y-yc).^2)/length(x);
    fprintf('   x          y          yc      \n');
    fprintf('---------------------------\n');
    for i = 1:length(x)
        fprintf('%5.1f %5.1f %8.3f \n',x(i),y(i),yc(i));
    end
    fprintf('\n\n');
    subplot(2,2,m-1),plot(x2,yc2,x,y,'o'),
    xlabel('x'), ylabel('y'), grid, axis([-10 10 -1500 500]);
    title(sprintf('Degree %d polynomial fit',m));
end
fprintf('  m    MSE \n')
fprintf('--------------------\n');
for m = 2:5
    fprintf(' %d %8.2f \n',m,MSE(m))
end
```

Program results

```
Output for m = 5 is only displayed here.
m = 5
   x          y          yc
---------------------------
-10.0     -980.0     -999.09
 -8.0     -620.0     -545.31
 -6.0      -70.0     -156.76
 -4.0       80.0       78.39
 -2.0      100.0      148.18
```

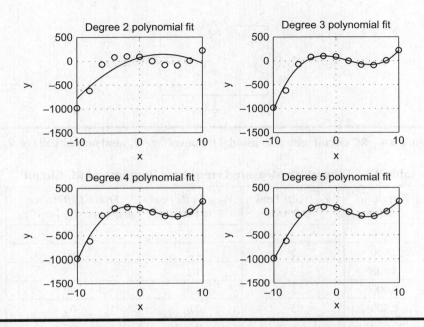

Figure 9.3   Approximating polynomial curves for data in Example 9.1.

```
 0.0        90.0        93.80
 2.0         0.0       -13.50
 4.0       -80.0       -95.45
 6.0       -90.0       -89.91
 8.0        10.0        26.15
10.0       220.0       213.50

m      MSE
-----------
2    32842.4
3     2660.0
4     2342.1
5     1502.9
>>
```

See Figure 9.3.

As would be expected, the MSE decreases as the order of the fitted polynomial is increased.

Example 9.2

Figure 9.4 shows an RC circuit and Table 9.1 shows a set of steady-state voltage amplitudes and phases, which were measured in the lab for V_{IN} and V_{OUT} at the given frequencies.

We wish to fit the data to the best possible first-order frequency response.

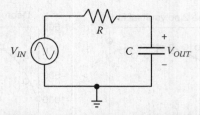

Figure 9.4 **RC circuit with a sinusoidal input voltage V_{IN} and output voltage V_{OUT}.**

Table 9.1 **Laboratory-Measured Frequency Response of RC Circuit**

Frequency (Hz)	V_{IN} (Volts Peak to Peak)	V_{OUT} (Volts Peak to Peak)	Phase Difference between V_{OUT} and V_{IN}
0	5	5	0
2,000	5	2.2	−60
4,000	5	1.2	−75
6,000	5	0.8	−80
8,000	5	0.6	−83
10,000	5	0.5	−85

The RC circuit can be considered a first-order low-pass filter with a frequency response of the form

$$H(f) = \frac{n_0}{j2\pi f + d_0} \tag{9.21}$$

where the *transfer function*, $H = V_{OUT}/V_{IN}$, is the frequency-dependent output-to-input ratio, n_0 and d_0 are the numerator and denominator coefficients to be determined.

Because Equation 9.21 has the polynomial in the denominator, we start by rearranging it into a polynomial of the form of Equation 9.18 so that we can apply MATLAB's polyfit function. This gives

$$\frac{1}{H} = \frac{1}{n_0} j2\pi f + \frac{d_0}{n_0} \tag{9.22}$$

Matching terms between Equations 9.18 and 9.22 shows that $a_1 = j2\pi/n_0$ and $a_2 = d_0/n_0$, and thus, the inverse of the frequency response $1/H$ can be fitted by using A=polyfit(f,1./H,1) where f is the vector of measured frequencies and H is the vector containing measured values of V_{OUT}/V_{IN}. The MATLAB program follows.

```
% Example_9_2.m:
% Fit lab-measured frequency response to 1st order response
clear; clc;
```

```
% Lab-measured data from Table 9.1:
f = [0 2e3 4e3 6e3 8e3 10e3];
V_in = [5 5 5 5 5 5];
V_out = [5 2.1 1.3 .8 .6 .5];
Phi_degrees = [0 -60 -75 -80 -83 -85];
% Ratio of V_out to V_in is the magnitude of the frequency
% response:
H_mag = V_out./V_in;
% The phase is the measured phase difference (converted to
% radians)
H_phase = Phi_degrees * pi/180;
% Express the complex freq response H as: mag * exp(j*phase)
H = H_mag.* exp(j*H_phase);
% Do polynomial fit to 1/H
A = polyfit(f,1./H,1);
% Compute coefficients of fitted freq response based polyfit
% results
n0 = 2*pi*j/A(1);
d0 = 2*pi*j*A(2)/A(1);
% Print fitted frequency response:
fprintf('Fitted frequency response:\n');
fprintf(' H(f) = %.0f/(j*2*pi*f + %.0f)\n',n0,d0);
fprintf('Pole location: %.0f rad/sec (%.0f Hz)\n',...
        d0, d0/(2*pi));
% For comparison purposes, calculate the fitted curve with
% fine precision. Use 100 log-spaced points from 1Hz to 1MHz
f_fine = logspace(0,6,100);
H_fit = n0./(j*2*pi*f_fine + d0);
H_fit_mag = abs(H_fit);
H_fit_phase = angle(H_fit);
% Create Bode plot:
subplot(2,1,1);
loglog(f,H_mag,'o',f_fine,H_fit_mag);
axis([1e0 1e6 1e-3 2]);
legend('measured','fitted function');
ylabel('|H(f)|');
title('Fit of measured data to 1st-order frequency response');
subplot(2,1,2);
semilogx(f,(180/pi)*H_phase,'o',f_fine, (180/ pi)*H_fit_phase);
axis([1e0 1e6 -100 10]);
legend('measured','fitted function');
ylabel('\angleH(f) (degrees)'); xlabel('frequency (Hz)');
```

Program results

```
Fitted frequency response:
    H(f) = 6204/(j*2*pi*f + 6842)
Pole location: 6842 rad/sec (1089 Hz)
```

The resulting plot is shown in Figure 9.5.

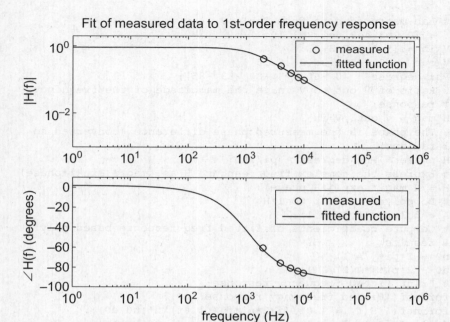

Figure 9.5   Fitted frequency response for a first-order RC circuit.

9.5 Cubic Splines

Given a set of n data points, suppose that an mth-degree polynomial is selected as the approximating curve and that this approximating curve produces curve values that are not allowed. For example, suppose it is known that a particular property represented by the approximating curve (such as absolute pressure or absolute temperature) must be positive and the approximating function produces values that are negative. In this case, the approximating function produces values that are not allowed and is therefore not satisfactory. The method of cubic splines eliminates this problem.

Given a set of $(n + 1)$ data points (x_i, y_i), $i = 1, 2, \ldots, (n + 1)$, the method of cubic splines develops a set of n cubic functions, such that $y(x)$ is represented by a different cubic in each of the n intervals and the set of cubics passes through the $(n + 1)$ data points.

This is accomplished by forcing the slopes and curvatures to be the same for each pair of cubics that join at a data point.

Note: Curvature, $K = \dfrac{\pm(d^2y/dx^2)}{[1 + (dy/dx)^2]^{3/2}}$ (9.23)

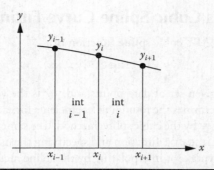

Figure 9.6 Two adjacent intervals in a cubic spline curve fitting scheme.

This is accomplished by the following equations:

$$[y(x_i)]_{\text{int}\,i-1} = [y(x_i)]_{\text{int}\,i}$$

$$[y'(x_i)]_{\text{int}\,i-1} = [y'(x_i)]_{\text{int}\,i} \tag{9.24}$$

$$[y''(x_i)]_{\text{int}\,i-1} = [y''(x_i)]_{\text{int}\,i}$$

In interval $(i-1)$, $(x_{i-1} \leq x \leq x_i)$ (see Figure 9.6):

$$y(x) = A_{i-1} + B_{i-1}(x - x_{i-1}) + C_{i-1}(x - x_{i-1})^2 + D_{i-1}(x - x_{i-1})^3 \tag{9.25}$$

In interval i, $(x_i \leq x \leq x_{i+1})$:

$$y(x) = A_i + B_i(x - x_i) + C_i(x - x_i)^2 + D_i(x - x_i)^3 \tag{9.26}$$

This gives fewer equations than the number of unknowns, and as a result, additional assumptions must be made. Values for d^2y/dx^2 at x_1 and x_{n+1} must be assumed.
    Several alternatives exist:

1. Assume $y''(x_1) = y''(x_{n+1}) = 0$.
    This is widely used and forces splines to approach straight lines at end points
2. Assume $y''(x_{n+1}) = y''(x_n)$ and $y''(x_1) = y''(x_2)$.
    This forces the splines to approach parabolas at the end points.

9.6 MATLAB®'s Cubic Spline Curve Fitting Function

The syntax for MATLAB's cubic spline function is

$$yy = \text{spline}(xi, yi, xx)$$

where (xi,yi) is a given set of data points and yy is the value of y at xx. The spline function determines the four cubic coefficients for each section in the given data and will evaluate yy by the cubic spline method. The same result can be obtained by using MATLAB's interp1 function and specifying the use of the spline method of interpolation. The syntax for interpolating by the spline method is

$$yi = \text{interp1}(x, y, xi, \text{'spline'})$$

Example 9.3

The following example involves a measured increase in air pressure at distances from a blast. The data specifies the pressure above normal atmospheric pressure and is designated as over-pressure. The program demonstrates the use of the MATLAB's spline function as well as MATLAB's interp1 function with the spline option to determine the pressure at distances not in the data. We see that the two methods produce the same results.

```
% Example_9_3.m
% This program uses both MATLAB's spline function and
% MATLAB's interp1 function with the cubic spline option to
% determine the over-pressure resulting from a blast. The
% over-pressure is in kPa and the distance from the blast
% in km.
clear; clc;
dist = 0.52:0.3:4.12;
press=[165.5 96.5 69.0 52.4 37.2 27.6 21.4 17.2 13.8 11.7 ...
       10.3 9.0 7.2];
d=0.52:0.1:4.12;
p1=spline(dist,press,d);
p2=interp1(dist,press,d,'spline');
d = 0.52:0.1:4.12;
p1 = spline(dist,press,d);
p2 = interp1(dist,press,d,'spline');
fo = fopen('output.txt','w');
fprintf(fo,'PEAK OVERPRESSURE VS. DISTANCE FROM BLAST \n');
fprintf(fo,'CUBIC SPLINE FIT \n');
fprintf(fo,' dist(km)  over-press(kPa)  over-press(kPa)\n');
fprintf(fo,'          by spline function      by interp1 \n');
fprintf(fo,'--------------------------------------------------- \n');
for n = 1:length(d)
    fprintf(fo,'%5.2f   %12.2f   %15.2f \n',d(n),p1(n),p2(n));
end
plot(d,p1,d,p2,'o'), xlabel('km from ground zero'),
```

```
ylabel('overpressure(kPa)'), grid,
title('peak over-pressure vs. distance from blast')
fclose(fo);
```

Program results

```
PEAK OVERPRESSURE VS. DISTANCE FROM BLAST
CUBIC SPLINE FIT
                    over-press(kPa)           over-press(kPa)
dist(km)            by spline function        by interp1
-----------------------------------------------------------
0.52                    165.50                    165.50
0.62                    135.72                    135.72
0.72                    113.15                    113.15
0.82                     96.50                     96.50
0.92                     84.46                     84.46
1.02                     75.72                     75.72
1.12                     69.00                     69.00
1.22                     63.15                     63.15
1.32                     57.71                     57.71
1.42                     52.40                     52.40
1.52                     47.02                     47.02
  :                        :                         :
3.12                     12.28                     12.28
3.22                     11.70                     11.70
3.32                     11.19                     11.19
3.42                     10.73                     10.73
3.52                     10.30                     10.30
3.62                      9.88                      9.88
3.72                      9.46                      9.46
3.82                      9.00                      9.00
3.92                      8.49                      8.49
4.02                      7.89                      7.89
4.12                      7.20                      7.20
```

See Figure 9.7.

Example 9.4

In this example, we use the spline option in MATLAB's interp1 function to add additional data to an audio signal, $y(t)$.

Program

```
% Example_9_4.m
% This program uses interpolation by cubic splines to
% upsample an audio signal y(t), time in microsec.
clear; clc;
% Define original datapoints for y(t) (time is in microsec)
```

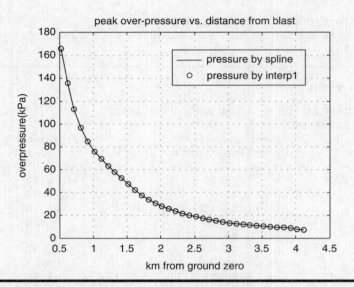

Figure 9.7   Peak overpressure vs. distance from the blast.

```
orig_t = [0 4 8 12 16 20 24 28 32 36 40 44 48 52 56];
orig_y = [.7 .9 .9 .7 .3 0 -.3 -.7 -.7 -.3 0 .3 .7 .7 .3];
% Define upsampled time points
upsample_t = 0:60;
% Calculate interpolated data points using cubic splines
upsample_y = interp1(orig_t,orig_y,upsample_t,'spline');
% Print output to screen
fprintf('UPSAMPLING VIA CUBIC SPLINE FIT\n');
fprintf('time (microsec)    upsample_y \n');
fprintf('-------------------------------\n');
for i = 1:length(upsample_t)
    fprintf('%8.2f      %18.3f \n',...
            upsample_t(i),upsample_y(i));
end
plot(orig_t,orig_y,'o',upsample_t,upsample_y);
xlabel('t (microsec)'); ylabel('y(t)'); grid;
title('Upsampling with Cubic Spline Interpolation');
legend('original','upsampled','Location','SouthWest');
```

Program results

```
UPSAMPLING VIA CUBIC SPLINE FIT
time (microsec)        upsample_y
-------------------------------
 0.00                    0.700
 1.00                    0.770
 2.00                    0.827
 3.00                    0.870
```

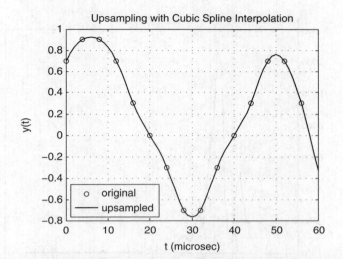

Figure 9.8   Upsampling with cubic spline interpolation.

4.00	0.900
5.00	0.918
6.00	0.923
7.00	0.917
8.00	0.900
9.00	0.872
10.00	0.830
11.00	0.774
12.00	0.700
⋮	⋮
52.00	0.700
53.00	0.631
54.00	0.539
55.00	0.428
56.00	0.300
57.00	0.158
58.00	0.004
59.00	-0.158
60.00	-0.326

See Figure 9.8.

9.7 Curve Fitting with Fourier Series

Example 9.5

The following example involves measured turbulent wind velocity as a function of time at a fixed point as a helicopter approaches and leaves the region of interest. The

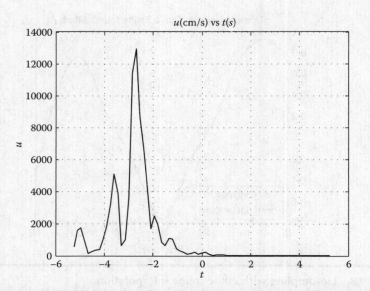

Figure 9.9   Experimental data of *u* vs. *t*.

experimental data set produced a plot as shown in Figure 9.9 and it is desired to obtain an analytical expression (approximating curve) that comes close to fitting the data. An attempt to fit a polynomial approximating curve to this data was not successful. However, the use of a Fourier series gave a reasonable analytical expression for approximating the data. The original time domain, t', ranged from 0 to 10.5 s. However, to use the Fourier series method, we needed to shift the time domain by letting $t = t' - 5.25$. Thus, the new time domain was from -5.25 to 5.25 s. The time domain, t, was subdivided into 70 equal spaces, with $\Delta t = 10.5/70 = 0.15$ s. Thus, $(t_{i+1} - t_i)$ was uniform over the entire time domain. The measured data that is shown in Table 9.2 represents turbulent wind velocity at a specific point as a helicopter approaches and leaves the region of interest.

If u_c is the approximating curve, then by a Fourier series,

$$u_c(t) = a_0 + \sum_{m=1}^{\infty} \left(a_m \cos\left(\frac{m\pi t}{L}\right) + b_m \sin\left(\frac{m\pi t}{L}\right) \right) \qquad (9.27)$$

where

$$a_0 = \frac{1}{2L} \int_{-L}^{L} u(t)\,dt \qquad (9.28)$$

$$a_m = \frac{1}{L} \int_{-L}^{L} u(t)\cos\left(\frac{m\pi t}{L}\right) dx \qquad (9.29)$$

Table 9.2   Shifted Velocity Data as a Function of Time

t	u	t	u	t	u	t	u
−5.25	557.78	−2.55	8778.24	0.15	233.17	2.85	3.35
−5.1	1,557.53	−2.4	6644.64	0.3	112.47	3	6.71
−4.95	1,737.36	−2.25	4511.04	0.45	34.75	3.15	2.74
−4.8	880.87	−2.1	1679.45	0.6	45.42	3.3	1.83
−4.65	144.48	−1.95	2493.26	0.75	69.49	3.45	2.74
−4.5	272.80	−1.8	1972.06	0.9	44.81	3.6	2.74
−4.35	338.33	−1.65	847.34	1.05	26.82	3.75	7.92
−4.2	408.43	−1.5	649.22	1.2	25.60	3.9	7.62
−4.05	984.50	−1.35	1097.28	1.35	26.82	4.05	6.40
−3.9	1,792.22	−1.2	1024.13	1.5	16.46	4.2	3.05
−3.75	3,200.40	−1.05	448.06	1.65	19.20	4.35	1.22
−3.6	5,090.16	−0.9	316.99	1.8	28.04	4.5	2.44
−3.45	3,901.44	−0.75	217.02	1.95	22.25	4.65	4.27
−3.3	637.03	−0.6	109.12	2.1	3.66	4.8	2.74
−3.15	987.55	−0.45	126.49	2.25	5.79	4.95	4.27
−3	3,596.64	−0.3	224.33	2.4	7.92	5.1	3.05
−2.85	11,460.48	−0.15	106.07	2.55	8.53	5.25	1.52
−2.7	12,954.00	0	194.46	2.7	8.23	—	—

$$b_m = \frac{1}{L} \int_{-L}^{L} u(t) \sin\left(\frac{m\pi t}{L}\right) dt \tag{9.30}$$

Using 30 terms in the series and Simpson's rule on integration and replacing t' with t, the following program calculates the approximating curve. The data in Table 9.2 were entered into the program as two column vectors.

The $a_m \cos\left(\dfrac{m\pi t}{L}\right) + b_m \sin\left(\dfrac{m\pi t}{L}\right)$ terms can be put into the following form by the trigonometric identity $a \cos\beta + b \sin\beta = c \sin(\beta - \phi)$, where c represents the amplitude. The amplitude, c_m, as a function of $m\pi/L$ is given by

$$c_m = \sqrt{\left(a_m^2 + b_m^2\right)}$$

The program follows:

```
% Example_9_5.m
% This program determines an approximating curve to the data in
% in awake3.txt by Fourier series.
clear; clc;
load awake3.txt
dt = 0.15;
t = awake3(:,1);
u = awake3(:,2);
plot(t,u), title('u vs t'), xlabel('t'), ylabel('u'),grid;
figure;
L = 5.25;
for n = 1:30
    i = 1;
    for j = 1:35
        arg1 = n*pi*t(i)/L;
        arg2 = n*pi*t(i+1)/L;
        arg3 = n*pi*t(i+2)/L;
        f(i) = u(i)*cos(arg1);
        f(i+1) = u(i+1)*cos(arg2);
        f(i+2) = u(i+2)*cos(arg3);
        A(j) = dt/3*(f(i)+4*f(i+1)+f(i+2));
        i = i+2;
    end
    a(n) = 1.0/L*sum(A);
end
for n = 1:30
    i = 1;
    for j = 1:35
        arg1 = n*pi*t(i)/L;
        arg2 = n*pi*t(i+1)/L;
        arg3 = n*pi*t(i+2)/L;
        f(i) = u(i)*sin(arg1);
        f(i+1) = u(i+1)*sin(arg2);
        f(i+2) = u(i+2)*sin(arg3);
        A(j) = dt/3*(f(i)+4*f(i+1)+f(i+2));
        i = i+2;
    end
    b(n) = 1.0/L*sum(A);
end
i = 1;
for j = 1:35
    f(i) = u(i);
    f(i+1) = u(i+1);
        f(i+2) = u(i+2);
        A(j) = dt/3*(f(i)+4*f(i+1)+f(i+2));
        i = i+2;
end
```

```
a0 = 0.5/L*sum(A);
for i = 1:71
    for n = 1:30
        arg = n*pi*t(i)/L;
        term(n) = a(n)*cos(arg)+b(n)*sin(arg);
    end
    uc(i) = a0+sum(term);
end
uc = uc';
plot(t,uc,t,u,'.'),xlabel('t'), ylabel('u,uc)'), grid,
title('uc(cm/s) and u(cm/s) vs. t(s)'), legend('uc','u');
figure;
for m = 1:30
    c(m) = sqrt(a(m)^2+b(m)^2);
    x(m) = m*pi/L;
end
plot(x,c), xlabel('m\pi/L'), ylabel('c'),
title('c(cm/s) vs. m\pi/L'),grid;
```

Program results

See Figures 9.10 and 9.11.

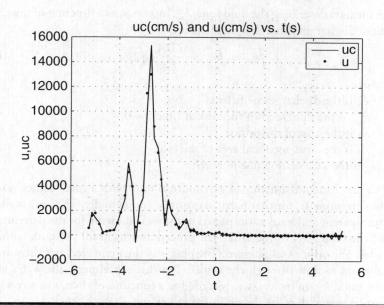

Figure 9.10   Fourier series curve fit to experimental data shown in Figure 9.9.

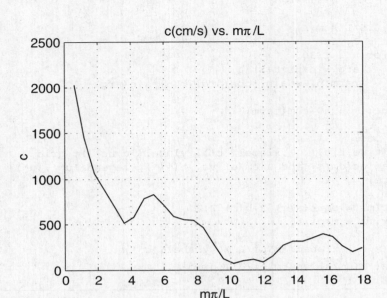

Figure 9.11   Amplitude of Fourier coefficients, c_m, as a function of $m\pi/L$.

Projects

P9.1 A formula describing the fluid level, h_{eq}, in a tank, as a function of time, as the fluid discharges through a small orifice, is

$$\sqrt{h_{eq}} = \sqrt{h_{eq,o}} - \frac{C_d A_0}{2 A_T} \sqrt{2g}\, t \qquad \text{(P9.1a)}$$

where

C_d is the discharge coefficient

$h_{eq,o}$ is the fluid level in the tank at time, $t = 0$

A_0 is the area of the orifice

A_T is the cross-sectional area of the tank

g is the acceleration due to gravity

An experiment consisting of a cylindrical tank with a small orifice was used to determine C_d for that particular orifice and cylinder. The tank walls were transparent and a ruler was pasted to the wall allowing for the determination of the fluid level in the tank. The procedure was to fill the tank with water while the orifice was plugged. The plug was then removed and the water was allowed to flow through the orifice. For this experiment, the water level in the tank, h_{exp} in meters, was recorded as a function of time, t in seconds. The experimental data are shown in the following:

$h_{exp} = [0.288\ 0.258\ 0.234\ 0.215\ 0.196\ 0.178\ 0.160\ 0.142\ 0.125\ 0.110$
$\phantom{h_{exp} = [}0.095\ 0.080\ 0.065\ 0.053\ 0.041\ 0.031\ 0.022\ 0.013\ 0.006\ 0.002\ 0.000]$

$t = [0\ 10\ 20\ 30\ 40\ 50\ 60\ 70\ 80\ 90\ 100\ 110\ 120\ 130\ 140\ 150\ 160\ 170$
$180\ 190\ 200]$

The diameters of the orifice and the tank are $d_o = 0.0055$ m and $D_t = 0.146$ m, respectively. The free surface elevation, $h_{eq,o}$, at $t = 0$ is 0.288 m. The gravitational constant $g = 9.81$ m/s².

Use the MSE as defined by Equation P9.1b to determine the value for C_d that best fits the data. Vary C_d from 0.3 to 0.9 in steps of 0.01 and evaluate the MSE for each C_d selected, where

$$\text{MSE} = \frac{1}{N} \sum_{i=1}^{N} [h_{eq}(t_i) - h_{exp}(t_i)]^2 \qquad \text{(P9.1b)}$$

where

N is the number of data points
$h_{eq}(t_i)$ is the water level in the tank at t_i as determined by Equation P9.1a
$h_{exp}(t_i)$ is the water level in the tank at t_i as determined by experiment

For the C_d with the lowest MSE, create a plot of h_{eq} vs. t (solid line) and superimpose h_{exp} vs. t as little x's onto the plot of h_{eq} vs. t. Also print out the value of C_d that gives the lowest MSE.

P9.2 Table P9.1 gives the head, H, vs. the flow rate, Q, developed by a particular pump. The (H vs. Q) data was obtained experimentally. The project involves determining the best fit polynomial approximating curve to the experimental (H vs. Q) data shown in Table P9.1.

**Table P9.1   Experimental
H vs. Q Data**

Q (m³/h)	H (m)	Q (m³/h)	H (m)
3.3	43.3	61.6	40.8
6.9	43.4	68.5	39.6
13.7	43.6	75.3	38.7
20.5	43.6	82.2	37.2
27.4	43.3	89	36.3
34.2	43.0	95.8	34.4
41.1	42.7	102.7	32.6

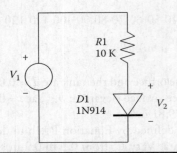

Figure P9.3 **Diode–resistor circuit for laboratory measurement of diode *I-V* curve.**

Try degree polynomials of 2 through 4 to determine which degree polynomial will give the smallest MSE. For each degree polynomial:

a. Use MATLAB's `polyfit` function to determine the coefficients of the polynomial approximating curve.

b. Use MATLAB's `polyval` function to determine the head, H_c, where H_c are the values of the approximating curve at the flow rates, Q.

c. Create a table containing Q, H, and H_c.

d. Create a plot containing both H_c vs. Q (solid line) and H vs. Q (small circles).

e. Print out the MSE.

P9.3 Figure P9.3 shows a resistor–diode circuit using a type 1N914 silicon diode (*D1*) and a 10 kΩ resistor (*R1*). Table P9.3 shows a list of laboratory

Table P9.3 Laboratory Measurements of Resistor–Diode Circuit

V_1 (V)	V_2 (V)	V_1 (V)	V_2 (V)
0.189	0.189	6.005	0.588
0.333	0.317	6.933	0.595
0.393	0.356	7.934	0.602
0.819	0.464	9.014	0.607
1.067	0.487	10.040	0.613
1.289	0.501	11.009	0.619
1.656	0.518	15.045	0.634
1.808	0.522	19.865	0.647
2.442	0.541	24.64	0.657
3.949	0.566	29.79	0.666
4.971	0.579		

measurements of V_2 for various applied voltage levels of V_1 at room temperature (300 K).

1. Use Kirchhoff's voltage law to calculate the diode current i_D in terms of V_1 and V_2.

2. Use the technique described in Section 9.3 to find the least-squares fit of i_D and v_D ($=V_2$) to the formula

$$i_D = I_S \exp\left(\frac{v_D}{V_T}\right) \tag{P9.3a}$$

In this case, you are using the raw data to find best-fit values for I_S and V_T. Plot both the lab data and your fitted curve on the same axes.

3. $V_T = kT/q$ is known as the *thermal voltage* (where k is the Boltzmann constant and q is the unit electric charge). For your best-fit value for V_T, what is the corresponding temperature value T (in kelvins)? Does this value seem reasonable?

4. A more accurate equation to model the diode is

$$i_D = I_S \exp\left(\frac{v_D}{nV_T}\right) \tag{P9.3b}$$

where n is the *ideality factor*. Find the best-fit value for n when $T = 300$ K.

5. The complete I-V model for the diode is

$$i_D = I_S \left(\exp\left(\frac{v_D}{nV_T}\right) - 1 \right)$$

Is it reasonable that we neglected the –1 term in our curve fit? Why?

P9.4 Reuse the data for `orig _ t` and `orig _ y` from Example 9.4 to perform a curve fit using Fourier series. Begin by shifting t so that the data are symmetric around the origin such that $-L \leq t \leq L$ (where $L = 28$ μs). Then, proceed as follows:

1. Write a MATLAB program to calculate the first m Fourier coefficients. Use Simpson's rule (as described Chapter 6) to calculate the coefficients $a_0, a_1, a_2, \ldots, a_m$, and $b_1, b_2, \ldots, b_m$ as in Equations 9.28 through 9.30.

2. Generate Fourier coefficients for $m = 10$.

3. Using 50 subdivisions on the t-axis for $-L \leq t \leq L$, determine values for the approximating curve (y_c vs. t) by substituting your coefficients back into Equation 9.27.

4. Plot on the same set of axes y_c vs. t (solid line) and `orig _ y` vs. `orig _ t` ("x" symbol).

Chapter 10

Simulink®

10.1 Introduction

Simulink® is used with MATLAB® to model, simulate, and analyze dynamic systems. Common uses are for solving differential equations, modeling feedback systems, and signal processing.

With Simulink, models can be built from scratch or additions can be made to existing models. Simulations can be made interactive, so a change in parameters can be made while running the simulation. Simulink supports linear and nonlinear systems, modeled in continuous time, sample time, or a combination of the two.

Simulink provides a graphical user interface (GUI) for building models as block diagrams, using click-and-drag mouse operations. By using Scopes and other Display blocks, simulation results can be seen quickly. The program includes a comprehensive library of components ("blocks") from which to construct models. Additional application-specific "toolboxes" are also available.

10.2 Creating a Model in Simulink®

1. In the MATLAB desktop, click on the Simulink icon in the Toolstrip (see Figure 10.1) or type `simulink` in the MATLAB command window. This brings up the Simulink Library Browser window (see Figure 10.2).
2. Click on *File* in the Simulink Library Browser window. Select a "*New*, then *Model*" (for a new model) or "*Open*" for an existing model. This will bring up an untitled model window (for the case of a new model—see Figure 10.3) or an existing model window.
3. Next click on one of the Library Browser groups to view the contents in that group. For example, clicking on the Commonly Used Groups category

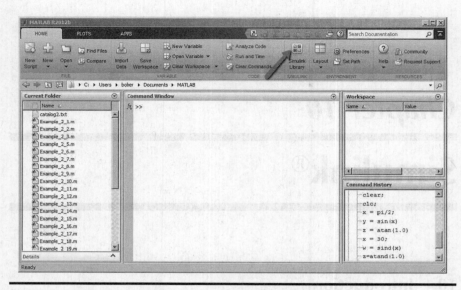

Figure 10.1 Simulink program icon in MATLAB®'s desktop.

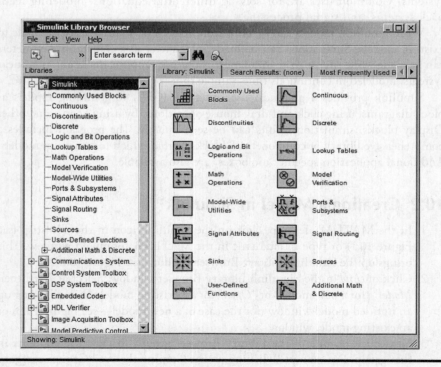

Figure 10.2 Simulink Library Browser.

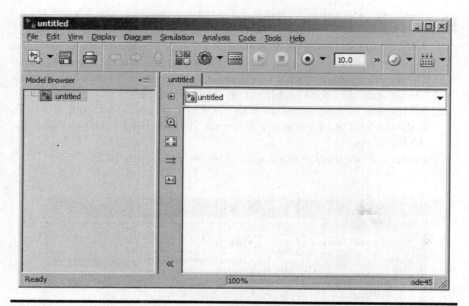

Figure 10.3   Untitled Model Window.

(see Figure 10.4a) brings up a window containing the available blocks in that category. To create a new model, you need to copy blocks from one of the Library Browser category windows into the new model window. This can be done by highlighting a particular block and dragging it into the model window (both the new model window and the Simulink Browser Window need to be open. See Figure 10.4b). To simplify the connections of blocks, you may need to rotate a block 90° or 180°. To do this, highlight the block and click on *Format* in the menu bar, and then select *Rotate Block* (for 90°) or *Flip Block* (for 180°).

4. To change the default values of various blocks, you can double click on the block or right click on the block and select the Block Parameters option in the window that opens. In either case, a window will open giving you the option of changing the default value of the block.

Simulink has many categories for displaying the library blocks; those of interest for this chapter are Commonly Used Blocks, Continuous, Discontinuities, Math Operations, Ports and Subsystems, Signal Routing, Sinks, Sources, and User-Defined Functions.

Common blocks that will be used repeatedly in this chapter are Constant and Clock (from the Sources library); Product, Gain, and Sum (from the Math Operations library); Integrator (from the Continuous library); Scope, Display, and To Workspace (from the Sink library); Relay (from the Discontinuities library); Switch and Mux (from the Signal Routing library); and Fcn (from the User-Defined Functions library).

10.3 Typical Building Blocks in Constructing a Model

1. Addition of two constants with displayed output (see Figure 10.5). To set the value for a constant, double-click on the block and edit the constant value. To run the program, click on *Simulation* in the menu bar and select *Start*, or alternatively click the Play button (▶) in the menu bar.
2. Subtraction of two constants with displayed output (see Figure 10.6). To make the Sum block perform a difference, double-click on the block and edit the list of signs.
3. Product of two blocks with displayed output (see Figure 10.7).

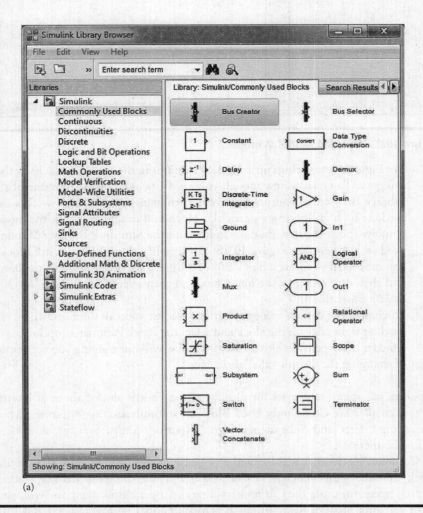

(a)

Figure 10.4 **(a) Commonly Used Blocks in Simulink's Library Browser.**

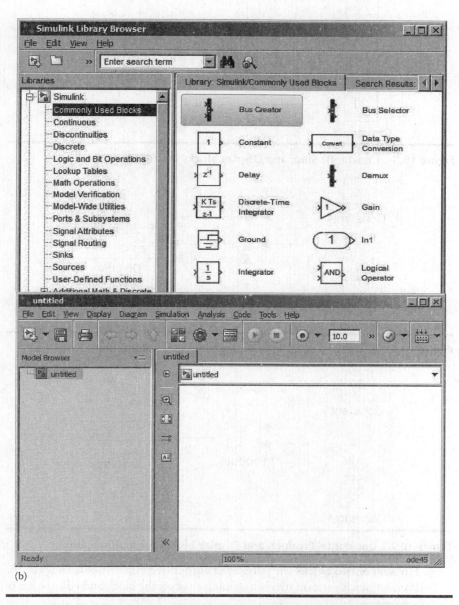

(b)

Figure 10.4 (continued) **(b) Overlap of the Commonly Used Blocks Library Window and the Untitled Model Browser Window.**

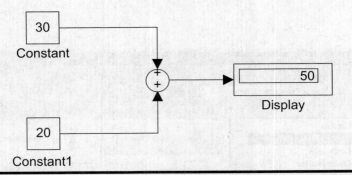

Figure 10.5   Constants, Sum, and Display blocks for addition.

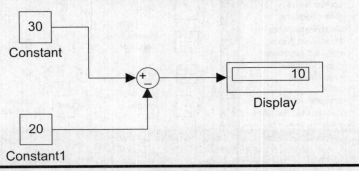

Figure 10.6   Constants, Sum, and Display blocks for subtraction.

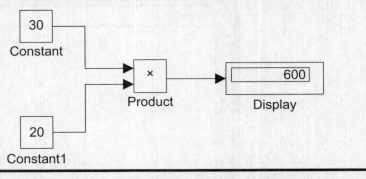

Figure 10.7   Constants, Product, and Display blocks for multiplication.

4. Division of two blocks with displayed output (see Figure 10.8). To make the Product block perform division, double-click on the block and edit the list of operations.
5. To construct a model involving a differential equation, rewrite the differential equation with the highest derivative term on the left side and all other terms on the right side. Construct a model consisting of all terms on the right side; the output will represent the derivative term, which you can then integrate

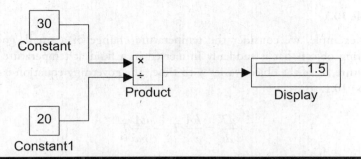

Figure 10.8   Constants, Product, and Display blocks for division.

with Simulink's Integrator block, giving the solution to the differential equation. If the right-hand side of the rewritten equation contains the independent variable, the output of the Integrator block will need to go to the input of some or all of the blocks representing the right-hand side of the differential equation. The method of solution is illustrated in Example 10.1.

10.4 Tips for Constructing and Running Models

1. To connect lines from the output of a block to the input of a second block, place the cursor on the output of the first block, right-click on the mouse, and drag the line to the input of the second block.
2. To connect a point on a line to the input of a block, place the cursor on the line, right-click on the mouse, and drag the line to the input of the block.
3. To add alphanumeric information above a line, double-click above the line and a text box will appear. Type in the desired label and click outside the box to complete.
4. To view the results on a Scope, double-click on the Scope to make the graph appear. To select the graph axis, right-click on the graph and select "axis properties" or "autoscale." In most cases, selecting autoscale is sufficient. You may also click on the binoculars icon to autoscale the graph.
5. To set initial conditions for an integrator, double-click on the block and edit the initial condition box.
6. By default, Simulink runs over a time interval of zero to 10 s. These times are inappropriate for many models (e.g., high-frequency circuits). To edit the start and stop times, click on Simulation in the menu bar, select Model Configuration Parameters, and edit the start and stop time boxes. Alternatively, you can adjust the stop time in the menu bar (the start time defaults to zero). Another simulation parameter option is the method of solution, which includes the ode4 (Runge–Kutta) method.
7. To run the simulation, click on Simulation in the menu bar and choose Start. Alternatively, click the Play button (▶) in the menu bar.

The solution of a simple ordinary first-order differential equation is illustrated in the following example.

Example 10.1

In this example, we consider the temperature change of a small, good heat-conducting object that is suddenly immersed in a fluid at temperature T_∞. The temperature, T, of the object varies with time. The governing equation is given by Equation 10.1.

$$\frac{dT}{dt} = \frac{hA_s}{mc}T_\infty - \frac{hA_s}{mc}T \tag{10.1}$$

where
- m is the mass of the object
- A_s is the surface area
- c is the specific heat of the object
- h is the convective heat transfer coefficient

Note that Equation 10.1 is in the form $\dfrac{dT}{dt} = f(T,t)$.

We will assume:

$$\frac{hA_s}{mc} = 8.7 \times 10^{-4}\,\text{s}^{-1}$$

$$T_\infty = 10°\text{C}, \quad T(0) = 100°\text{C}$$

The block diagram for Equation 10.1 is shown in Figure 10.9. The simulation stop time was changed to 500 s.

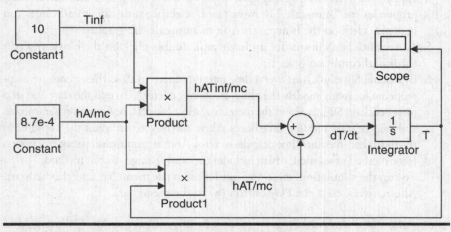

Figure 10.9   Block diagram for solving Equation 10.1.

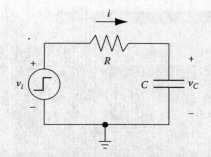

Figure 10.10   First-order *RC* circuit.

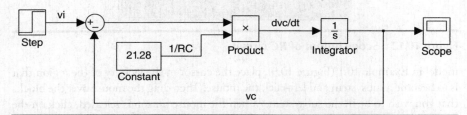

Figure 10.11   Block diagram for solving Equation 10.2.

Example 10.2

For the series RC circuit of Figure 10.10, we will examine the capacitor voltage v_C with respect to time in response to a unit-step input voltage. The governing equation (as derived in Exercise E7.1, with v_i replacing V_D) is

$$\frac{dv_C}{dt} = \frac{1}{RC}(v_i - v_C) \tag{10.2}$$

where
$R = 10 \text{ k}\Omega$
$C = 4.7 \text{ μF}$
$v_i = \begin{cases} 0 \text{ V} & \text{for } t \leq 0 \\ 1 \text{ V} & \text{for } t > 0 \end{cases}$

The block diagram for Equation 10.2 is shown in Figure 10.11, and the resulting scope output is shown in Figure 10.12.

10.5 Constructing a Subsystem

Suppose we build a large system consisting of many blocks and we wish to reduce the number of blocks appearing in the overall block diagram. This can be done by creating a subsystem. The subsystem will appear as a single block. To create a subsystem of the

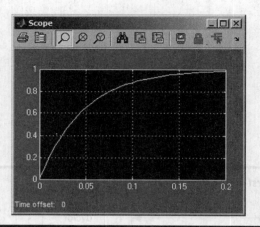

Figure 10.12   Scope output of *RC* circuit.

model in Example 10.1 (Figure 10.9), place the cursor in the vicinity of the region that is to become a subsystem and left-click the mouse. Then drag the mouse over the blocks that you wish to be in the subsystem. When the mouse button is released, click on the *Diagram* option in the menu bar and select *Subsystem and Model Reference*. Then select *Create Subsystem* from the drop down menu. This will result in the selected multiple blocks to be replaced with a single subsystem block as shown in Figure 10.13. In that figure the constants and the product blocks have been combined into the subsystem. This particular subsystem has one input and two outputs, but in general a subsystem may have multiple inputs and outputs. By double-clicking on the subsystem, you may view its components as shown in Figure 10.14. Blocks and connecting lines in any view can be moved to create a model flow to your liking.

Similarly, in Example 10.2 (see Figure 10.11), the Constant, Product, and Integrator blocks can be combined into a subsystem (see Figures 10.15 and 10.16). This particular subsystem has one input and one output. By double-clicking on the subsystem, you may view its components (see Figure 10.17).

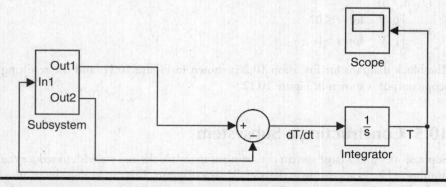

Figure 10.13   Block diagram containing a subsystem for solving Equation 10.1.

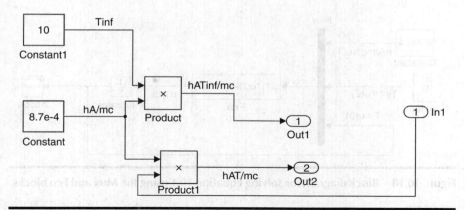

Figure 10.14   Components of subsystem used in solving Equation 10.1.

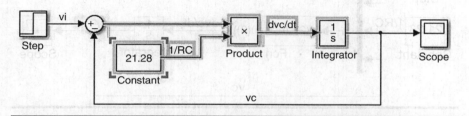

Figure 10.15   Creating a subsystem in block diagram for solving Equation 10.2.

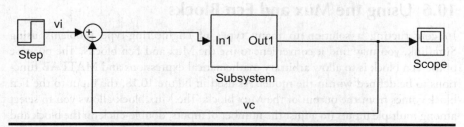

Figure 10.16   Block diagram with subsystem for solving Equation 10.2.

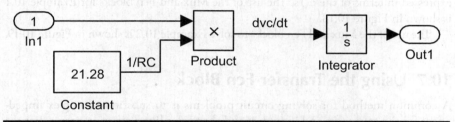

Figure 10.17   Components of subsystem used in Figure 10.16.

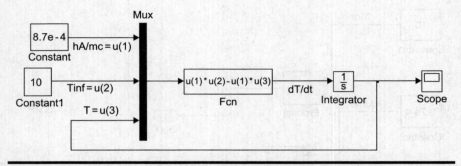

Figure 10.18 Block diagram for solving Equation 10.1 using the Mux and Fcn blocks.

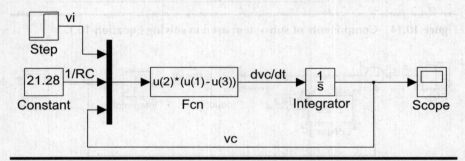

Figure 10.19 Block diagram for solving Equation 10.2 using the Mux and Fcn blocks.

10.6 Using the Mux and Fcn Blocks

In constructing a solution to many types of engineering-type problems using Simulink, you may find it convenient to use the Mux and Fcn blocks. The purpose of the Fcn block is to allow arbitrary mathematical expressions and MATLAB functions to be defined within the model. As used in Figure 10.18, the input to the Fcn block comes from the output of the Mux block. The Mux block allows you to select among multiple inputs (to adjust the number of inputs, double-click on the block and edit the number of inputs). The uppermost input is designated as $u(1)$, the next one below is designated as $u(2)$, etc. The mathematical expressions in the Fcn block are expressed in terms of the $u()$'s. The use of the Mux and Fcn blocks for Example 10.1 is shown in Figure 10.18.

The use of the Mux and Fcn blocks to solve Example 10.2 is shown in Figure 10.19.

10.7 Using the Transfer Fcn Block

A common method for solving circuit problems is to substitute complex impedances for the capacitors and inductors and then solve like a resistive circuit. For the RC circuit, the impedance of the resistor is simply $Z_R = R$ and the impedance of

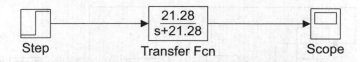

Figure 10.20   Block diagram using the Transfer Fcn block.

the capacitor is $Z_C = 1/Cs$ (where $s = j\omega$). Then, the capacitor voltage is simply the output of a resistive divider:

$$
\begin{aligned}
v_C &= \frac{Z_C}{Z_R + Z_C} v_i \\
&= \frac{\dfrac{1}{Cs}}{R + \dfrac{1}{Cs}} v_i \\
&= \frac{\dfrac{1}{RC}}{s + \dfrac{1}{RC}} v_i \\
&= H(s) v_i
\end{aligned}
$$

where $H(s) = (1/RC)/(s + 1/RC)$ is commonly referred to as the *transfer function*. In Simulink, the Transfer Fcn block allows direct entry of a transfer function into your model as shown in Figure 10.20.

10.8 Using the Relay and Switch Blocks

Relays and switches are used in designs to enable a low-power device (e.g., an electronic controller) to control a high-power system (e.g., a boiler or furnace). Simulink has Relay and Switch blocks, which can be used to simulate these types of systems.

Example 10.3

In a home heating system, a temperature sensor is used to switch the boiler on and off in order to heat the house to a comfortable temperature. However, because most boilers do not turn on and off instantaneously (i.e., they take a few minutes to heat up after turning on and also take time to cool after turning off), the control of the room temperature requires some hysteresis in order to avoid cycling the boiler on and off too often (which causes excessive wear on the boiler). This concept can be represented by a simple differential equation in which the temperature T is set to fluctuate at a constant rate between 20°C and 22°C:

$$
\frac{dT}{dt} = c \quad \text{where } c = \begin{cases} 30 & \text{if } T \leq 20 \\ -30 & \text{if } T \geq 22 \end{cases} \tag{10.3}
$$

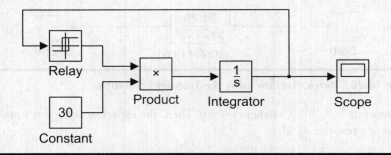

Figure 10.21   Block diagram using the Relay block from Example 10.3.

The block diagram for this system consists of an integrator, a constant, a relay, a product, and a scope (Figure 10.21). The relay is used to invert the sign on c depending on whether the relay is on the on or off position. The relay parameters may be edited by double-clicking on the Relay block and setting the parameters to the following values:

Switch point on = 22
Switch point off = 20
Output when on = –1
Output when off = +1

The "switch point on" value must be greater than the value of "switch point off". We assume that the initial room temperature is $T = 18°C$, and this initial value is entered into the parameters for the Integrator block. At the start of the simulation, $T \leq 20$ and thus the relay switch is off and the output of the relay is +1, causing T to increase. The relay output will remain +1 until T reaches 22, at which point the relay will turn on and its output will be –1, causing T to decrease. The relay output will remain –1 until T reaches 20, at which point the relay will turn off and its output becomes +1 again. This process will continue until the simulation end time is reached. The output of the scope is shown in Figure 10.22.

Example 10.4

Some problems may involve a function that varies in time for $0 \leq t \leq t_1$ and is constant for $t_1 < t \leq t_2$. This type of function can best be modeled with the Switch block, which implements a double-pole single-throw (DPST) switch with an additional terminal to control the opening and closing of the switch.

Suppose

$$y = \begin{cases} 5t & \text{for } 0 \leq t \leq 10 \\ 50 & \text{for } 10 < t \leq 20 \end{cases} \tag{10.4}$$

The Simulink model for this problem is shown in Figure 10.23.

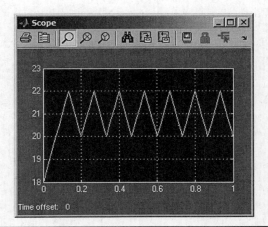

Figure 10.22   Scope output of Example 10.3.

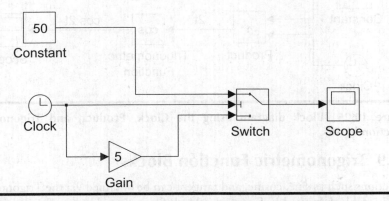

Figure 10.23   Block diagram using the Switch block from Example 10.4.

The parameters for the switch are as follows (note that terminal "$u1$" is the top switch input, "$u2$" is the middle (control) input, and "$u3$" is the bottom switch input):

Criteria for passing first input: $u2 \geq$ Threshold

Threshold: 10

The Gain block multiplies the input by a constant value (gain). The input and the gain can each be a scalar, vector, or matrix.

The resulting output is shown in Figure 10.24. Note that we used the Clock block to generate the independent variable, which, in this case, is the time, t.

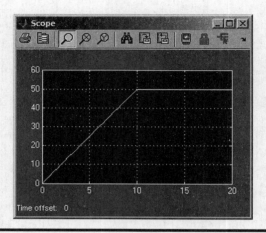

Figure 10.24   Scope output of Example 10.4.

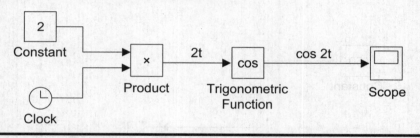

Figure 10.25   Block diagram using the Clock, Product, and Trigonometric Function blocks.

10.9 Trigonometric Function Blocks

Functions such as sine, cosine, and tangent can be obtained via the Trigonometric Function block located in Simulink's Math Operations library. The input to this block is the argument to the desired trig function. If the argument involves the independent variable t (as in sin ωt), then we can use the Clock block to obtain the value of t. This is shown in Figure 10.25 where we compute the value of cos $2t$.

Example 10.5: Simulation of a spring–dashpot system

The governing equation for a simple spring–dashpot system subjected to an oscillatory force is

$$\ddot{x} + \frac{c}{m}\dot{x} + \frac{k}{m}x = \frac{F}{m}\sin(\omega t), \quad x(0) = 5 \text{ m}, \quad \dot{x}(0) = 0 \text{ m/s} \quad (10.5)$$

The Simulink program in Figure 10.26 gives the solution. The values used are

$$m = 10 \text{ kg}, \quad k = 5 \text{ N/m}, \quad c = 0.5 \text{ kg/s}, \quad \omega = 20 \text{ rad/s}, \quad F = 1 \text{ N}$$

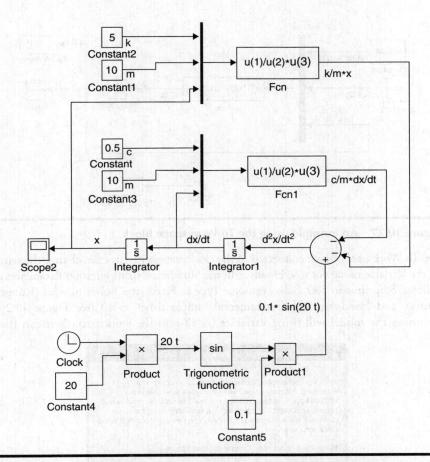

Figure 10.26 Block diagram for solving the second-order differential equation of Equation 10.5.

10.10 To Workspace Block

There may be occasions when you may wish to have an output from a model go to the MATLAB workspace for further manipulation. For example, you may wish to create a table or to construct a MATLAB plot that is not available from the model's Scope. This can be done with Simulink's To Workspace block, which is found in the Sinks section. We will modify Example 10.1 model (shown in Figure 10.18) to include the To Workspace block (see Figure 10.27) and then write a MATLAB program to print a table and create a MATLAB plot. To accomplish this, double click on the To Workspace block and make the following changes: in the Variable name slot, change simout (default) to *T*, and in the Save format slot, change Structure (default) to array. Do the same to

Mux
8.7e-4
Constant
hA/mc = u (1)
10
Constant1
Tinf = u (2)
T = u (3)
u(1) * u(2) - u(1) * u(3)
Fcn
dT/dt
1
s
Integrator
T
To Workspace
Scope
Clock
t
To Workspace1

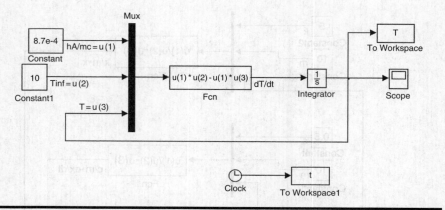

Figure 10.27 An example using the To Workspace block.

the To Workspace1 block connected to the clock, except in this case, change the name in the Variable name slot to *t*. Finally, edit the Simulation Configuration Parameters as follows: Stop time to 500, Solver options: Type to Fixed-step, Solver to ode4 (Runge–Kutta), and Fixed-step size (fundamental sample time) to 10 (see Figure 10.29). Running the model will bring variables (*t*, *T*) into the workspace. You can then

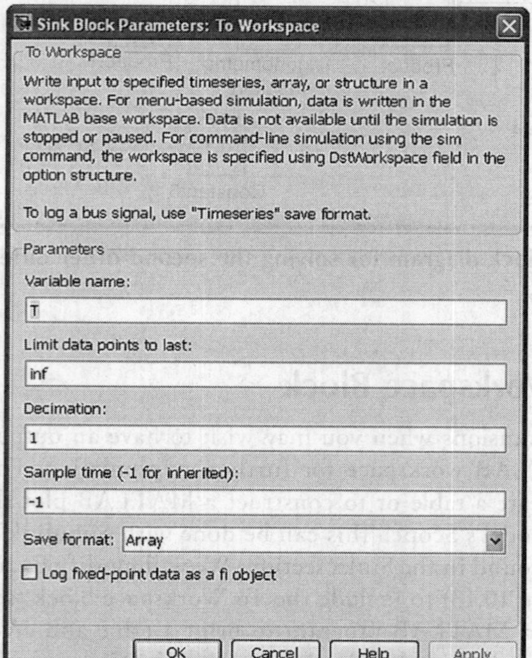

Sink Block Parameters: To Workspace
To Workspace
Write input to specified timeseries, array, or structure in a workspace. For menu-based simulation, data is written in the MATLAB base workspace. Data is not available until the simulation is stopped or paused. For command-line simulation using the sim command, the workspace is specified using DstWorkspace field in the option structure.
To log a bus signal, use "Timeseries" save format.
Parameters
Variable name:
T
Limit data points to last:
inf
Decimation:
1
Sample time (-1 for inherited):
-1
Save format: Array
Log fixed-point data as a fi object
OK
Cancel
Help
Apply

Figure 10.28 To Workspace parameter box.

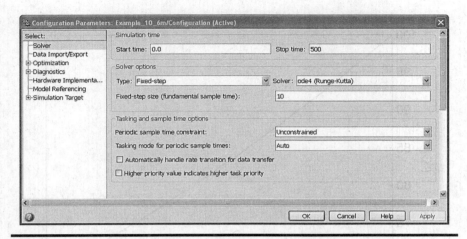

Figure 10.29   Simulation Configuration Parameter box.

run a standard MATLAB program to print out a table and create a MATLAB plot. The MATLAB program follows.

Example 10.6

```
% Example_10_6.m
% This program takes data from the output of a Simulink model
% and creates a table and a MATLAB plot.
% Do not use the clear statement in this program.
clc;
fprintf('        t(s)    T(C) \n');
fprintf('--------------------------\n');
for i = 1:length(t)
    fprintf('%4.1f       %8.2f \n',t(i),T(i));
end
plot(t,T), xlabel('t(s)'), ylabel('T(\circC)'),
title('Temperature vs. time'), grid;
```

Program results

```
    t(s)    T(C)
--------------------------

     0.0    100.00
    10.0     99.22
    20.0     98.45
    30.0     97.68
    40.0     96.92
    50.0     96.17
     ⋮        ⋮
```

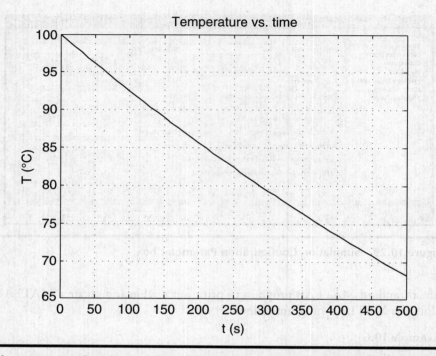

Figure 10.30 Plot of *T* vs. *t* using data from the workspace with MATLAB®'s plot command.

```
        450.0   70.84
        460.0   70.32
        470.0   69.79
        480.0   69.28
        490.0   68.76
        500.0   68.25
>>
```
See Figure 10.30.

Example ■ 10.7: Simulation of an *RLC* Circuit

In Projects P2.7 and P2.8, we studied the parallel RLC circuit. If we use a sinusoidal input current to the circuit in Figure P2.7 and assume that the switch closes at $t = 0$, the governing equation is

$$\frac{d^2 i_L}{dt^2} + \frac{1}{RC}\frac{di_L}{dt} + \frac{1}{LC}i_L = \frac{1}{LC}I_o \sin \omega t \qquad (10.6)$$

where the circuit is "driven" by a sinusoidal current of frequency ω and magnitude I_o. The Simulink model that solves this second-order differential equation is shown in Figure 10.31.

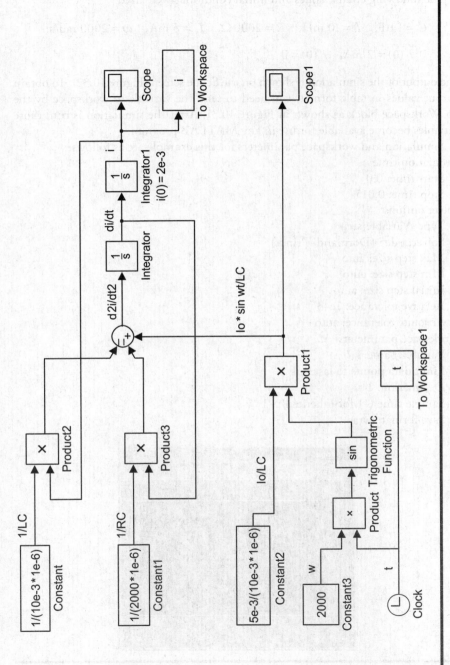

Figure 10.31 Block diagram for solving Equation 10.6 including To Workspace block.

The following circuit values and initial conditions were used:

$$C = 1 \; \mu F, \quad L = 10 \; mH, \quad R = 2000 \; \Omega, \quad I_o = 5 \; mA, \quad \omega = 2000 \; rad/s,$$

$$i_L(0) = 2 \; mA, \quad i'_L(0) = 0$$

The output of the simulation is shown on the Scope screen (Figure 10.32). To obtain output values in table format, you need to send the variable to workspace by the To Workspace block as shown in Figure 10.31. After the simulation is run, those variables become available for use in any MATLAB program.

Simulation and workspace parameters for this example are as follows:
Simulation time
    Start time: 0.0
    Stop time: 0.015
Solver options
    Type: Variable-step
    Solver: ode5 (Dormand–Prince)
    Max step size: auto
    Min step size: auto
    Initial step size: auto
    Relative tolerance: 1e–3
    Absolute tolerance: auto
Workspace parameters
    Variable name: i
    Limit data points to last: inf
    Decimation: 1
    Sample time (–1 for inherited): –1
    Save format: array

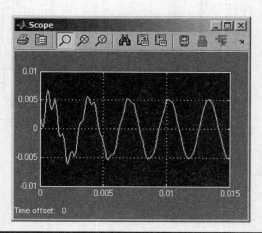

Figure 10.32   Scope output of Example 10.7.

Exercises

E10.1 Implement the following function in Simulink for $0 \leq t \leq 5$ and run the model:

$$y = 5t^2 + 4t + 2$$

1. Using one Clock block, three Constants, two product blocks, and a Sum block.
2. Using one Clock block, three Constants, a Mux block, an Fcn block, a Scope block, and two To Workspace blocks. The Scope block should give a plot of y vs. t. Also create a MATLAB program that constructs a table of y vs. t in the Command window and plot the result using MATLAB's plot command.

E10.2 In Exercise E2.5, the governing equations describing the motion of a ball bearing that is dropped in a vat of oil were given. The obtained governing equations were

$$W - B - D = \frac{W}{g}\frac{dV}{dt} = \frac{W}{g}\frac{d^2y}{dt^2} \qquad (10.7)$$

$$W = \rho_{steel} g \times \frac{4}{3}\pi R^3 \qquad (10.8)$$

$$D = 6\pi R\mu V \qquad (10.9)$$

$$B = \rho_{oil} g \times \frac{4}{3}\pi R^3 \qquad (10.10)$$

Develop a Simulink model to determine the position, y, and velocity, V, as a function of time, t. Run the model and send the results to the workspace and print in the Command window a table containing (t, y, V). Also, create plots of y vs. t and V vs. t using MATLAB's plot command.

Use the following parameters for your model: $\mu = 3.85$ (N-s)/m², $R = 0.01$ m, $\rho_{steel} = 7910$ kg/m³, $\rho_{oil} = 899$ kg/m³, and $g = 9.81$ m/s². Use a fixed time step of 0.005 s and run the model for 0.3 s. Assume $y(0) = 0$ and $\frac{dy}{dt}(0) = 0$.

Projects

P10.1 In Project P7.4, the governing equations describing the motion of a small rocket that is fired from a rocket launcher were developed. The governing equations are

$$\frac{d^2x}{dt^2} = \frac{v_x T}{m\sqrt{v_x^2 + v_y^2}} - \frac{v_x K\sqrt{v_x^2 + v_y^2}}{m} \qquad (P10.1a)$$

$$\frac{d^2 y}{dt^2} = \frac{v_y T}{m\sqrt{v_x^2 + v_y^2}} - \frac{v_y K \sqrt{v_x^2 + v_y^2}}{m} - g \qquad \text{(P10.1b)}$$

$$v^2 = v_x^2 + v_y^2$$

where
  m is the mass of the rocket (varies with time)
  (v_x, v_y) are the x and y components of the rocket's velocity relative to the
    ground
  K is the drag coefficient
  g is the gravitational constant
  (x, y) are the position of the rocket relative to the ground
  t is the time of rocket flight

The initial mass of the rocket is 350 kg, which includes a mass of 100 kg
of fuel. The rocket leaves the launcher at velocity v_o and at an angle of θ_o
with the horizontal. Neglect the fuel consumed inside the rocket launcher.
The rocket burns fuel at the rate of 10 kg/s and develops a thrust,
$T = 6000$ N, which lasts for 10 s. In developing the governing equations,
it is assumed that the thrust acts in the axial direction along the rocket. It
is also assumed that the drag force acts in the axial direction and is proportional to the square of the rocket velocity:

a. Develop a Simulink model that will solve for x, y, v_x, v_y for $0 \leq t \leq 60$ s.
    Use a fixed time step of 0.1 s. Run the program for $t \leq 60$ s.
    Take $x(0) = 0$, $y(0) = 0$, $v_x(0) = v_o\cos\theta_o$, $v_y(0) = v_o\sin\theta_o$, $v_o = 150$ m/s,
    $\theta_o = 60°$, $K = 0.045$ N-s^2/m^2, and $g = 9.81$ m/s^2. Send the results to the
    workspace.
b. Develop a MATLAB program that utilizes the data sent to the work-
    space; prints out a table for t, x, y, v_x, v_y every 1 s; and creates plots of (x and
    y vs. t) and (v_x and v_y vs. t).

P10.2 We wish to examine the time temperature variation of a fluid, T_f, enclosed
in a container with a heating element and a thermostat. The walls of the
container are pure copper. The fluid is engine oil, which has a temperature
T_f that varies with time. The thermostat is set to cut off power to the heating element when T_f reaches 65°C and to resume supplying power when T_f
reaches 55°C.

◆ Wall properties:

$$k_w = 386.0 \text{ W/m-C}, \quad c_w = 0.3831 \times 10^3 \text{ J/kg-C}, \quad \rho_w = 8954 \text{ kg/m}^3$$

Engine oil properties:

$$k_f = 0.137 \text{ W/m-C}, \quad c_f = 2.219 \times 10^3 \text{ J/kg-C}, \quad \rho_f = 840 \text{ kg/m}^3$$

The inside size of the container is 0.5 m × 0.5 m × 0.5 m.

The wall thickness is 0.01 m. Thus, the inside surface area $A_{s,i}$ = 1.5 m², outside surface area $A_{s,o}$ = 1.5606 m², engine oil volume V_{oil} = 0.125 m³, and wall volume V_{wall} = 0.0153 m³.

The power, Q, of the heating element = 10,000 W.

The inside convective heat transfer coefficient h_i = 560 W/m²-C.

The outside convective heat transfer coefficient h_o = 110 W/m²-C.

Using a lump parameter analysis (assuming that the engine oil is well mixed) and the first law of thermodynamics, the governing equations describing the time temperature variation of both materials are as follows:

$$\frac{d\theta_f}{dt} = -a_1(\theta_f - \theta_w) + a_5 \tag{P10.2a}$$

$$\frac{d\theta_w}{dt} = a_2(\theta_f - \theta_w) - a_3\theta_w = a_2\theta_f - (a_2 + a_3)\theta_w \tag{P10.2b}$$

where

$$\theta_f = T_f - T_\infty$$

$$\theta_w = T_w - T_\infty$$

$$a_1 = \frac{h_i A_{s,i}}{m_f c_f}, a_2 = \frac{h_i A_{s,i}}{m_w c_w}, a_3 = \frac{h_o A_{s,o}}{m_w c_w}, a_4 = a_2 + a_3, a_5 = \frac{Q}{m_f c_f}$$

Initial conditions

$$T_f(0) = T_w(0) = 15°\text{C}$$

$$T_\infty = 15°\text{C}$$

Construct a model in Simulink that simulates this system. Run the model for 3600 s.

Send $T_f(t)$ and $T_w(t)$ to the workspace and construct a MATLAB program that will plot both $T_f(t)$ and $T_w(t)$ vs. t.

P10.3 A *decision circuit* is used in a communication system in order to decide whether a received digital signal is a logical one or zero. It then outputs its decision using onboard logic levels. For example, in the RS-232 serial communication protocol [1], the nominal logic levels to represent a logical one or zero are ±12 V. However, typical logic levels on a computer mother-board are +5 V and 0 V. We can use a Relay block in Simulink to examine the behavior of a decision circuit, which implements these requirements in the presence of noise that is induced in the RS-232 cable due to fluorescent lights, radio waves, etc.:

1. Use Simulink's Signal Generator block (from the Sources library) to generate a binary signal of alternating ones and zeros using RS-232 logic levels at a rate of 1200 bits/s. (Note: 1200 bits/s is *not* the same as 1200 Hz.)

2. Add white noise to your RS-232 signal with a Band-Limited White Noise block (from the Sources library) and a Sum block. Set the parameters on the white noise to be 2×10^{-4} for the noise power and 1×10^{-5} for the sample time. View the resulting noisy binary signal with a Scope block and confirm that it still looks like a binary signal (albeit with noise).

3. Use a Relay block to convert your RS-232 binary signal from +12 V (for a one) and –12 V (for a zero) into +5 V (for a one) and 0 V (for a zero). Use 0 V for both the switch-on and switch-off points for the relay. Run the simulation for 0.01 s and view the resulting output from the relay with a scope. You should see some errors on the relay output due to the noise on the input.

4. We will now quantify the bit errors on the relay output as follows:

   a. First, we need to generate a "perfect" (i.e., noiseless) version of the relay output for comparison purposes. Thus, make a copy of your existing blocks (including Signal Generator, Relay, and Scope) within the current model, but omit the noise generator in the copy. Then, run the simulation again for 0.01 s and confirm that you have both perfect and noisy versions at the outputs of your two relays.

   b. Next, create an error signal by subtracting the two relay outputs with a Sum block.

   c. Next, generate the magnitude of the error signal with an Abs (absolute value) block (from the Math Operations library). The purpose of this is to make sure that the error signal is always positive so that it may be easily counted.

   d. Finally, send the output of your Abs block to an Integrator block (with initial condition of zero), and use a Display block to monitor the output of your integrator. The purpose of the integrator is to provide a running total of all of the detected errors.

5. Run the simulation for 0.01 s on your complete model containing dual relays with cumulative error detection. At the end of the simulation, you should wind up with a positive value at the output of the integrator. Also try running the integration for 1 s and see that you have even more errors.

Table P10.3   Cumulative Errors for the Decision Circuit of Project 10.3 for Various Amounts of Hysteresis

Switch-On and Switch-Off Values	Cumulative Error Detected at Integrator Output (per Second)
0, 0	—
+0.5, −0.5	—
+2, −2	—
+5, −5	—
+8, −8	—

6. One method of improving the error performance of this decision circuit is to insert hysteresis into the relay in order to increase the noise margins. To observe this, modify the relay switch-on and switch-off values to +0.5 and −0.5 and rerun the simulation for a 1 s interval and confirm that you see fewer errors. Try the various sets of switch-on and switch-off values as listed in Table P10.3 and complete the table. What happens if you use too much hysteresis?

P10.4 Repeat Project P7.6 for a Sallen–Key circuit with a step and impulse input, but this time use Simulink to construct a simulation of the system. Scope output should be for v_{out} and v_1 vs. t. Set the end time to $t = 0.0001$ s and print out the block diagram, impulse response, and step response.

Reference

1. McNamara, J., *Technical Aspects of Data Communication*, 3rd ed., Butterworth-Heinemann, Woburn, MA, 1988.

Chapter 11

Optimization

11.1 Introduction

The objective of optimization is to maximize or minimize some function f (called the *object function*). You are probably familiar with determining the maxima and minima of functions by using techniques from differential calculus. However, the problem becomes more complicated when we place *constraints* on the allowable solutions to f. As an example, suppose there is an electronics company that manufactures several different types of circuit boards. Each circuit board must pass through several different departments (such as drilling, pick-and-place, testing) before shipping. The time required for each circuit board to pass through the various departments is also known. There is a minimum production quantity per month that the company must produce. However, the company is capable of producing more than the minimum production requirement for each type of circuit board each month. The profit the company will make on each circuit board it produces is known. The problem is to determine the production amount of each type of circuit board per month that will result in the maximum profit. A similar type of problem may be one in which the object is to minimize the cost of producing a particular product. These types of optimization problems are discussed in greater detail later in this chapter.

In most optimization problems, the object function f will usually depend on several variables, $x_1, x_2, x_3, \ldots, x_n$. These are called the *control variables* because their values can be selected. Optimization theory develops methods for selecting optimal values for the control variables $x_1, x_2, x_3, \ldots, x_n$ that either maximize (or minimize) the objective function f. In many cases, the choice of values for $x_1, x_2, x_3, \ldots, x_n$ is not entirely free but is subject to some constraints.

11.2 Unconstrained Optimization Problems

In calculus, it is shown that a necessary (but not sufficient) condition for f to have a maximum or minimum at point P is that each of the first partial derivatives at P be zero, that is,

$$\frac{\partial f}{\partial x_1}(P) = \frac{\partial f}{\partial x_2}(P) = \cdots = \frac{\partial f}{\partial x_n}(P) = 0 \tag{11.1}$$

where the notation $\frac{\partial f}{\partial x_i}(P)$ indicates the partial derivative with respect to x_i evaluated at point P, that is, $\left.\frac{\partial f}{\partial x_i}\right|_{x=P}$. If $n = 1$ and the object function is $y = f(x)$, then a necessary condition for an extremum (maximum or minimum) at x_0 is for $y'(x_0) = 0$.

For y to have a local minimum at x_0, $y'(x_0) = 0$ and $y''(x_0) > 0$.

For y to have a local maximum at x_0, $y'(x_0) = 0$ and $y''(x_0) < 0$.

For f involving several variables, the condition for f to have a relative minimum is more complicated. First, Equation 11.1 must be satisfied. Second, the quadratic form

$$Q = \sum_{i=1}^{n} \sum_{j=1}^{n} \frac{\partial^2 f}{\partial x_i \partial x_j}(P)\left(x_i - x_i(P)\right)\left(x_j - x_j(P)\right) \tag{11.2}$$

must be positive for all choices of x_i and x_j in the vicinity of point P, and $Q = 0$ only when $x_i = x_i(P)$ for $i = 1, 2, \ldots, n$. This condition comes from a Taylor series expansion of $f(x_1, x_2, \ldots, x_n)$ about point P using only the terms up to $\frac{\partial^2 f}{\partial x_i \partial x_j}(P)$.

This gives

$$f(x_1, x_2, \ldots, x_n) = f(P) + \sum_{i} \frac{\partial f}{\partial x_i}(P)\left(x_i - x_i(P)\right)$$

$$+ \sum_{i=1}^{n} \sum_{j=1}^{n} \frac{\partial^2 f}{\partial x_i \partial x_j}(P)\left(x_i - x_i(P)\right)\left(x_j - x_j(P)\right) \tag{11.3}$$

If $f(x_1, x_2, \ldots, x_n)$ has a relative minimum at point P, then $\frac{\partial f}{\partial x_i}(P) = 0$ for $i = 1, 2, \ldots, n$ and $f(x_1, x_2, \ldots, x_n) - f(P) > 0$ for all $(x_1, x_2, \ldots, x_n)$ in the vicinity of point P. But $f(x_1, x_2, \ldots, x_n) - f(P) = Q$. Thus, for $f(x_1, x_2, \ldots, x_n)$ to have a relative minimum at point P, Q must be positive for all choices of x_i and x_j in the vicinity of point P.

Example 11.1

As an example, let us consider the function

$$f(x_1, x_2) = 4 + 4.5x_1 - 4x_2 + x_1^2 + 2x_2^2 - 2x_1x_2 + x_1^4 - 2x_1^2 x_2 \qquad (11.4)$$

A relative minimum exists at $(x_1, x_2) = (1.941, 3.854)$. Thus, point P has coordinates $(1.941, 3.854)$. Let us evaluate Q in the vicinity of point P. For f as a function of (x_1, x_2), the equation for Q is

$$Q = \frac{\partial^2 f}{\partial x_1^2}(P)\big(x_1 - x_1(P)\big)^2 + 2\frac{\partial^2 f}{\partial x_1 \partial x_2}(P)\big(x_1 - x_1(P)\big)\big(x_2 - x_2(P)\big)$$

$$+ \frac{\partial^2 f}{\partial x_2^2}(P)\big(x_2 - x_2(P)\big)^2 \qquad (11.5)$$

Taking the partial derivatives of f and substituting these expressions into Equation 11.2 gives

$$\frac{\partial^2 f}{\partial x_1^2} = 2 + 12x_1^2 - 4x_2; \quad \frac{\partial^2 f}{\partial x_1 \partial x_2} = -2 - 4x_1; \quad \frac{\partial^2 f}{\partial x_2^2} = 4$$

To evaluate Q in the vicinity of point P, we will take points on a small circle around point P, that is, $\big(x_1 - x_1(P)\big) = \Delta s \cos \vartheta$ and $\big(x_2 - x_2(P)\big) = \Delta s \sin \vartheta$ with $0 \leq \vartheta \leq 360°$.

The program follows:

```
% Example_11_1.m
% Quadratic_form
% This program determines Q in the vicinity of a relative
% minimum for the following function:
% f(x1,x2) = 4+4.5*x1-4*x2+x1^2+2*x2^2-2x1*x2+x1^4-2x1^2*x2.
% Q = d2fdx1^2(P)[x1-x1(P)]^2 +2*d2fdx1dx2(P)[x1-x1(P)]
% [x2-x2(p)]+ d2fdx2^2(P)[x2-x2(P)]^2
clear; clc;
x1p = 1.941; x2p = 3.854;
f=@(x1,x2) ···
    (4+4.5*x1-4*x2+x1^2+2*x2^2-2*x1*x2+x1^4-2*x1^2*x2);
d2fdx1dx1 = @(x1,x2) (2+12*x1^2-4*x2);
d2fdx2dx2 = 4;
d2fdx1dx2 = @(x1) (-2-4*x1);
ds = 0.1;
theta = 0:18:360;
fmin = f(1.941,3.854);
fprintf('fmin =%10.4f \n',fmin);
fprintf('Determining Q around minimum value of f \n');
fprintf('Minimum value of f occurs at ');
```

```
fprintf('(x1,x2)=(%5.3f,%5.3f) \n',x1p,x2p);
fprintf(' theta        Q \n');
fprintf('---------------\n');
for i = 1:length(theta)
    theta1 = theta(i);
    Q=d2fdx1dx1(x1p,x2p)*(ds*cosd(theta1))^2+...
        4*(ds*sind(theta1))^2+...
        d2fdx1dx2(x1p)*(ds*cosd(theta1)*ds*sind(theta1));
    fprintf('%5.0f %10.4f \n',theta1,Q);
end
```

Program results

```
    fmin = 0.9856
    Determining Q around minimum value of f
    Minimum value of f occurs at (x1, x2) = (1.941, 3.854)
     theta     Q
     -------------
        0    0.3179
       18    0.2627
       36    0.1755
       54    0.0896
       72    0.0378
       90    0.0400
      108    0.0952
      126    0.1825
      144    0.2683
      162    0.3201
      180    0.3179
      198    0.2627
      216    0.1755
      234    0.0896
      252    0.0378
      270    0.0400
      288    0.0952
      306    0.1825
      324    0.2683
      342    0.3201
      360    0.3179
    >>
```

We see that Q is positive for all of the test points around point P.

Since the aforementioned analysis is quite complicated when f is a function of several variables, an iterative scheme is frequently used as a method of solution. One such method is the method of steepest descent. In this method, we first need to guess for a point where an extremum exists. Using a grid to evaluate the function at different values of the control variables can be helpful in establishing a good starting point for the iteration process.

11.3 Method of Steepest Descent

Consider a function f of three variables x, y, and z. From calculus, we know that the gradient of f, written as ∇f, is given by

$$\nabla f = \frac{\partial f}{\partial x}\hat{\mathbf{e}}_x + \frac{\partial f}{\partial y}\hat{\mathbf{e}}_y + \frac{\partial f}{\partial z}\hat{\mathbf{e}}_z \tag{11.6}$$

where $\hat{\mathbf{e}}_x$, $\hat{\mathbf{e}}_y$, and $\hat{\mathbf{e}}_z$ are unit vectors in the x, y, and z directions, respectively.

At (x_0, y_0, z_0), we also know that $\nabla f(x_0, y_0, z_0)$ points in the direction of the maximum rate of change of f with respect to distance. A unit vector $\hat{\mathbf{e}}_g$ that points in this direction is

$$\hat{\mathbf{e}}_g = \frac{\nabla f}{|\nabla f|} \tag{11.7}$$

where

$$|\nabla f| = \sqrt{\left(\frac{\partial f}{\partial x}\right)^2 + \left(\frac{\partial f}{\partial y}\right)^2 + \left(\frac{\partial f}{\partial z}\right)^2}$$

To find a relative minimum via the method of steepest descent, we start at some initial point and move in small steps in the direction of steepest descent, which is $-\hat{\mathbf{e}}_g$. Let $(x_{n+1}, y_{n+1}, z_{n+1})$ be the new position on the nth iteration and (x_n, y_n, z_n) the previous position. Then,

$$x_{n+1} = x_n - \frac{\frac{\partial f}{\partial x}(x_n, y_n, z_n)}{|\nabla f(x_n, y_n, z_n)|}\Delta s$$

$$y_{n+1} = y_n - \frac{\frac{\partial f}{\partial y}(x_n, y_n, z_n)}{|\nabla f(x_n, y_n, z_n)|}\Delta s \tag{11.8}$$

$$z_{n+1} = z_n - \frac{\frac{\partial f}{\partial z}(x_n, y_n, z_n)}{|\nabla f(x_n, y_n, z_n)|}\Delta s$$

where Δs is some small length.

Example 11.2

Given

$$f(x_1, x_2) = 4 + 4.5x_1 - 4x_2 + x_1^2 + 2x_2^2 - 2x_1x_2 + x_1^4 - 2x_1^2x_2$$

determine the minimum of f by the method of steepest descent, starting at point $(x_1, x_2) = (6, 10)$. Use a $\Delta s = 0.1$ and a maximum of 100 iterations.

```
% Example_11_2.m
% This program determines a relative minimum by the method
% of steepest decent.
% The function is:
% f(x1,x2) = 4+4.5*x1-4*x2+x1^2+2*x2^2-2*x1*x2+x1^4-2*x1^2*x2
% Note: this function has known minima at (x1,x2) =
% (1.941,3.854) and (-1.053,1.028). The functional values at
% the minima points are 0.9855 and -0.5134 respectively.
clear; clc;
% Define function and its partial derivatives
fx_func = @(x1,x2) 4+4.5*x1-4*x2+x1^2+2*x2^2-2*x1*x2+···
          x1^4-2*x1^2*x2;
dfx1_func = @(x1,x2) 4.5+2*x1-2*x2+4*x1^3-4*x1*x2;
dfx2_func = @(x1,x2) -4+4*x2-2*x1-2*x1^2;
% Define stepping parameters
ds = 0.1;
max_iterations = 100;
% First guess
x1 = 6;
x2 = 10;
fx = fx_func(x1,x2);
% Print headings and initial guess to screen
n = 0
fprintf(' n        x1        x2           fx \n');
fprintf('-------------------------------------------\n');
fprintf(' %2d %7.4f %7.4f %10.4f \n',n,x1,x2,fx);
for n = 1:max_iterations
    % compute partial derivatives
    dfx1 = dfx1_func(x1,x2);
    dfx2 = dfx2_func(x1,x2);
    % compute magnitude of gradient
    gradf_mag = sqrt(dfx1^2+dfx2^2);
    % compute next values of x1 and x2 as per Equation 11.8
    x1n = x1-dfx1/gradf_mag*ds;
    x2n = x2-dfx2/gradf_mag*ds;
    fxn = fx_func(x1n,x2n);
    fprintf('%2d %7.4f %7.4f %10.4f \n',n,x1n,x2n,fxn);
```

```
    % if new value is larger than previous, then minimum has
    % been passed
    if(fxn > fx)
        fprintf('A minimum has been passed after ');
        fprintf(' %d iterations \n\n',n);
        break;
    % otherwise, store new values over current ones and
    % continue
    else
        x1 = x1n;
        x2 = x2n;
        fx = fxn;
    end
end
if n >= max_iterations
    fprintf('Error: no solution found after %d iterations\n',n);
else
    fmin = fx_func(x1,x2);
    fprintf('The relative minimum occurs at approximately \n');
    fprintf(' x1 = %.4f x2 = %.4f \n',x1,x2);
    fprintf('The minimum value for f = %.4f \n',fmin);
end
```

Program results

```
n       x1        x2          fx
---------------------------------
0      6.0000    10.0000     683.0000
1      5.9003    10.0077     622.7192
2      5.8006    10.0155     566.2528
3      5.7009    10.0233     513.4609
       ⋮          ⋮           ⋮
89     1.9936     4.0234      1.0001
90     1.9638     3.9279      0.9883
91     1.9495     3.8289      0.9901

A minimum has been passed after 91 iterations
The relative minimum occurs at approximately
x1 = 1.9638 x2 = 3.9279
The minimum value for f = 0.9883.
```

To obtain a more accurate result, we could rerun the program with revised starting values (from iteration 90 earlier) and use a smaller Δs.

Example 11.3

In the previous example, instead of starting the method of steepest decent at some arbitrary point $(x1, x2) = (6, 10)$, we could have first used a grid program to establish a good starting point. In addition, a grid program might also have indicated that there was more than one relative minimum point in the range of interest. The following program demonstrates the grid program that could be used.

```
% Example_11_3.m
% This program calculates the values of a specified function
% f(x1,x2) of 2 variables for determining a good starting
% point for the method of steepest decent. The range of
% interest is from -5.0 <= x1 <= 5.0 and
% -10.0 <= x2 <= 10.0.
clear; clc;
% Define function of interest:
fx_func = @(x1,x2) 4+4.5*x1-4*x2+x1^2+2*x2^2-2*x1*x2+···
          x1^4-2*x1^2*x2;
% Define grid endpoints and step size
x1min = -5.0; x1max = 5.0; dx1 = 2.0;
x2min = -10.0; x2max = 10.0; dx2 = 2.0;
% Define grid and calculate f(x1,x2) at each point
x1 = x1min:dx1:x1max;
x2 = x2min:dx2:x2max;
for i = 1:length(x1)
    for j = 1:length(x2)
        f(i,j) = fx_func(x1(i),x2(j));
    end
end
% Print heading
fprintf('-----------------------------------------------\n');
fprintf('   x2 |                   x1 \n');
fprintf('-----------------------------------------------\n');
fprintf('      |');
for i = 1:length(x1)
   fprintf('%7.1f ',x1(i));
end
fprintf('\n');
fprintf('-----------------------------------------------\n');
% Print values of f(x1,x2)
for j = 1:length(x2)
    fprintf('%6.1f |',x2(j));
    for i = 1:length(x1)
        fprintf('%7.1f ',f(i,j));
    end
    fprintf('\n');
end
```

Program results

```
---------------------------------------------------------------
x2    |                              x1
---------------------------------------------------------------
      |   -5.0     -3.0     -1.0     1.0      3.0      5.0
---------------------------------------------------------------
-10.0 |  1271.5   440.5    241.5    290.5    587.5    1516.5
 -8.0 |  1111.5   336.5    161.5    202.5    459.5    1316.5
 -6.0 |   967.5   248.5     97.5    130.5    347.5    1132.5
 -4.0 |   839.5   176.5     49.5     74.5    251.5     964.5
 -2.0 |   727.5   120.5     17.5     34.5    171.5     812.5
  0.0 |   631.5    80.5      1.5     10.5    107.5     676.5
  2.0 |   551.5    56.5      1.5      2.5     59.5     556.5
  4.0 |   487.5    48.5     17.5     10.5     27.5     452.5
  6.0 |   439.5    56.5     49.5     34.5     11.5     364.5
  8.0 |   407.5    80.5     97.5     74.5     11.5     292.5
 10.0 |   391.5   120.5    161.5    130.5     27.5     236.5
>>
```

Since the functional value at $(x1, x2) = (0, 2)$ is lower than the functional values at surrounding points, we suspect that there is a relative minimum in the vicinity of $(x1, x2) = (0, 2)$. A second relative minimum appears to be in the vicinity of $(x1, x2) = (2, 6)$.

11.4 MATLAB®'s `fminbnd` and `fminsearch` Functions

■ MATLAB's `fminbnd` function

MATLAB provides the `fminbnd` function to determine the relative minimum of a single variable function in the interval $x1 < x < x2$. The syntax for the function is:

```
[X,FVAL] = fminbnd(FUN,x1,x2)
```

The three arguments to `fminbnd` are the function, `FUN`, that contains a relative minimum that we wish to determine, and the positions `x1` and `x2`, which gives the interval in which the relative minimum may lie. `FUN` can be a function defined in a separate *.m* file or may be defined anonymously within the script. The outputs of `fminbnd` are `X` and `FVAL`, where `X` is the position of the relative minimum and the `FVAL` is the minimum value of `FUN`.

Note that MATLAB does not have a separate function to find a relative maximum. In order to find a maximum, redefine your function to return the negative value of the function and then use `fminbnd` to find the minimum (see Example 11.4).

Example 11.4

Given: $y(x) = x^3 + 5.7x^2 - 35.1x + 85.176$.
Determine: the relative minima and maxima.

```
% Example_11_4.m
% Find the minima and maxima of
% y = x^3 + 5.7 x^2 - 35.1x + 85.176
clc; clear;
% First, plot the function so that we can determine the
% x range to use in fminbnd is x1=-10, x2=10;
x = -10:0.1:10;
fprintf('This output is from MATLAB fminbnd function \n');
for i=1:length(x)
    y(i) = x(i)^3 + 5.7*x(i)^2 - 35.1*x(i) + 85.176;
end
plot(x,y), xlabel('x'),ylabel('y'), grid, title('y vs x');
% Examine the plot to see if a relative minimum and a
% relative maximum exists in the specified range.
% Next, find the position of the relative minimum and the
% minimum value.
[xmin, Fmin] = fminbnd( @(x) x^3+5.7*x^2-35.1*x+85.176,-10,10 );
% Print results.
fprintf('xmin=%.3f minvalue=%.1f \n',xmin,Fmin)
% Next, find the position of the relative maximum and the
% maximum value.
% Note: To find a maximum, find the minimum of the negative
% of the function.
[xmax, Fmax]= fminbnd( @(x) -(x^3+5.7*x^2-35.1*x+85.176),-10,10);
% Print results.
fprintf('xmax=%.3f maxvalue=%.1f \n',xmax,-Fmax);
```

Program results:

```
This output is from MATLAB fminbnd function
xmin=2.013 minvalue=45.8
xmax=-5.813 maxvalue=285.4
>>
```

See Figure 11.1

■ MATLAB's fminsearch function

MATLAB provides the fminsearch(FUN,X0)function to determine the relative minimum of a multidimensional unconstrained function in the vicinity of X0. The syntax for the function is:

```
[X,FVAL] = fminsearch(FUN,X0)
```

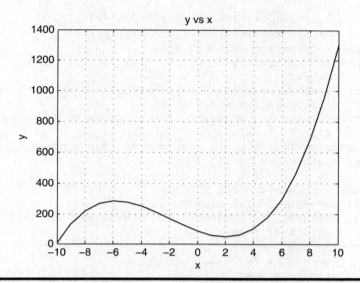

Figure 11.1   Plot of *y* vs. *x* for Example 11.4.

The two arguments to fminsearch are the function, FUN, whose relative minimum we wish to determine and X0, which is the starting coordinates in the search for the relative minimum of FUN. FUN can be a function defined in a separate *.m* file or may be defined anonymously within the script. The output of fminsearch is X and FVAL, where X is the position of the relative minimum and the FVAL is the minimum value of FUN.

Note that MATLAB does not have a separate function to find a relative maximum. In order to find a maximum, redefine your function to return the negative value of the function and then use fminsearch to find the minimum (as in Example 11.4).

Example 11.5

```
% Example_11_5.m
% This program determines the minimum value of the function in
% Example 11.2 by MATLAB's fminsearch function. The function is
% f(x1,x2)=4+4.5*x(1)-4*x(2)+x(1)^2+2*x(2)^2-2*x(1)*x(2)+...
%          x(1)^4-2*x(1)^2*x(2);
% To use MATLAB's fminsearch function, we need to make an
% initial guess for the position of the relative minimum.
clear; clc;
% Initial guess for the position of the relative minimum.
% From Example 11.3 output, a minimum appears to be
% in the vicinity of [0 2]
xo=[0 2];
```

```
fun_x12 = @(x)  4+4.5*x(1)-4*x(2)+x(1)^2+2*x(2)^2-...
          2*x(1)*x(2)+x(1)^4-2*x(1)^2*x(2);
[X, Fmin] = fminsearch(fun_x12,xo);
% x1min=X(1); x2min=X(2);
% Print results.
fprintf('Results from MATLAB fminsearch function \n');
fprintf('First relative minimum \n');
fprintf('x1min=%6.4f  x2min=%6.4f   minvalue=%8.4f \n',...
          X(1),X(2),Fmin);
% From Example 11.3 output, a second minimum appears to be
% in the vicinity of [2 6]
xo=[2 6];
[X, Fmin] = fminsearch(fun_x12,xo);
fprintf('Second relative minimum \n');
fprintf('x1min=%6.4f  x2min=%6.4f   minvalue=%8.4f \n',...
          X(1),X(2),Fmin);
```

```
Results from MATLAB fminsearch function
First relative minimum
x1min=-1.0527  x2min=1.0278   minvalue= -0.5134
Second relative minimum
x1min=1.9410   x2min=3.8543   minvalue=  0.9856
>>
```

11.5 Optimization with Constraints

In many optimization problems, the variables in the function to be maximized or minimized are not all independent but are related by one or more conditions or constraints. A simple example that illustrates the constraint concept follows.

Example 11.6

Suppose we wish to determine the maximum and minimum values of the objective function

$$f(x,y) = 2x + 3y \tag{11.10}$$

with the following constraints:

$$\text{Lower bounds (LB): } x \geq 0, \ y \geq 0 \tag{11.11}$$

$$\text{Upper bounds (UB): } x \leq 3, \ y \leq 3 \tag{11.12}$$

$$\text{Constraint 1 (L1): } \quad x + y \leq 4 \tag{11.13}$$

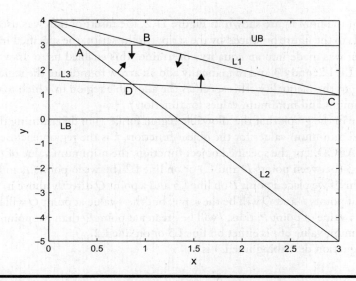

Figure 11.2   Allowable region for control variables in determining maximum and minimum points of the object function.

$$\text{Constraint 2 (L2):} \quad 6x + 2y \geq 8 \tag{11.14}$$

$$\text{Constraint 3 (L3):} \quad x + 5y \geq 8 \tag{11.15}$$

The bounds are graphed in Figure 11.2. The MATLAB code that produces Figure 11.2 follows.

```
% Example_11_6.m
% To draw straight lines, specify two points on the line
% upper bound on y
x(1) = 0; y(1) = 3; x(2) = 3; y(2) = 3;
% upper bound on x
x1(1) = 3; y1(1) = 0; x1(2) = 3; y1(2) = 3;
% lower bound on y
x2(1) = 0; y2(1) = 0; x2(2) = 3; y2(2) = 0;
% lower bound on x
x3(1) = 0; y3(1) = 0; x3(2) = 0; y3(2) = 3;
% line L1
x4(1) = 0; y4(1) = 4; x4(2) = 3; y4(2) = 1;
% line L2
x5(1) = 0; y5(1) = 4; x5(2) = 3; y5(2) = -5;
% line L3
x6(1) = 0; y6(1) = 8/5; x6(2) = 3; y6(2) = 1;
% plot the allowable region
plot(x,y,x1,y1,x2,y2,x3,y3,x4,y4,x5,y5,x6,y6)
xlabel('x'), ylabel('y'), title('y vs x');
```

To obtain the final figure shown in Figure 11.2, the insert option was used to add text labels to the figure produced by the script. In the script, the specified inequality constraint was made into an equality constraint. This enabled us to draw lines for LB, UB, L1, L2, and L3 and to manually add an arrow to indicate the region corresponding to the inequality. This specifies the allowable region in which to consider the maximum and minimum values for function f.

We see in the graph that the allowable region for (x, y) in determining the maximum and minimum values for the object function, f, is the region enclosed by the polygon ABCD. For the specified object function, the minimum value of f is either on line L3 (between points D and C) or on line L2 (between points A and D). We can see this if we place a point P on line L3 and a point Q directly above it, then the x values at points P and Q will be the same, but the y value at point Q will be larger than the y value of point P; thus, f will be greater at point Q than f at point P. Thus, the minimum value of f is either on line L3 or on line L2.

The equation describing line L3 is

$$y = \frac{8}{5} - \frac{x}{5} \qquad \text{(line L3)}$$

Substituting this expression for y into Equation 11.10 gives the value of f on line L3, which is

$$f = \frac{24}{5} + \frac{7x}{5} \qquad \text{(on L3)}$$

The minimum value for f occurs at the lowest allowable x value on L3, which is at point D. Point D results from the intersection of line L3 and line L2. The equation describing line L2 is

$$y = 4 - 3x \qquad \text{(line L2)}$$

The x-coordinate at point D can be obtained by equating the y expressions for line L1 and line L2 giving the x value at point D, or $x_D = 12/14 = 0.8571$. At point D, the value of f is 6.00.

Substituting the expression for y on line L2 into Equation 11.10 gives the equation for f on line L2, which is

$$f = 12 - 7x \qquad \text{(on line L2)}$$

The minimum value for f on line AD occurs at the maximum allowable value for x on AD, which is point D. Thus, in the allowable region, $f_{min} = 6.000$.

From Figure 11.2, we see that the maximum value of f will be either at point B or somewhere else on line L1. The equation describing line L1 is

$$y = 4 - x \qquad \text{(on line L1)}$$

Substituting this expression for y into Equation 11.10 gives

$$f = 12 - x \qquad \text{(on line L1)}$$

The maximum f occurs where x is a minimum on line L1, which is at point B. At point B, $x = 1$. Thus, $f_{max} = 11.0$.

11.6 Lagrange Multipliers

A more general and mathematical discussion of the optimization problem with constraints follows. Suppose we are given the object function $f(x_1, x_2, x_3, \ldots, x_n)$ in which the variables $x_1, x_2, x_3, \ldots, x_n$ are subject to N constraints, say,

$$\Phi_1(x_1, x_2, x_3, \ldots, x_n) = 0$$

$$\Phi_2(x_1, x_2, x_3, \ldots, x_n) = 0 \qquad (11.16)$$

$$\vdots$$

$$\Phi_N(x_1, x_2, x_3, \ldots, x_n) = 0$$

Theoretically, the N x's can be solved in terms of the remaining x's. Then these N variables can be eliminated from the objective function f by substitution, and the extrema problem can be solved as if there were no constraints. This method is referred to as the implicit method and in most cases is impractical.

The method of *Lagrange multipliers* provides the means for solving an extrema problem with constraints analytically. Suppose $f(x_1, x_2, x_3, \ldots, x_n)$ is to be maximized subject to constraints $\Phi_1(x_1, x_2, \ldots, x_n) = 0, \Phi_2(x_1, x_2, \ldots, x_n) = 0, \ldots, \Phi_n(x_1, x_2, \ldots, x_n) = 0$ as in Equation 11.16. We define the *Lagrange* function F as

$$F(x_1, x_2, x_3, \ldots, x_n) = f(x_1, x_2, x_3, \ldots, x_n) + \lambda_1 \Phi_1(x_1, x_2, x_3, \ldots, x_n)$$

$$+ \lambda_2 \Phi_2(x_1, x_2, x_3, \ldots, x_n) + \cdots + \lambda_N \Phi_N(x_1, x_2, x_3, \ldots, x_n)$$

where λ_i are the unknown Lagrange multipliers to be determined. We now set

$$\frac{\partial F}{\partial x_1} = 0, \ \frac{\partial F}{\partial x_2} = 0, \ \frac{dF}{dx_3} = 0, \ \cdots, \frac{\partial F}{\partial x_n} = 0$$

$$\Phi_1 = 0, \ \Phi_2 = 0, \ \Phi_3 = 0, \ \cdots, \Phi_N = 0 \qquad (11.17)$$

(Note: $\dfrac{\partial F}{\partial \lambda_j} = 0.$ implies $\Phi_j = 0.$)

This set of $n + N$ equations gives all possible extrema of f [1].

Example 11.7

A silo is to consist of a right circular cylinder of radius R and length L, with a hemispherical roof (see Figure 11.3). Assume that the silo is to have a specified volume $V = 8400$ m³. Find the dimensions, R and L, which makes its surface area S a minimum. Assume that the silo has a floor of the same material. Note: $V_{sphere} = (4/3)\pi R^3$ and $S_{sphere} = 4\pi R^2$.

Solution

$$V = \frac{2}{3}\pi R^3 + \pi R^2 L \qquad (11.18)$$

$$\Phi = \frac{2}{3}\pi R^3 + \pi R^2 L - V \qquad (11.19)$$

$$S = 2\pi RL + \pi R^2 + 2\pi R^2$$

$$= 2\pi RL + 3\pi R^2 \qquad (11.20)$$

In this case, the surface area S is the function to be minimized and the volume V is the constraint. The Lagrange function is

$$F = S + \lambda\Phi$$

$$= 2\pi RL + 3\pi R^2 + \lambda\left(\frac{2}{3}\pi R^3 + \pi R^2 L - V\right) \qquad (11.21)$$

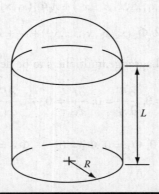

Figure 11.3   Silo consisting of a right circular cylinder topped by a hemisphere.

The variables are R, L, and λ. Taking partial derivatives with respect to R and L gives

$$\frac{\partial F}{\partial R} = 2\pi L + 6\pi R + \lambda(2\pi R^2 + 2\pi RL) = 0 \qquad (11.22)$$

$$\frac{\partial F}{\partial L} = 2\pi R + \lambda\pi R^2 = 0 \;\rightarrow\; R = -2 \quad \text{or} \quad \lambda = -\frac{2}{R} \qquad (11.23)$$

Substituting the value of λ from Equations 11.23 into 11.22 gives

$$2\pi L + 6\pi R - \frac{2}{R}(2\pi R^2 + 2\pi RL) = 0 \qquad (11.24)$$

The above equation reduces to $R - L = 0$ or $R = L$.
   Substituting this result into Equation 11.18 gives

$$V = \frac{2}{3}\pi R^3 + \pi R^3$$

$$= \frac{5}{3}\pi R^3$$

For $V = 8400 \text{ m}^3$,

$$R = \left(\frac{8400 \times 3}{5\pi}\right)^{1/3} = 11.7065 \text{ m}$$

Substituting the values for R and L into Equation 11.20 gives $S = 2152.6 \text{ m}^2$.

11.7 MATLAB®'s `fmincon` Function

MATLAB's function for solving optimization problems with constraints is `fmincon`, which is in the optimization Toolbox. `fmincon` takes as arguments a user-defined function `FUN`, an initial guess, `X0`, plus additional arguments that depend on the type of constraints defined in the problem.

The function FUN defines the object function to be minimized. The vector X0 is an initial guess of the control variables that minimizes the object function. MATLAB's fmincon handles four types of constraints:

1. Inequality constraints: constraints that involve one or more inequalities of the form $\mathbf{AX} \leq \mathbf{B}$. Suppose the problem specifies three linear inequality constraints of the form

$$a_{11}x_1 + a_{12}x_2 + \cdots + a_{1n}x_n \leq b_1$$

$$a_{21}x_1 + a_{22}x_2 + \cdots + a_{2n}x_n \leq b_2 \qquad (11.25)$$

$$a_{31}x_1 + a_{32}x_2 + \cdots + a_{3n}x_n \leq b_3$$

These constraints would be specified in MATLAB for use with fmincon as

$$\mathrm{A} = [a_{11}\, a_{12} \ldots a_{1n};\quad a_{21}\, a_{22} \ldots a_{2n};\quad a_{31}\, a_{32} \ldots a_{3n}]$$

$$\mathrm{B} = [b_1;\, b_2;\, b_3]$$

2. Equality constraints: the constraints involve one or more equalities of the form $\mathbf{Aeq\, X} = \mathbf{Beq}$. Suppose the problem specifies two linear equality constraints of the form

$$\tilde{a}_{11}x_1 + \tilde{a}_{12}x_2 + \cdots + \tilde{a}_{1n}x_n = \tilde{b}_1$$

$$\tilde{a}_{21}x_1 + \tilde{a}_{22}x_2 + \cdots + \tilde{a}_{2n}x_n = \tilde{b}_2 \qquad (11.26)$$

where $\tilde{a}_{ij}$ and $\tilde{b}_i$ are the elements of **Aeq** and **Beq**, respectively. These constraints would be specified in MATLAB for use with fmincon as

$$\mathrm{Aeq} = [\tilde{a}_{11}\, \tilde{a}_{12} \cdots \tilde{a}_{1n};\quad \tilde{a}_{21}\, \tilde{a}_{22} \cdots \tilde{a}_{2n}]$$

$$\mathrm{Beq} = [\tilde{b}_1;\, \tilde{b}_2]$$

3. Bounds: the constraints involve lower or upper limits of the form $\mathbf{X} \geq \mathbf{LB}$ and $\mathbf{X} \leq \mathbf{UB}$. Then you would specify the bounds in MATLAB for use with fmincon as

$$\mathrm{LB} = [l_1\, l_2 \cdots l_n]$$

$$\mathrm{UB} = [u_1\, u_2 \cdots u_n]$$

where $x_1 \geq l_1, x_2 \geq l_2, \ldots, x_n \geq l_n$ and $x_1 \leq u_1, x_2 \leq u_2, \ldots, x_n \leq u_n$.

4. Nonlinear constraints: the constraints are defined by the function statement. [C,Ceq] = NONLCON(X), where NONLCON is a user-defined MATLAB function that specifies the nonlinear inequality constraints C and the

nonlinear equality constraints Ceq. Suppose the problem specifies two non-linear inequality constraints and one nonlinear equality constraint. The nonlinear inequality and equality constraint functions, f_1, f_2, and f_3, need to be set up such that

$$f_1(x_1, x_2,\ldots,x_n) \leq 0, \quad f_2(x_1,x_2,\ldots,x_n) \leq 0, \quad \text{and} \quad f_3(x_1,x_2,\ldots,x_n) = 0$$

Then, in the function NONLCON, set C(1) = $f_1(x_1, x_2, \ldots, x_n)$, C(2) = $f_2(x_1, x_2, \ldots, x_n)$, and Ceq(1) = $f_3(x_1, x_2, \ldots, x_n)$.

A description of the many invocations of fmincon can be obtained by typing **help fmincon** in the Command Window, some of which are described as follows.

Usage 1

```
X = FMINCON(FUN,X0,A,B) starts at X0 and finds a minimum X to the
function FUN, subject to the linear inequalities A*X <= B.
FUN accepts input X and returns a scalar function value F
evaluated at X. X0 may be a scalar, vector, or matrix.
```

This first version is used only if all the constraints are linear inequalities; that is, A*X <= B. In this case, the script starts at X0 and finds control variables X that minimizes the object function contained in FUN and returns the X values to the calling program. The initial guess X0 is a column vector, and A,B describe a system of equations defining the inequality constraints.

Usage 2

```
X = FMINCON(FUN,X0,A,B,Aeq,Beq) minimizes FUN subject to th
linear equalities Aeq*X = Beq as well as A*X <= B. (Set A =
[] and B = [] if no inequalities exist.)
```

This version is used if there are also linear equality constraints, that is, Aeq*X = Beq, where Aeq and Beq describe a system of equations describing the equality constraints. If there are no inequality constraints, then use [] for A and B.

Usage 3

```
X = FMINCON(FUN,X0,A,B,Aeq,Beq,LB,UB) defines a set of lower
and upper bounds on the design variables, X, so that a
solution is found in the range LB <= X <= UB. Use empty
matrices for LB and UB if no bounds exist. Set LB(i) = -Inf if
X(i) is unbounded below; set UB(i) = Inf if X(i) is unbounded
above.
```

This version is used if you wish to set lower and upper limits LB and UB to the control variables, where LB and UB are vectors containing the bounds on X.

Usage 4

`X = FMINCON(FUN,X0,A,B,Aeq,Beq,LB,UB,NONLCON)` subjects the minimization to the constraints defined in function NONLCON. The function NONLCON accepts X and returns the vectors C and Ceq, representing the nonlinear inequalities and equalities respectively. FMINCON minimizes FUN such that C(X) < = 0 and Ceq(X) = 0. (Set LB = [] and/or UB = [] if no bounds exist, do the same for A,B,Aeq,Beq if there are no linear inequality or linear equality constraints.)

This version is used if there are nonlinear constraints as defined in the function NONLCON as demonstrated in Example 11.8.

Note that in all usages earlier, we have invoked fmincon as returning a single vector, that is, the form X = fmincon(FUN,X0,...). However, for convenience, we can also use the form [X,FVAL] = fmincon(FUN,X0,...), which also returns FVAL, which is the minimum value of the object function.

Example 11.8

In this example, we use fmincon to determine the maximum and minimum values of the object function of Example 11.6 subject to the constraints specified in that example.

The program follows:

```
% Example_11_8.m
% Linear Programming Example using MATLAB's fmincon function
% for optimization
% Object function: 2*x(1)+3*x(2)
% Lower bounds
%       x(1)>= 0 x(2)>= 0
% Upper bounds
%       x(1)<= 3 x(2)<= 3
% Inequality constraints
%       x(1)+x(2)<= 4
%       6*x(1)+2*x(2)>= 8
%       x(1)+5*x(2)>= 8
clear; clc;
lb = [0; 0];                %Lower bound
ub = [3; 3];                %upper bound
x0 = [0; 0];                %Initial guess
Aeq = [];                   %No linear equality constraints
Beq = [];                   %No linear equality constraints
A = [1 1; -6 -2; -1 -5];    %Linear inequality constraints
B = [4; -8; -8];            %Linear inequality constraints
```

```
object_fmin = @(x) (2*x(1)+3*x(2));
object_fmax = @(x) (-(2*x(1)+3*x(2)));
[xmin,fvalmin] = fmincon(object_fmin,x0,A,B,Aeq,Beq,lb,ub);
[xmax,fvalmax] = fmincon(object_fmax,x0,A,B,Aeq,Beq,lb,ub);
% Print min and max values:
fprintf('MIN \n');
fprintf('-------------------------------------------------\n');
fprintf('xmin(1) =%8.4f xmin(2) =%8.4f \n',xmin(1),xmin(2));
fprintf('fmin =%8.4f \n\n',fvalmin);
fprintf('\nMAX \n');
fprintf('-------------------------------------------------\n');
fprintf('xmax(1) =%8.4f xmax(2) =%8.4f \n',xmax(1),xmax(2));
fprintf('fmax =%8.4f \n',-fvalmax);
```

Program results

MATLAB gives a warning which can be ignored.

```
MIN
-----------------------------------
xmin(1) = 0.8571 xmin(2) = 1.4286
fmin = 6.0000

MAX
-----------------------------------
xmax(1) = 1.0000 xmax(2) = 3.0000
fmax = 11.0000
>>
```

Example 11.9

Solve for the minimum surface area of the silo in Example 11.7 by using MATLAB's fmincon function.

The program follows:

```
% Example_11_9.m
% This program minimizes the material surface area of a silo.
% The silo consists of a right cylinder topped by a hemisphere.
% Solve for a silo volume of 8400 m^3.
% Define the vector X = [R,L] where R is radius and L is length
% (as drawn in Figure 11.3)
% The function objfun_silo calculates the silo surface area and
% is the objective function to be minimized. The function
% confun_silo defines the constraint that the silo volume be
% fixed at 8400.
clear; clc;
global V;
V = 8400;
% Define bounds: radius and length must be positive.
```

```
LB = [0,0];
UB = [];
% Initial guesses for R and L:
Xo = [10.0 20.0];
% Run the optimization. We have no linear constraints, so
% pass [] for those arguments:
objfun_silo = @(X) (2.0*pi*X(1)*X(2)+3.0*pi*X(1)^2);
[X,fval] = fmincon(objfun_silo,Xo,[],[],[],[],LB,UB,···
  @confun_silo);
% Print results:
fprintf('Optimization Problem:\n');
fprintf('This program minimizes the surface area of a silo\n');
fprintf('consisting of a right cylinder topped by a ');
fprintf('hemisphere.\n');
fprintf('The silo volume is set at %.0f m^3 \n',V);
fprintf('Minimum surface area:%.3f m^2\n',fval);
fprintf('Optimum radius:%.3f m\n',X(1));
fprintf('Optimum length:%.3f m\n',X(2));
```

```
% confun_silo.m (constraint function for Example 11.9)
function [c, ceq] = confun_silo(X)
% Variables are: radius R = X(1), length of cylinder L = X(2).
global V;
% Nonlinear equality constraints:
ceq = pi*X(1)^2*X(2) + 2.0/3.0*pi*X(1)^3 - V;
% No nonlinear inequality constraints:
c = [];
```

Program results

```
Optimization Problem:
This program minimizes the surface area of a silo
consisting of a right cylinder topped by a hemisphere.
The silo volume is set at 8400 m^3
Minimum surface area: 2152.651 m^2
Optimum radius:        11.707 m
Optimum length:        11.706 m
```

Example 11.10

Suppose we now reverse Example 11.9 and determine the maximum volume of the silo described in Example 11.9 subjected to the constraint that the surface area of the silo is to be less than a specified amount, say, 2152.6 m².

The program follows:

```
% Example_11_10.m
% This program maximizes the volume of a silo.
```

```
% The silo consists of a right cylinder topped by a
% hemisphere.
% The variables to optimize are radius 'R' and cylinder
% height 'L'.
% Let X(1) = R and X(2) = L.
% Equation for surface area: S = 3*pi*X(1)^2+2*pi*X(1)*X(2);
% Constraint: 3*pi*X(1)^2+2*pi*X(1)*X(2) <= 2152.6 m^2.
% We need to set up the constraint in the form:
% 3*pi*X(1)^2+2*pi*X(1)*X(2)- 2152.6 <= 0
% We wish to maximize the volume under the constraint.
% The equation for the volume, V, of the silo is given by
% V = 2.0/3.0*pi*X(1)^3 + pi*X(1)^2*X(2).
% Since we wish to maximize V, we will need to place a minus
% sign in front of the expression for V and then minimize.
clear; clc;
global SAmax;
SAmax = 2152.6;
% Define bounds: radius and length must be positive.
LB = [0,0];
UB = [];
% Take an initial guess at the solution
Xo = [10.0 20.0];
objfun_silo2 = @(X) (-(2.0/3.0*pi*X(1)^3+pi*X(1)^2*X(2)));
% Run the optimization. We have no linear constraints, so
% pass [] for those arguments:
[X,fval] = fmincon(objfun_silo2,Xo,[],[],[],[],LB,UB,...
                   @confun_silo2);
SA = 3*pi*X(1)^2 + 2*pi*X(1)*X(2);
% Print results:
fprintf('Optimization Problem:\n');
fprintf('This program maximizes the volume of a silo \n');
fprintf('consisting of a right cylinder topped by a ');
fprintf('hemisphere.\n');
fprintf('The maximum surface area is set at %.1f m^3 \n', ...
        SAmax);
fprintf('Maximum surface area: %9.3f m^2\n',SA);
fprintf('Maximum volume:       %9.3f m^3\n',-fval);
fprintf('Optimum radius:       %9.3f m\n',X(1));
fprintf('Optimum length:       %9.3f m\n',X(2));
```

```
% confun_silo2.m (constraint function for Example 11.10)
function [c, ceq] = confun_silo2(X)
global SAmax;
% Variables are: radius R = X(1), length of cylinder L = X(2)
% Nonlinear inequality constraint:
c = 3*pi*X(1)^2 + 2*pi*X(1)*X(2) - SAmax;
% No nonlinear equality constraints:
ceq = [];
```

Program results

```
Optimization Problem:
This program maximizes the volume of a silo
consisting of a right cylinder topped by a hemisphere.
The maximum surface area is set at 2152.6 m^3
Maximum surface area: 2152.600 m^2
Maximum volume:        8399.701 m^3
Optimum radius:          11.706 m
Optimum length:          11.706 m
```

Example 11.11

Two machine shops, machine shop A and machine shop B, are to manufacture two types of motor shafts, shaft S1 and shaft S2. Each machine shop has two turning machines: turning machine T1 and turning machine T2. The following table lists the production time for each shaft type on each machine and at each location:

Time to Manufacture (Minutes)				
	Machine Shop A		Machine Shop B	
	Shaft S1	Shaft S2	Shaft S1	Shaft S2
Turning machine T1	4	9	5	8
Turning machine T2	2	6	3	5

Shaft S1 sells for \$35 and shaft S2 sells for \$85. Determine the number of S1 and S2 shafts that should be produced at each machine shop and on each machine that will maximize the revenue for 1 hour of shop time.

The program follows:

```
% Example_11_11.m
% Shaft Production Problem
% This program maximizes the revenue/hr for the production
% of two types of shafts, type S1 and type S2. There are two
% machine shops producing these shafts, shop A and shop B.
% Each shop has two types of turning machines, T1 and T2,
% capable of producing these shafts.
% Shop A:
% Machine T1 takes 4 minutes to produce type S1 shafts
% and 9 minutes to produce type S2 shafts.
```

```
% Machine T2 takes 2 minutes to produce type S1 shafts
% and 6 minutes to produce type S2 shafts.
% Shop B:
% Machine T1 takes 5 minutes to produce type S1 shafts
% and 8 minutes to produce type S2 shafts.
% Machine T2 takes 3 minutes to produce type S1 shaft
% and 5 minutes to produce type S2 shafts.
% Shaft S1 sells for $35 and shaft S2 sells for $85.
% We wish to determine the number of S1 & S2 shafts that
% should be produced at each shop and by each machine
% that will maximize the revenue/hr.
% Let:
% X(1) = number of S1 shafts produced/hr by machine T1 at
% shop A
% X(2) = number of S2 shafts produced/hr by machine T1 at
% shop A
% X(3) = number of S1 shafts produced/hr by machine T2 at
% shop A
% X(4) = number of S2 shafts produced/hr by machine T2 at
% shop A
% X(5) = number of S1 shafts produced/hr by machine T1 at
% shop B
% X(6) = number of S2 shafts produced/hr by machine T1 at
% shop B
% X(7) = number of S1 shafts produced/hr by machine T2 at
% shop B
% X(8) = number of S2 shafts produced/hr by machine T2 at
% shop B
% Let r = total revenue/hr for producing these shafts,
% then
% r = 35*(X(1)+X(3)+X(5)+X(7))+75*(X(2)+X(4)+X(6)+X(8))
clear; clc;
% Objective function: total revenue per hour for
% manufactured shafts.
% Have the function return a negative number because we are
% maximizing instead of minimizing.
revenue = @(x)  -(35*(x(1)+x(3)+x(5)+x(7))+85*(x(2)+x(4)+ ...
                x(6)+x(8)));
% Take a guess at the solution
Xo = [0 0 0 0 0 0 0 0];
% Lower and upper bounds:
LB = [0 0 0 0 0 0 0 0];
UB = [];
% We have linear inequality constraints. We require that
% each machine make an integral number of shafts per
% 60 minutes. However, we do allow some machines to remain
% out of production. The constraints are:
% 4*X(1)+9*X(2) <= 60
% 2*X(3)+6*X(4) <= 60
```

```
% 5*X(5)+8*X(6) <= 60
% 3*X(7)+5*X(8) <= 60
A = [4 9 0 0 0 0 0 0;
     0 0 2 6 0 0 0 0;
     0 0 0 0 5 8 0 0;
     0 0 0 0 0 0 3 5];
B = [60 60 60 60]';
% We have no linear equality constraints, so pass [] for those
% arguments.
[X, rmax] = fmincon(revenue,Xo,A,B,[],[],LB,UB);
fprintf('Optimization Results:\n');
fprintf('No. of S1 shafts produced at shop A');
fprintf('T1: %2.0f\n',X(1));
fprintf('No. of S2 shafts produced at shop A');
fprintf('T1: %2.0f\n',X(2));
fprintf('No. of S1 shafts produced at shop A');
fprintf('T2: %2.0f\n',X(3));
fprintf('No. of S2 shafts produced at shop A');
fprintf('T2: %2.0f\n',X(4));
fprintf('No. of S1 shafts produced at shop B');
fprintf('T1: %2.0f\n',X(5));
fprintf('No. of S2 shafts produced at shop B');
fprintf('T1: %2.0f\n',X(6));
fprintf('No. of S1 shafts produced at shop B');
fprintf('T2: %2.0f\n',X(7));
fprintf('No. of S2 shafts produced at shop B');
fprintf('T2: %2.0f\n',X(8));
fprintf('\n');
fprintf('The max revenue per hour: $%.0f/hour \n',-rmax);
```

Program results

```
Optimization Results:
No. of S1 shafts produced at shop A, T1:  0
No. of S2 shafts produced at shop A, T1:  7
No. of S1 shafts produced at shop A, T2: 30
No. of S2 shafts produced at shop A, T2:  0
No. of S1 shafts produced at shop B, T1:  0
No. of S2 shafts produced at shop B, T1:  8
No. of S1 shafts produced at shop B, T2:  0
No. of S2 shafts produced at shop B, T2: 12
The max revenue per hour: $3274/hour
```

Note: When the program is run, MATLAB gives diagnostic warnings to the screen that can be ignored. If a satisfactory solution is obtained, MATLAB will inform you by stating Local minimum found that satisfies the constraints.

Exercises

E11.1 Use MATLAB's `fminbnd()` function to find the maxima and minima of the following functions:

a. $f(x) = x^4 + 10x^3 - 20x - 15$

b. $f(x) = |x^2 - 20x + 15|$

c. $f(x) = e^{-0.5x} \sin 2x$

d. Use the method of steepest descent to obtain the approximate position that makes f a relative minimum where

$$f(x_1, x_2) = 8x_1^2 - 20x_1x_2 + 17x_2^2 - 32x_1 + 40x_2$$

Use $(x_1, x_2) = (6, 4)$ as the starting point, $\Delta s = 0.05$, and 200 steps.

e. Repeat part d, but this time use MATLAB's `fminsearch` function to find the relative minimum of $f(x_1, x_2)$. Print to the screen the position and value of the minimum of $f(x_1, x_2)$

E11.2 Use Lagrange multipliers to find the volume of the largest box that can be placed inside the ellipsoid

$$\frac{x^2}{a^2} + \frac{y^2}{b^2} + \frac{z^2}{c^2} = 1$$

so that the edges will be parallel to the coordinate axis.

Projects

P11.1 A silo consists of a right circular cylinder topped by a right circular cone as shown in Figure P11.1. The radius of the cylinder and the base of the cone is R. The length of the cylinder is L and the height of the cone is H.

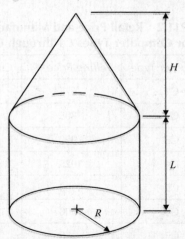

Figure P11.1 **Silo consisting of a right circular cylinder topped by a right circular cone.**

The cylinder, the cone, and the silo floor are all made of the same material. Write a program using MATLAB's fmincon function to determine the values of R, H, and L that will result in the minimum surface silo area for an internal silo volume of 7000 m³.

Note: For a right circular cone

$$V = \frac{\pi R^2 H}{3}$$

$$S = \pi R \sqrt{R^2 + H^2}$$

P11.2 A retail store sells computers to the public. There are eight different computer types that the store may carry. Table P11.2 lists the type of computer, the selling price, and the cost to the store. The store plans to spend $20,000 per month purchasing the computers.

The store plans to spend no more than 30% of its costs on computer types C1 and C2, no more than 30% on computer types C3 and C4, no more than 10% on computer types C5 and C6, and no more than 30% on computer types C7 and C8. The store estimates that it can sell 30% more of type C1 than C2, 20% more of type C3 than type C4, 20% more of type C5 than C6, and 60% more of type C7 than C8. Use the fmincon function in MATLAB to determine the number of each type of computer that will provide the store with the most profit. Print out the number of each type of computer the store should purchase per month, the total profit per month, and the total cost per month to the store.

Table P11.2   Retail Price and Manufacturing Cost for Computer Types C1 through C8

Computer Type	Selling Price ($)	Cost ($)
C1	675	637
C2	805	780
C3	900	874
C4	1025	990
C5	1300	1250
C6	1500	1435
C7	350	340
C8	1000	1030

P11.3 The Jones Electronics Corp. has a contract to manufacture four different computer circuit boards. The manufacturing process requires each of the boards to pass through four departments before shipping. These are: etching and lamination (etches circuits into board), drilling (drills holes to secure components), assembly (installs transistors, microprocessors, etc.), and testing. The time requirement in minutes for each unit produced and its corresponding profit value are summarized in Table P11.3a.

Each department is limited to 3 days/week to work on this contract. The minimum weekly production requirement to fulfill the contract is shown in Table P11.3b.

Write a MATLAB program that will

1. Determine the number of each type of circuit board for the coming week that will provide the maximum profit. Assume that there are 8 hours/day and 5 days/week available for factory operations. Note: Not all departments work on the same day.
2. Determine the total profit for the week.
3. Determine the total number of minutes it takes to produce all the boards.
4. Determine the total number of minutes spent in each of the four departments.
5. Print out to a file the requested information.

Table P11.3a   Manufacturing Time for Various Circuit Board Types in Each Department

Circuit Board	Etching and Lamination (min)	Drilling (min)	Assembly (min)	Testing (min)	Unit Profit ($)
Board A	15	10	8	15	12
Board B	12	8	10	12	10
Board C	18	12	12	17	15
Board D	13	9	4	13	10

Table P11.3b   Minimum Weekly Production Requirement for Various Circuit Board Types

Circuit Board	Minimum Production Count
Board A	10
Board B	10
Board C	10
Board D	10

P11.4 The XYZ oil company operates three oil wells (OW1, OW2, OW3) and supplies crude oil to four refineries (refinery A, refinery B, refinery C, refinery D). The cost of shipping the crude oil from each oil well to each of the refineries, the capacity of each of the three oil wells, and the demand (equality constraint) for gasoline at each refinery are tabulated in Table P11.4a. The crude oil at each refinery is distilled into six basic products: gasoline, lubricating oil, kerosene, jet fuel, heating oil, and plastics. The cost of distillation per 100 liter at each refinery from each of the oil wells is given in Table P11.4b. The percentage of each distilled product per liter is tabulated in Table P11.4c. The profit from each product is tabulated in Table P11.4d.

Table P11.4a   Cost of Shipping per 100 L ($)

Oil Well	Refinery A	Refinery B	Refinery C	Refinery D	Oil Well Capacity (L)
OW 1	9	7	10	11	7000
OW 2	7	10	8	10	6100
OW 3	10	11	6	7	6500
Demand (liters of gasoline)	2000	1800	2100	1900	

Table P11.4b   Cost of Distillation per 100 L ($)

Oil Well	Refinery A	Refinery B	Refinery C	Refinery D
OW 1	15	16	12	14
OW 2	17	12	14	10
OW 3	12	15	16	17

Table P11.4c   Distillation Products per Liter of Crude Oil

	Product Percentage per Liter from Distillation					
Oil Well	Gasoline	Lubricating Oil	Kerosene	Jet Fuel	Heating Oil	Plastics
OW 1	43	10	9	15	13	10
OW 2	38	12	5	14	16	15
OW 3	46	8	8	12	12	14

Table P11.4d   Product Revenue per Liter ($)

Gasoline	Lubricating Oil	Kerosene	Jet Fuel	Heating Oil	Plastics
0.40	0.20	0.20	0.50	0.25	0.15

Using the fmincon function in MATLAB, determine the liters of oil to be produced at each oil well and shipped to each of the four refineries that will satisfy the gasoline demand and that will produce the maximum profit. Print out the following items:
1. The liters produced at each oil well
2. The liters of gasoline received at each refinery
3. The total cost of shipping and distillation of all products
4. The total revenue from the sale of all of the products
5. The total profit from all of the products

Reference

1. Wylie, C.R., *Advanced Engineering Mathematics*, McGraw Hill, New York, 1955, p. 596.

Chapter 12

Iteration Method

12.1 Introduction

Some engineering problems are best solved by an iteration procedure. For example, an iteration method is often used in fluid mechanics to solve pipe flow problems. The Hardy–Cross iteration method may also be used for determining the flow rates and head losses throughout a pipe network, if the pipe sizes, lengths, and pipe roughness factors are known. In heat transfer, problems involving Laplace's equation can also be solved by the Gauss–Seidel iterative method. These methods are discussed in this chapter.

12.2 Iteration in Pipe Flow Analysis

Example 12.1

Consider the piping system shown in Figure 12.1.

The energy equation for the system is [1]

$$\left(\frac{p}{\gamma} + \frac{V^2}{2g} + z\right)_1 - \sum h_L = \left(\frac{p}{\gamma} + \frac{V^2}{2g} + z\right)_2 \qquad (12.1)$$

where
p is the pressure (in N/m²)
V is the average fluid velocity in the pipe (in m/s)
γ is the specific weight of the fluid (in N/m³)

389

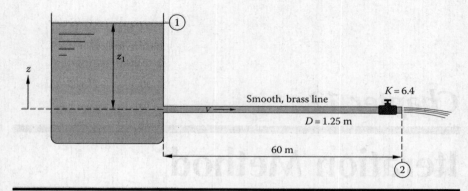

Figure 12.1 Piping system (no pump).

z is the elevation in (m)

g is the gravitational constant = 9.81 in (m/s²)

$\sum h_L$ is the sum of head losses in (m)

The subscripts indicate the conditions at points ① and ② as shown in Figure 12.1

For this system,

$p_1 = p_2 = p_{atm}$ (any surface open to the atmosphere is considered to be at atmospheric pressure)

$V_1 \approx 0$, $V_2 = V$, $z_1 - z_2$ is specified.

The sum of head losses consists of a head loss, h_f, in the pipe due to viscous or turbulent effects and minor head losses due to valves, elbows, and pipe entrance and exit losses.

The head loss in the pipe is given by

$$h_f = \frac{V^2}{2g} \frac{L}{D} f \tag{12.2}$$

where

L is the pipe length (known)

D is the pipe diameter (known)

f is the friction factor

The determination of the friction factor, f, for smooth pipes (an equation for f for rough pipes is given later in this chapter) can be approximated by the following formula [1]:

$$f = (1.82 \log_{10} \text{Re} - 1.64)^{-2} \tag{12.3}$$

where Re is the Reynolds number, which is given by

$$\text{Re} = \frac{VD}{\upsilon} = \frac{4Q}{\pi D \upsilon} \qquad (12.4)$$

where
  Q is the volume flow rate
  υ is the kinematic viscosity

The expression for V that was used in Equation 12.4 is

$$V = \frac{Q}{A} = \frac{4Q}{\pi D^2} \qquad (12.5)$$

where A is the cross-sectional area of a circular pipe. The minor head losses are expressed by the equation

$$h_{L,i} = K_i \frac{V^2}{2g} \qquad (12.6)$$

The K values for the minor head losses are $K_{entrance} = 0.5$ and $K_{valve} = 6.4$.
  Substituting these relationships into Equation 12.1 and rearranging gives

$$z_2 - z_1 + \frac{V^2}{2g}\left(\frac{L}{D}f + K_{entrance} + K_{valve} + 1\right) = 0 \qquad (12.7)$$

Substituting Equation 12.5 into Equation 12.7 and solving for Q^2 give

$$Q^2 = \frac{g\pi^2 D^4 (z_1 - z_2)}{8\left(\dfrac{L}{D}f + K_{entrance} + K_{valve} + 1\right)} \qquad (12.8)$$

If the unknown in the problem is either the flow rate, Q, or the pipe diameter, D, then an iteration scheme may be used to solve the problem. This is due to the fact that V is related to Q and f is related to the Re, which is related to Q. So, Equation 12.8 is an implicit function of Q. The following iteration scheme may be used to solve for Q.

1. Assume a value for f, say, $f_1 = 0.03$. (Experiment indicates that f ranges from 0.008 to 0.08.)
2. Solve Equation 12.8 for Q.
3. Solve Equation 12.4 for Re.
4. Solve Equation 12.3 for f, and call it f_2.
5. If $|f_2 - f_1| < \varepsilon$, then Q is the correct value, otherwise, set $f_1 = f_2$ and repeat process until condition of item 5 is satisfied.

We will now develop a MATLAB® program to determine the flow rate, Q, and the friction factor, f, for the pipe system shown in Figure 12.1 by the iteration method described above.

We will use the following values for the problem.

$$L = 60 \text{ m}, D = 1.25 \text{ cm and } (z_1 - z_2) = 15 \text{ m}, v = 1.141 \times 10^{-6} \text{ m}^2/\text{s},$$
$$g = 9.81 \text{ m/s}^2, \varepsilon = 10^{-5}$$

The program follows:

```
% Example_12_1.m
% This program determines the flow rate, Q, in a pipe flow
% system.
% An iteration scheme is used to solve the problem.
clear; clc;
L = 60.0; D = 0.0125; z1 = 15.0; z2 = 0;
% units are in meters.
g = 9.81; % units are m/s^2
nu = 1.141e-06; % units are m^2/s
Kent = 0.5; Kvalve = 6.4;
eps = 1.0e-5;
f1 = 0.03;
fprintf(' f1          Q(m^3/s)              Re          f2 \n');
fprintf('-----------------------------------------------------\n');
for i = 1:10
Kvalve+1
    Qsq = g/8*pi^2*D^4*(z1-z2)/(L/D*f1+Kent+Kvalve+1);
    Q = sqrt(Qsq);
    Re = 4*Q/(pi*D*nu);
    f2 = 1/(1.82*log10(Re)-1.64)^2;
    fprintf('%6.4f    %15.4e    %15e    %10.4f \n',f1,Q,Re,f2);
    if abs(f2-f1)< eps
        break;
    else
        f1 = f2;
    end
end
fprintf('Final answer \n');
fprintf('%6.4f    %15.4e    %15.4e    %10.4f \n',f1,Q,Re,f2);
```

Program results

f1	Q(m^3/s)	Re	f2
0.0300	1.7082e-04	1.5249e+04	0.0280
0.0280	1.7641e-04	1.5748e+04	0.0278
0.0278	1.7712e-04	1.5812e+04	0.0278
0.0278	1.7721e-04	1.5820e+04	0.0278
Final answer			
0.0278	1.7721e-04	1.5820e+04	0.0278

`>>`

Example 12.2

Consider the piping system shown in Figure 12.2.

The energy equation must now include the head developed by the pump, $(h_p)_{sys}$, as shown in Equation 12.9. The energy equation for this system is [1]

$$\left(\frac{p}{\gamma}+\frac{V^2}{2g}+z\right)_1 + (h_p)_{sys} - \sum h_L = \left(\frac{p}{\gamma}+\frac{V^2}{2g}+z\right)_2 \tag{12.9}$$

To determine the flow rate, Q, we need to know the pump characteristics, which are usually given as an (h_{pc}) vs. Q curve. A valid solution occurs when

$$(h_p)_{sys} = (h_{pc}) \tag{12.10}$$

Suppose

$$(h_{pc}) = 120 - 500Q^2 \tag{12.11}$$

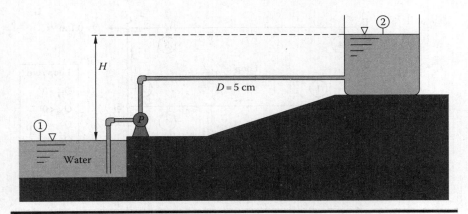

Figure 12.2   Piping system with pump.

For this configuration, $p_1 = p_2 = p_{atm}$, $V_1 \approx 0$, $V_2 \approx 0$. Substituting Equations 12.2, 12.5, 12.6, 12.10, and 12.11 into Equation 12.9 gives

$$Q^2 \left[\frac{8}{\pi^2 D^4 g} \left(\frac{L}{D} f + \sum_i K_i \right) + 500 \right] = 120 - (z_2 - z_1) \tag{12.12}$$

In the above equation, we have assumed that all pipe sections have the same diameter and that L is the total length of all pipe sections. The K_i are the minor loss coefficients and $z_2 - z_1 = H$.

Using Equations 12.3 and 12.4 and the iterative procedure described in Example 12.1, the flow rate can be determined. See Project P12.2.

12.3 Hardy–Cross Method

The Hardy–Cross method, which is an iterative method, provides the means for determining the flow rates and head losses throughout a pipe network, assuming that the pipe sizes, lengths, and pipe roughnesses are known. The description of the method is taken from Refs. [1] and [3].

The following two definitions are used in describing the method:

1. A *loop* is a series of pipes forming a closed path (see Figure 12.3). In Figure 12.3, the numbers inside the circles identify the line number within the loop. In the Q_{ij} expression, the i represents the loop number and the j represents the line number within that loop. A sign convention is used in

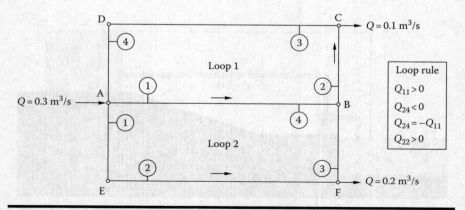

Figure 12.3   Loop rule for pipe network.

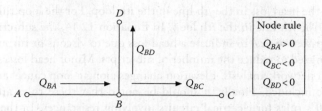

Figure 12.4   Node rule for pipe network.

describing the loop rules. The flow rate, Q, and the head loss, h_f, are considered positive if the flow is in the counterclockwise direction around the loop. It should be noted that two loops with a common pipe may have a positive Q in one loop and a negative Q in the other loop.

2. A *node* is a point where two or more lines are joined. A sign convention is also used for node rules (see Figure 12.4). A flow is considered as positive if the flow direction is toward the node, and negative if away from the node.

It should be noted that Q may be positive when the loop rule is applied and negative when the node rule is applied.

The Hardy–Cross method is based on two concepts:

1. The law of mass conservation
2. The fact that the total head at a node is single valued

Concept (1) leads to the node rule, which is applied at each node in the network. The node rule is

$$\sum_{\beta} Q_{\alpha\beta} = 0 \qquad (12.13)$$

where
   α indicates the node under consideration
   β indicates the node from which the flow is coming from

The sign convention gives the direction of flow. It should be noted that

$$Q_{\beta\alpha} = -Q_{\alpha\beta}$$

Concept (2) leads to the loop rule, which may be stated as follows: For loop i,

$$\sum^{i} h_{ij} = 0 \qquad (12.14)$$

where h_{ij} is the head loss in the jth line in the ith loop. For the loop rule, Q_{ij} is the flow rate in the jth line in the ith loop. In Equation 12.14, the subscript, f, which is usually written with h to indicate a head loss due to viscous or turbulent effects, has been omitted to reduce the number of subscripts. Minor head losses are usually neglected in network analysis. Elevation changes along a loop cancel and therefore need not be included. Finally, it should be noted that these rules are analogous to Kirchhoff's rules for electrical circuits involving resistances. In the analogy, Q corresponds to electrical current and pressure drop corresponds to voltage drop. A description of the method follows:

1. Subdivide the network into a number of loops, making sure that all lines are included in at least one loop.
2. Determine the zeroth estimate for the flow rates, $Q_{\alpha\beta}^{(0)}$, for each line according to the following procedure. Let s equal the total number of nodes in the network and r the total number of lines. Invariably, r will be greater than s. Writing the law of mass conservation at each node gives s equations in r unknowns. Therefore, one needs to assume $(r - s + 1)$ $Q_{\alpha\beta}^{(0)}$ values, which are consistent with the mass conservation rule. The remaining $Q_{\alpha\beta}^{(0)}$ are to be determined by applying the law of mass conservation at each node. This should give a set of linear equations in s unknowns that can readily be solved for the remaining $Q_{\alpha\beta}^{(0)}$ unknowns.
3. This initial guess will not satisfy Equation 12.14; as a result, one needs to apply a correction to each $Q_{\alpha\beta}^{(0)}$ value. This is done by applying a Taylor series expansion (using only two terms in the expansion) to the $h(Q)$ equation:

$$h(Q + \Delta Q) = h(Q) + \left(\frac{dh}{dQ}\right)_Q \Delta Q \tag{12.15}$$

Taking $h(Q + \Delta Q) = h^{(1)}$ and $h(Q) = h^{(0)}$, Equation 12.15 becomes

$$h^{(1)} = h^{(0)} + \left(\frac{dh}{dQ}\right)_{Q^{(0)}} \Delta Q \tag{12.16}$$

Applying Equation 12.16 to Equation 12.14 gives

$$\sum_j h_{ij}^{(1)} = \sum_j \left[h_{ij}^{(0)} + \left(\frac{dh}{dQ}\right)_{ij} \Delta Q_i \right] = 0 \tag{12.17}$$

For each loop, the ΔQ_i can be factored out, thus giving a correction factor equation for each loop. That is,

$$\Delta Q_i = -\frac{\sum_j h_{ij}^{(0)}}{\sum_j \left(\dfrac{dh}{dQ}\right)_{ij}} \tag{12.18}$$

where $\left(\dfrac{dh}{dQ}\right)_{ij}$ is evaluated at $Q_{ij}^{(0)}$.

The Darcy–Weisbach equation relates h to the friction factor f, which is

$$h = \frac{V^2}{2g}\frac{L}{D}f = \frac{8LQ^2}{\pi^2 gD^5}f = KQ^2 f \tag{12.19}$$

The Swamee–Jain formula [3] gives an explicit formula for f, which is

$$f = \frac{1.325}{\left[\ln\left(\dfrac{\varepsilon}{3.7D} + \dfrac{5.74}{\mathrm{Re}^{0.9}}\right)\right]^2} \tag{12.20}$$

where

Re is the Reynolds number $= \dfrac{4Q}{\pi D \upsilon}$

υ is the kinematic viscosity (m²/s)

D is the pipe diameter (m)

ε is the pipe roughness (m)

Equation 12.20 is valid for $10^{-6} \le \dfrac{\varepsilon}{D} \le 10^{-2}$ and $5 \times 10^3 \le \mathrm{Re} \le 10^8$.

In applying the loop rule, some of the lines will experience a head gain and not a head loss. This occurs when the flow direction is opposite to the positive loop direction. To account for this, take

$$h = \begin{cases} KQ^2 f & \text{if } Q \ge 0 \\[2mm] -KQ^2 f & \text{if } Q < 0 \end{cases} \tag{12.21}$$

and

$$\frac{dh}{dQ} = \pm\left(2KfQ + KQ^2\frac{df}{dQ}\right) \tag{12.22}$$

where the (+) sign is used if $Q > 0$ and the (–) sign is used if $Q < 0$.

The formula for df/dQ is

$$\frac{df}{dQ} = \frac{13.69\left(\dfrac{\varepsilon}{3.7D} + \dfrac{5.74}{|\mathrm{Re}|^{0.9}}\right)^{-1}}{|\mathrm{Re}|^{0.9}Q\left[\ln\dfrac{\varepsilon}{3.7D} + \dfrac{5.74}{|\mathrm{Re}|^{0.9}}\right]^3} \tag{12.23}$$

Lines that are in common in two loops need to be treated as follows. If line j in loop i is in common with line m in loop k, then $Q_{km} = -Q_{ij}$ and for each iteration take

$$Q_{ij}^{(n+1)} = Q_{ij}^{(n)} + \Delta Q_i - \Delta Q_k$$

For example, referring to Figure 12.3, for the first iteration,

$$Q_{11}^{(1)} = Q_{11}^{(0)} + \Delta Q_1 - \Delta Q_2 \tag{12.24}$$

For lines that are not in common with a line in another loop, take

$$Q_{ij}^{(n+1)} = Q_{ij}^{(n)} + \Delta Q_i \tag{12.25}$$

We can assign an identification number, *ID*, for each line in the network to determine if that line is to have one ΔQ corrections or two ΔQ corrections. Use $ID(i,j) = 0$, if the line is not a common line. Here i represents the loop number and j represents the line number in that loop. Set $ID(i,j) = $ the loop number that the line is in common with, where j is the line number in the ith loop.

The formulation for the Hardy–Cross method is now complete. The following example illustrates the method, and Project P12.3 involves the Hardy–Cross method in determining the flow rate distribution in a three-loop network.

Example 12.3

Let us consider the network displayed in Figure 12.3. The following table lists the pipe lengths and diameters.

Loop Number	Pipe Number	Pipe Length (m)	Pipe Diameter (cm)
1	1	4000	40
	2	1000	40
	3	4000	30
	4	1000	30
2	1	1000	30
	2	4000	40
	3	1000	30
	4	4000	40

The first step is to write the governing equations at each node based on the node rule (see Figures 12.5a through 12.5f). The number of lines, $r = 7$, the number of nodes, $s = 6$, thus, $(r - s + 1) = 2$. This means that we need to assume two Q values. Suppose we assume $Q_{AB} = -0.15$ m³/s and $Q_{FB} = 0.1$ m³/s.

Initial guess for Q distribution (all unknown Q values are assumed positive):

Node A

$$Q_{AD} + Q_{AE} + 0.3 - 0.15 = 0$$

Node B

$$Q_{BC} + Q_{BF} + 0.15 = 0$$

Node C

$$Q_{CB} + Q_{CD} - 0.1 = 0$$

Node D

$$Q_{DC} + Q_{DA} = 0$$

Node E

$$Q_{EA} + Q_{EF} = 0$$

Node F

$$Q_{FB} + Q_{FE} - 0.2 = 0.1 + Q_{FE} - 0.2 = 0$$

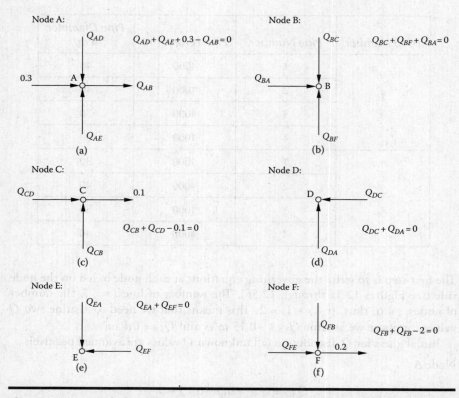

Figure 12.5 (a) Node A Qs, (b) Node B Qs, (c) Node C Qs, (d) Node D Qs, (e) Node E Qs, and (f) Node F Qs.

From Node F, we see that $Q_{FE} = 0.1$ and $Q_{EF} = -0.1$. From Node E, $Q_{EA} = 0.1$ and $Q_{AE} = -0.1$.

From Node A, $Q_{AD} = -0.05$ and $Q_{DA} = 0.05$. From Node D, $Q_{DC} = -0.05$ and $Q_{CD} = 0.05$.

From Node C, $Q_{CB} = 0.05$ and $Q_{BC} = -0.05$. We now have all the $Q_{\alpha\beta}$ terms to establish our initial guess based on the loop rules. These are as follows:

$$Q_{11}^{(0)} = 0.15, \quad Q_{12}^{(0)} = 0.05, \quad Q_{13}^{(0)} = -0.05, \quad Q_{14}^{(0)} = -0.05$$

$$Q_{21}^{(0)} = 0.1, \quad Q_{22}^{(0)} = 0.1, \quad Q_{23}^{(0)} = -0.1, \quad Q_{24}^{(0)} = -0.15$$

See Figure 12.6.

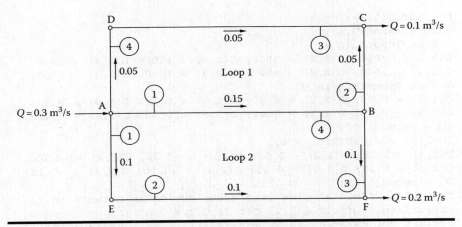

Figure 12.6 Initial network Q distribution based on loop rules.

The program follows:

```
% Example_12_3.m
% Hardy_Cross.m
% This program uses the Hardy-Cross method to solve for the
% flow rate distribution in a 2 loop network. The pipe
% lengths and diameters are given in the table below.
% Loop      Line    Length   Diameter   Initial Guess
% Number    Number   (m)      (cm)        (m3/s)
%-----------------------------------------------------------
%            1       4000      40          0.15
%     1      2       1000      40          0.05
%            3       4000      30         -0.05
%            4       1000      30         -0.05
%-----------------------------------------------------------
%            1       1000      30          0.10
%     2      2       4000      40          0.10
%            3       1000      30         -0.10
%            4       4000      40         -0.15
%-----------------------------------------------------------
clear; clc;
fo = fopen('output.txt','w');
fprintf(fo,'Hardy-Cross method for a two loop network. \n');
% viscosity nu at 20 degree C is 1.005e-6 m^2/s
nu = 1.308e-6; g = 9.81;
% roughness of cast iron pipe, e = 0.026 cm;
e = 0.00026;
eps = 1.0e-6;
Q = zeros(2,4);
Q2 = zeros(2,4);
L = zeros(2,4);
D = zeros(2,4);
```

```
h = zeros(2,4);
dhdQ = zeros(2,4);
% Pipe length in m
L(1,1) = 4000; L(1,2) = 1000; L(1,3) = 4000; L(1,4) = 1000;
L(2,1) = 1000; L(2,2) = 4000; L(2,3) = 1000; L(2,4) = 4000;
% Pipe Diameters in m
D(1,1) = 0.40; D(1,2) = 0.30; D(1,3) = 0.40; D(1,4) = 0.30;
D(2,1) = 0.30; D(2,2) = 0.40; D(2,3) = 0.30; D(2,4) = 0.40;
ID = zeros(2,4);
ID(1,1) = 2; ID(2,4) = 1;
Q(1,1) = 0.15; Q(1,2) = 0.05; Q(1,3) = -0.05; Q(1,4) = -0.05;
Q(2,1) = 0.10; Q(2,2) = 0.10; Q(2,3) = -0.10; Q(2,4) = -0.15;
DQ = zeros(1,2);
fprintf(fo,'Initial flow rate distribution in network \n');
fprintf(fo,' Q(i,1) Q(i,2) Q(i,3) Q(i,4) \n');
fprintf(fo,'------------------------------------------- \n');
for i = 1:2
    fprintf(fo,'Loop # %2i \n',i);
    fprintf(fo,'%10.4f %10.4f %10.4f %10.4f \n',...
            Q(i,1),Q(i,2),Q(i,3), Q(i,4));
end
% Iteration for Q
for k = 1:50
    % determine f,h, dhdQ for each line in each loop
    for i = 1:2
        for j = 1:4
            eod = e/D(i,j);
            Re = 4*abs(Q(i,j))/(pi*D(i,j)*nu);
            arg = eod/3.7+5.74/Re^0.9;
            f = 1.325/(log(arg))^2;
            num = 13.69/arg;
            den = Re^0.9*Q(i,j)*(log(arg))^3;
            dfdQ = num/den;
            alpha = 8*L(i,j)/(pi^2*g*D(i,j)^5);
            if Q(i,j) >= 0.0
                h(i,j) = alpha*Q(i,j)^2*f;
                dhdQ(i,j) = 2*alpha*f*Q(i,j)+ ...
                            alpha*Q(i,j)^2*dfdQ;
            end
            if Q(i,j) < 0.0
                h(i,j) = -alpha*Q(i,j)^2*f;
                dhdQ(i,j) = -(2*alpha*f*Q(i,j)+ ...
                            alpha*Q(i,j)^2*dfdQ);
            end
        end
    end
    for i = 1:2
        sumh(i) = 0.0; sumdhdQ(i) = 0.0;
```

```
    for j = 1:4
        sumh(i) = sumh(i)+h(i,j);
        sumdhdQ(i) = sumdhdQ(i)+dhdQ(i,j);
    end
end
for i = 1:2
    dQ(i) = -sumh(i)/sumdhdQ(i);
    %fprintf(fo,'dQ =%10.6f \n',dQ(i));
end
for i = 1:2
    for j = 1:4
        if ID(i,j) == 0
            Q2(i,j) = Q(i,j)+dQ(i);
        else
            n = ID(i,j);
            Q2(i,j) = Q(i,j)+dQ(i)-dQ(n);
        end
    end
end
fprintf(fo,'\n Iteration number %4.0f \n',k);
fprintf(fo,'flow rate distribution in network \n');
fprintf(fo,'  i    j        Q2        Q \n');
fprintf(fo,'----------------------------------- \n');
for i = 1:2
    for j = 1:4
        fprintf(fo,'%3i %3i %10.6f %10.6f \n',...
                i,j,Q2(i,j),Q(i,j));
    end
    fprintf(fo,'\n');
end
% test1(i,j) equals the difference between new Q's & old
% Q's in each line
for i = 1:2
    for j = 1:4
        test1(i,j) = abs(Q2(i,j)-Q(i,j));
    end
    testm=max(test1);
    % textm equals the maximum difference in Q's for
    % each loop
end
    testmax=max(testm);
    % testmax = maximum difference in Q's for entire
    % network
    fprintf(fo,'maximum difference in Q''s in the');
    fprintf(fo,'network=%10.6f \n', testmax);
if testmax > eps
    for i = 1:2
        for j = 1:4
            Q(i,j) = Q2(i,j);
```

```
            end
        end
    end
    if testmax < eps
        break;
    end
end
fprintf(fo,'---------------------------------------- \n');
fprintf(fo,'System has converged, ');
fprintf(fo,'number of iterations=%4i \n',k);
fprintf(fo,'Maximum difference in Q''s = %8.6f \n',testmax);
fprintf(fo,'flow rate distribution in network \n');
fprintf(fo,' i        j       Q2(i,j)        Q(i,j) \n');
fprintf(fo,'---------------------------------------- \n');
for i = 1:2
    for j = 1:4
        fprintf(fo,'%3i %3i %10.6f  %10.6f \n',...
            i,j,Q2(i,j),Q(i,j));
    end
fprintf(fo,'\n');
end
```

Program results

```
Hardy-Cross method for a two loop network.
Initial flow rate distribution in network
Q(i,1) Q(i,2) Q(i,3) Q(i,4)
----------------------------------------

Loop # 1
      0.1500 0.0500 -0.0500        -0.0500
Loop # 2
      0.1000 0.1000 -0.1000        -0.1500


Iteration number 1
Flow rate distribution in network
  i    j       Q2            Q
----------------------------------------

  1    1     0.104861      0.150000
  1    2     0.018079      0.050000
  1    3    -0.081921     -0.050000
  1    4    -0.081921     -0.050000

  2    1     0.113218      0.100000
  2    2     0.113218      0.100000
  2    3    -0.086782     -0.100000
  2    4    -0.104861     -0.150000

Maximum difference in Q's in the network = 0.045139
Iteration number 2
```

Flow rate distribution in network

i	j	Q2	Q
1	1	0.118482	0.104861
1	2	0.022965	0.018079
1	3	-0.077035	-0.081921
1	4	-0.077035	-0.081921
2	1	0.104483	0.113218
2	2	0.104483	0.113218
2	3	-0.095517	-0.086782
2	4	-0.118482	-0.104861

Maximum difference in Q's in the network = 0.013621
Flow rate distribution in network

⋮

Iteration number 14
Flow rate distribution in network

i	j	Q2	Q
1	1	0.114970	0.114970
1	2	0.019934	0.019934
1	3	-0.080066	-0.080066
1	4	-0.080066	-0.080066
2	1	0.104964	0.104964
2	2	0.104964	0.104964
2	3	-0.095036	-0.095036
2	4	-0.114970	-0.114970

maximum difference in Q's in the network = 0.000000

System has converged, number of iterations = 11
Maximum difference in Q's = 0.000000
Flow rate distribution in network

i	j	Q2(i,j)	Q(i,j)
1	1	0.114970	0.114970
1	2	0.019934	0.019934
1	3	-0.080066	-0.080066
1	4	-0.080066	-0.080066
2	1	0.104964	0.104964
2	2	0.104964	0.104964
2	3	-0.095036	-0.095036
2	4	-0.114970	-0.114970

Projects

P12.1 Determine the flow rate, Q, and the friction factor, f, for the pipe system shown in Figure P12.1 using the following values:

$$L = 30 \text{ m}, \quad D = 15 \text{ cm}, \quad \gamma = 9790 \text{ N/m}^3, \quad \upsilon = 1.005 \times 10^{-6} \text{ m}^2/\text{s}$$

$$(z_1 - z_2) = 22 \text{ m}, \quad g = 9.81 \text{ m/s}^2, \quad K_{entrance} = 0.1, \quad K_{elbow} = 1.5, \quad K_{valve} = 6.4.$$

Assume a starting value for $f = 0.03$.

Print to the screen a table of assumed f, Q, Re, and calculated f by Equation 12.3.

P12.2 Solve the problem described in Example 12.2. Take

$$L = 3000 \text{ m}, \quad D = 5 \text{ cm}, \quad \gamma = 9790 \text{ N/m}^3, \quad \upsilon = 1.005 \times 10^{-6} \text{ m}^2/\text{s}$$

$$(z_2 - z_1) = 50 \text{ m}, \quad g = 9.81 \text{ m/s}^2, \quad K_{entrance} = 0.5, \quad K_{elbow} = 1.5, \quad \text{and} \quad K_{exit} = 1.0.$$

Assume a starting value for $f = 0.03$.

Print to the screen a table of assumed f, Q, Re, and calculated f by Equation 12.3. Also print out $(h_p)_{sys}$.

P12.3 This project involves determining the volume flow rate that a pump will deliver to the upper open tank as shown in Figure P12.3. The pump characteristic curve (h_{pc} vs. Q) is approximated as a cubic as given in Equation P12.3a:

$$h_{pc} = -0.85 \, Q^3 - 3.25 \, Q^2 + 1.75 \, Q + 43.30 \tag{P12.3a}$$

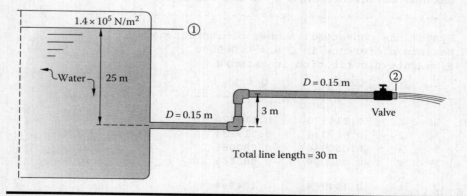

Figure P12.1   Piping system for Project P12.1.

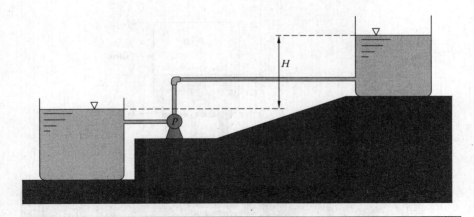

Figure P12.3   Piping system with pump for Project P12.3.

Using the iterative method described in Section 12.1 and the following parameters, determine the volume flow rate, Q, that the pump will deliver to the upper open tank. Print out a table containing the assumed f_1, the iterated Q, the iterated Re, and the determined f_2.

Total pipe length, $L = 150$ m, $H = 10$ m, $D = 0.25$ m, $\sum K_i = 2.25$,

$\dfrac{\varepsilon}{D} = 0.001$, $g = 9.81$ m/s^2, $v = 1.141 \times 10^{-6}$ m^2/s

P12.4 This project involves determining the volume flow rate a pump will deliver to a closed tank as a function of time. The pump characteristic curve (H vs. Q) was taken from a pump manufacturer's catalog. The configuration for this project is shown in Figure P12.4.

Assume that the tank receiving water is closed. Thus, as water fills up in the tank, the air in the tank is compressed. Isothermal compression is to be assumed. The problem is to determine the time it takes to raise the water level in the tank by a specified amount. Data points of the (H vs. Q) curve provided by a pump manufacturer are shown in Table P12.4.

Determine the best-fit third-degree polynomial by the method of least squares using MATLAB's function `polyfit`. The `polyfit` function returns the coefficients for the third-degree polynomial, a_1, a_2, a_3, a_4, that best fits the data. The approximating function as described by Equation P12.4a is used in the analysis:

$$H = a_1 Q^3 + a_2 Q^2 + a_3 Q + a_4 \qquad \text{(P12.4a)}$$

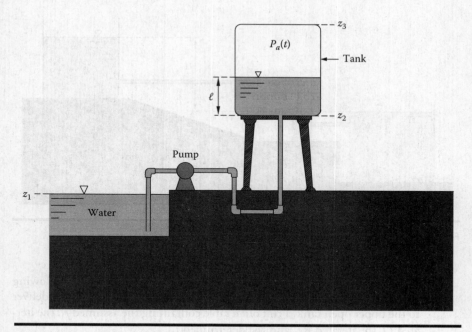

Figure P12.4   Piping system with pump for Project P12.4.

Table P12.4   *H* vs. *Q* Data from Pump Manufacturer

Q (m³/h)	H (m)	Q (m³/h)	H (m)
3.3	43.3	61.6	40.8
6.9	43.4	68.5	39.6
13.7	43.6	75.3	38.7
20.5	43.6	82.2	37.2
27.4	43.3	89	36.3
34.2	43	95.8	34.4
41.1	42.7	102.7	32.6
47.9	42.4	109.6	30.5
54.8	41.8		

Application of the energy equation [4] to the system shown in Figure P12.4 gives

$$(h_p)_{sys} = h_f + (z_2 - z_1 + \ell) + \frac{p_a - p_{atm}}{\gamma} + \sum_i h_i \qquad \text{(P12.4b)}$$

where

$(h_p)_{sys}$ is the head the pump delivers to the water flowing through the pipe

h_f is the viscous head loss in the system

z is the elevation

p_a is the absolute air pressure in the tank

γ is the specific weight of water

P_{atm} is the surrounding atmospheric pressure

ℓ is the water level above the bottom of the tank

The flow rate, developed by the pump must satisfy Q, the manufactures H vs. Q data i.e., for a particular Q,

$$(h_p)_{sys} = H \qquad \text{(P12.4c)}$$

Viscous head loss, h_f, in a pipe is given by [4]

$$h_f = \frac{8Q^2 L}{\pi^2 g D^5} f \qquad \text{(P12.4d)}$$

where

Q is the volume flow rate through the pipe

L is the total pipe length

g is the gravitational constant

f is the friction factor

D is the pipe diameter

Minor losses due to elbows, entrance losses, etc., are expressed by [4]

$$\sum_i h_{minor\ losses} = \frac{8Q^2}{\pi^2 g D^4} \sum_i K_i \qquad \text{(P12.4e)}$$

where K_i is the ith minor head loss coefficient in the system.

Substituting Equations P12.3a, P12.3b, P12.3d, and P12.3e into Equation P12.3c gives

$$a_1 Q^3 + a_2 Q^2 + a_3 Q + a_4 = \frac{8Q^2}{\pi^2 g D^4}\left(\frac{L}{D}f + \sum_i K_i\right) + \frac{p_a - p_{atm}}{\gamma} + (z_2 - z_1 + \ell)$$

(P12.4f)

Rearranging terms gives the following cubic equation:

$$a_1 Q^3 + \left(a_2 - \frac{8}{g\pi^2 D^4}\left(\frac{L}{D}f + \sum_i K_i\right)\right)Q^2 + a_3 Q$$

$$+ \left(a_4 - \frac{p_a - p_{atm}}{\gamma} - (z_2 - z_1 + \ell)\right) = 0$$

(P12.4g)

For a smooth pipe line, the friction factor, f, can be approximated by [2]

$$f = (1.82 \log_{10} \text{Re} - 1.64)^{-2}$$

(P12.4h)

where

$$\text{Re} = \text{Reynolds number} = \frac{4Q}{\pi D \nu}$$

(P12.4i)

where ν is the kinematic viscosity of water.

The air pressure, p_a, at time t, is determined from the ideal gas law for an isothermal process, that is,

$$p_a(t) = \frac{p_{a,i}\mathbb{V}_i}{\mathbb{V}(t)} = p_{a,i}\frac{A_T(z_3 - z_2 - \ell_i)}{A_T(z_3 - z_2 - \ell(t))} = p_{a,i}\frac{(z_3 - z_2 - \ell_i)}{(z_3 - z_2 - \ell(t))}$$

(P12.4j)

where

$p_{a,i}$ is the initial air pressure
A_T is the cross-sectional area of tank $= \pi D_T^2/4$
$\mathbb{V}$ is the air volume in tank
D_T is the diameter of tank

An iterative method of solution for determining Q is as follows:

1. Assume a value for f, say, f_1. Start with $f_1 = 0.03$.
2. Solve Equation P12.4g for Q by the MATLAB function `roots`, selecting the maximum positive root.

3. Solve for Re by Equation P12.4i.
4. Solve for f, say, f_2, by Equation P12.4h.
5. If $|f_2 - f_1| < 1.0 \times 10^{-6}$, then $f = f_2$ and Q = the value obtained in step 2. If $|f_2 - f_1| > 1.0 \times 10^{-6}$, then set $f_1 = f_2$ and repeat process (step 1 through step 5).
6. Continue iteration until $|f_2 - f_1| < 1.0 \times 10^{-6}$.

The governing equation for the time it takes to raise the water level in the tank is as follows:

$$\frac{d(A_T \ell)}{dt} = Q(\ell, p_a) \quad \text{or} \quad \frac{d\ell}{dt} = \frac{Q(\ell, p_a)}{A_T} \tag{P12.4k}$$

Separating the variables and integrating from the starting water level in the tank, ℓ_i, to the final water level in the tank, ℓ_f gives

$$t_f = A_T \int_{\ell_i}^{\ell_f} \frac{d\ell}{Q(\ell, p_a)} = A_T \int_{\ell_i}^{\ell_f} F(\ell, p_a) d\ell \tag{P12.4l}$$

Use Simpson's rule to obtain t_f. A review of Simpson's rule follows.

$$\text{Given } I = \int_a^b F(x) dx$$

subdivide the x domain into N even subdivisions giving $x_1, x_2, x_3, \ldots, x_{N+1}$. Let the function values at $x_1, x_2, x_3, \ldots, x_{N+1}$ be $F_1, F_2, F_3, \ldots, F_{N+1}$, then

$$I = \frac{\Delta x}{3}(F_1 + 4F_2 + 2F_3 + 4F_4 + 2F_5 + \cdots + 2F_N + F_{N+1}) \tag{P12.4m}$$

For this project, $\ell \sim x$ and $1/Q \sim F$.

Procedure
1. At each ℓ_j, determine $p_{a,j}$ by Equation P12.4j.
2. For each ℓ_j, iterate for Q_j by the iteration procedure described earlier, obtaining all the Q_j's.
3. Then determine all the F_j's.
4. Apply Equation P12.4m to obtain the time t_f.

Use the following values for the variables:

$$(z_2 - z_1) = 30 \text{ m}, \quad \ell_i = 1.2 \text{ m}, \quad \ell_f = 4.2 \text{ m}, \quad v = 1.0 \times 10^{-6} \text{ m}^2/\text{s},$$

$$p_{atm} = 1.0132 \times 10^5 \text{ N/m}^2$$

$$\gamma = 9790 \text{ N/m}^3, \quad \sum_i K_i = 4.5, \quad D = 15 \text{ cm}, \quad L = 60 \text{ m}, \quad p_a(0) = p_{atm},$$

$$N = 100, \quad D_T = 1.5 \text{ m}.$$

1. Create a plot of the approximating curve of H (m) vs. Q (m³/h) (solid line) and on the same plot include the data points as circles. Note: Except for the plots, Q needs to be in (m³/s).
2. Create plots of Q (m³/h) vs. $\ell(m)$, $t(s)$ vs. $\ell(m)$, p_a(N/m² gage) vs. $\ell(m)$, and f vs. $\ell(m)$.
3. Print out the time, $t_f(s)$, it takes to raise the water level in the tank by 3 m.
4. Print out the air pressure (N/m², gage) in the tank at time t_f.
5. Print out the initial flow rate, Q_i, and final flow rate, Q_f, in (m³/h).

P12.5 Use the Hardy–Cross method to determine the flow rate distribution (Q's) in the network shown in Figure P12.5. Print out a table indicating the loop

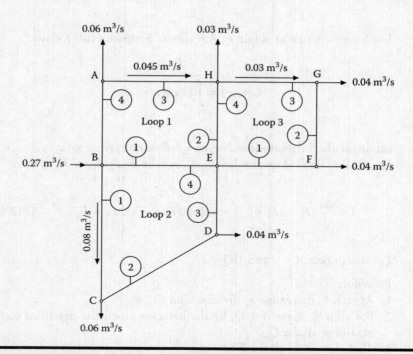

Figure P12.5 Pipe network to be used with the Hardy–Cross method.

Table P12.5   Network Parameters

Loop Number	Line Number	Length (m)	Diameter (cm)	Initial Guess Q (m³/s)
1	1	3220	40	—
	2	4830	30	—
	3	3200	35	−0.45
	4	4830	40	—
2	1	5630	40	0.1
	2	4020	35	—
	3	3200	30	—
	4	3200	40	—
3	1	4830	30	—
	2	4830	25	—
	3	4830	30	−0.03
	4	4830	30	—

number, the line number, and the flow rate in that line. The network parameters are described in Table P12.5.

Hint: To apply the ΔQ corrections to lines that are common in two loops, use an ID matrix for every line identifying the loop number of the common line. Set the ID element to zero if the line is not a common line. For example, suppose line $(1, 1)$ is in common with line $(2, 4)$, then $ID(1, 1) = 2$, and suppose line $(1, 3)$ is not a common line, then $ID(1, 3) = 0$.

Take $\nu = 1.308 \times 10^{-6}$ m²/s, $\varepsilon = 0.026$ cm, $g = 9.81$ m/s².

References

1. Bober, W. and Kenyon, R.A., *Fluid Mechanics*, John Wiley, New York, 1980.
2. Holman, J.P., *Heat Transfer*, 9th Edn., McGraw-Hill, New York, 2002.
3. Bober, W., The use of the Swamee-Jain Formula in pipe network problems, *Journal of Pipelines*, 4, 315–317, 1984.
4. White, F., *Fluid Mechanics*, 6th Edn., McGraw-Hill, New York, 2008.

Chapter 13

Partial Differential Equations

13.1 Classification of Partial Differential Equations

The mathematical modeling of many types of engineering problems involves partial differential equations (PDEs). Many PDEs fall into the general form of Equation 13.1, where A, B, and C are constants:

$$A\frac{\partial^2 u}{\partial x^2} + B\frac{\partial^2 u}{\partial x \partial y} + C\frac{\partial^2 u}{\partial y^2} = f\left(x, y, u, \frac{\partial u}{\partial x}, \frac{\partial u}{\partial y}\right) \tag{13.1}$$

Depending on the values of A, B, and C, these PDEs may be classified as follows:

If $B^2 - 4AC < 0$, the equation is said to be *elliptic*.
If $B^2 - 4AC = 0$, the equation is said to be *parabolic*.
If $B^2 - 4AC > 0$, the equation is said to be *hyperbolic*.

The steady-state heat conduction problem in two dimensions is an example of an elliptic PDE. Laplace's PDE also falls into this category. The parabolic PDE is also called the diffusion equation. The unsteady heat conduction problem is an example of a parabolic PDE. The hyperbolic PDE is also called the wave equation. Sound waves and vibration problems, such as the vibrating string, fall into this category. How a PDE is treated depends on the category that it falls into. However, there are cases in all three categories where a closed-form solution can be obtained by a method called *separation of variables*. This solution method is discussed in the next section.

13.2 Solution by Separation of Variables

13.2.1 Vibrating String

The first problem to be considered is the vibrating string, such as a violin or a viola string (see Figure 13.1). We will assume that

1. The string is elastic.
2. The string motion is vertical.
3. The gravitational forces are negligible compared to the tension in the string.
4. The displacement, $Y(x, t)$, from the horizontal is small and that the angle that the string makes with the horizontal is small. Then, $\partial Y/\partial t$ is the vertical velocity of the string and $\partial^2 Y/\partial t^2$ is the acceleration of the string at position x.

To obtain the governing equation, select an arbitrary element of the string as shown in Figure 13.2. Taking the sum of the forces in the y direction and applying Newton's second law to this element gives

$$M \frac{\partial^2 Y}{\partial t^2} = (T \sin \theta)_{x+\Delta x} - (T \sin \theta)_x \qquad (13.2)$$

Since there is no horizontal movement in the string,

$$(T \cos \theta)_{x+\Delta x} = (T \cos \theta)_x = T_0 \qquad (13.3)$$

Figure 13.1   Vibrating string.

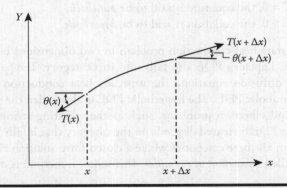

Figure 13.2   Arbitrary string section.

Dividing both sides of Equation 13.2 by T_0 (but using the appropriate expression from Equation 13.3) gives

$$\frac{M}{T_0}\frac{\partial^2 Y}{\partial t^2} = \left(\frac{T\sin\theta}{T\cos\theta}\right)_{x+\Delta x} - \left(\frac{T\sin\theta}{T\cos\theta}\right)_x = (\tan\theta)_{x+\Delta x} - (\tan\theta)_x \qquad (13.4)$$

Let

$$M = \rho\Delta x \qquad (13.5)$$

where ρ is the mass per unit length. Substituting Equation 13.5 into Equation 13.4 and dividing both sides by Δx and taking the limit as $\Delta x \to 0$ on both sides of Equation 13.4 gives

$$\frac{\rho}{T_0}\frac{\partial^2 Y}{\partial t^2} = \frac{\partial(\tan\theta)}{\partial x} \qquad (13.6)$$

But $\tan\theta$ is the slope of the string, which is $\partial Y/\partial x$. Thus, Equation 13.6 becomes

$$\frac{1}{c^2}\frac{\partial^2 Y}{\partial t^2} = \frac{\partial^2 Y}{\partial x^2} \qquad (13.7)$$

where

$$c^2 = \frac{T_o}{\rho} \qquad (13.8)$$

Comparing Equation 13.7 with Equation 13.1, we see that Equation 13.7 is a wave equation.

Since θ is very small, $\cos\theta \approx 1$, and thus, T_0 is essentially the force in the string. To complete the formulation, two initial conditions are needed (because the PDE is second order in t) and two boundary conditions are needed (because the PDE is second order in x). We will assume that the string is deflected at position x_0 and then released from rest. Then, the initial conditions become

$$\frac{\partial Y}{\partial t}(x,0) = 0 \qquad (13.9)$$

$$Y(x,0) = f(x) \qquad (13.10)$$

We will assume that the deflection at the endpoints are zero giving the following boundary conditions

$$Y(0,t) = 0 \qquad (13.11)$$

$$Y(L,t) = 0 \qquad (13.12)$$

We seek a function Y that satisfies the PDE and initial and boundary conditions. Let us examine the possibility that Y is a product of a pure function of x and a pure function of t, that is,

$$Y = F(x)G(t)$$

Substituting this expression into the Equation 13.7 gives

$$\frac{F(x)}{c^2}\frac{d^2G}{dt^2} = G(t)\frac{d^2F}{dx^2}$$

Dividing both sides by GF gives

$$\frac{1}{c^2G}\frac{d^2G}{dt^2} = \frac{1}{F}\frac{d^2F}{dx^2} \tag{13.13}$$

The left-hand side is a pure function of t and the right-hand side is a pure function of x. Since x and t are independent variables, Equation 13.13 can only be true if both sides equal the same constant, say, $(-\lambda^2)$. The minus sign is selected so that Y does not blow up as $t \to \infty$. Then, Equation 13.13 reduces to two ODEs, which are

$$\frac{d^2F}{dx^2} + \lambda^2 F = 0 \tag{13.14}$$

and

$$\frac{d^2G}{dt^2} + c^2\lambda^2 G = 0 \tag{13.15}$$

The general solution to Equation 13.14 is

$$F = a_1 \cos \lambda x + a_2 \sin \lambda x \tag{13.16}$$

The boundary condition given by Equation 13.11 reduces to

$$F(0)G(t) = 0 \rightarrow F(0) = 0 \rightarrow a_1 = 0$$

or

$$F = a_2 \sin \lambda x \tag{13.17}$$

The boundary condition given by Equation 13.12 is

$$F(L)G(t) = 0 \rightarrow F(L) = 0$$

or

$$a_2 \sin \lambda L = 0 \tag{13.18}$$

There are an infinite number of solutions to Equation 13.18, that is,

$$\lambda L = n\pi, \quad \text{where } n = 0, 1, 2, \dots, \infty$$

Then,

$$\lambda = \frac{n\pi}{L} \tag{13.19}$$

The solution to Equation 13.15 is

$$G = b_1 \cos(c\lambda t) + b_2 \sin(c\lambda t) \tag{13.20}$$

and

$$\frac{dG}{dt} = -c\lambda b_1 \sin(c\lambda t) + c\lambda b_2 \cos(c\lambda t)$$

Applying the initial condition given by Equation 13.9 gives

$$F(x)\frac{dG}{dt}(0) = 0 \rightarrow \frac{dG}{dt}(0) = 0 \rightarrow b_2 = 0$$

Then,

$$G(t) = b_1 \cos(c\lambda t) \tag{13.21}$$

The complete solution for Y is obtained from the product of Equations 13.18 and 13.21. The b's can be absorbed into the a_n constants giving

$$Y(x,t) = \sum_{n=1}^{\infty} a_n \sin\frac{n\pi x}{L} \cos\frac{n\pi ct}{L} \tag{13.22}$$

Applying initial condition given by Equation 13.10 gives

$$Y(x,0) = f(x) = \sum_{n=1}^{\infty} a_n \sin\frac{n\pi x}{L} \tag{13.23}$$

The coefficients a_n can be determined because the $\sin\dfrac{n\pi x}{L}$ functions are orthogonal, that is,

$$\int_0^L \sin\frac{n\pi x}{L} \sin\frac{m\pi x}{L} dx = \begin{cases} 0 & \text{if } m \neq n \\ \dfrac{L}{2} & \text{if } m = n \end{cases} \tag{13.24}$$

Multiplying Equation 13.23 by $\sin\dfrac{m\pi x}{L} dx$ and integrating from 0 to L gives

$$\int_0^L f(x)\sin\frac{m\pi x}{L} dx = \sum_{n=0}^{\infty} a_n \int_0^L \sin\frac{n\pi x}{L} \sin\frac{m\pi x}{L} dx = a_m \frac{L}{2} \tag{13.25}$$

Since m is an index from 1 to ∞, we can replace m with n and write

$$a_n = \frac{2}{L}\int_0^L f(x)\sin\frac{n\pi x}{L}\,dx \qquad (13.26)$$

Having now determined the values for a_n, Equation 13.22 gives the solution for $Y(x, t)$. See Project P13.1.

13.2.2 Unsteady Heat Transfer I

Consider a thick plate of width $2L$, as shown in Figure 13.3, that is initially at a uniform temperature, T_0, that is suddenly immersed in a large bath at temperature, T_∞. We wish to determine the temperature time history of the bar, $T(x, t)$, and the amount of heat transferred to the bath (see Appendix B for derivation of the heat conduction equation). Due to symmetry, we only need to consider $T(x, t)$ for $0 \le x \le L$.

The PDE is

$$\frac{1}{a}\frac{\partial T}{\partial t} = \frac{\partial^2 T}{\partial x^2} \qquad (13.27)$$

The initial condition is

$$T(x,0) = T_0 \qquad (13.28)$$

The boundary conditions are

$$\frac{\partial T}{\partial x}(0,t) = 0 \qquad (13.29)$$

$$\frac{\partial T}{\partial x}(L,t) + \frac{h}{k}(T(L,t) - T_\infty) = 0 \qquad (13.30)$$

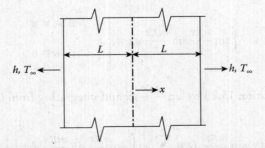

Figure 13.3 A thick bar suddenly immersed in a liquid.

where
t is the time
a is thermal diffusivity of the bar material
h is the convective heat transfer coefficient
k is the thermal conductivity of the bar material

Equation 13.29 is a statement that there is no heat transfer in any direction at $x = 0$ (this is due to problem symmetry). Equation 13.30 is a statement that the rate that heat leaves the bar at $x = L$ by conduction is equal to the rate that heat is carried away by convection in the bath. To make the boundary condition at $x = L$ homogeneous, let

$$\vartheta(x,t) = T(x,t) - T_\infty$$

Then the PDE and the initial and the boundary conditions become

$$\frac{1}{a}\frac{\partial \vartheta}{\partial t} = \frac{\partial^2 \vartheta}{\partial x^2} \tag{13.31}$$

$$\vartheta(x,0) = T_0 - T_\infty \tag{13.32}$$

$$\frac{\partial \vartheta}{\partial x}(0,t) = 0 \tag{13.33}$$

$$\frac{\partial \vartheta}{\partial x}(L,t) + \frac{h}{k}\vartheta(L,t) = 0 \tag{13.34}$$

Comparing Equation 13.31 with Equation 13.1, we see that Equation 13.31 is a parabolic (diffusion) equation. We will assume that

$$\vartheta = F(x)G(t) \tag{13.35}$$

where F is a pure function of x and G is a pure function of t. Substituting Equation 13.35 into Equation 13.31 gives

$$\frac{1}{a}F(x)G'(t) = G(t)F''(x) \tag{13.36}$$

where

$$G'(t) = \frac{dG}{dt} \quad \text{and} \quad F''(x) = \frac{d^2 F}{dx^2}$$

Dividing both sides of Equation 13.36 by FG gives

$$\frac{1}{a}\frac{G'}{G} = \frac{F''}{F} \tag{13.37}$$

The left-hand side of Equation 13.37 is a pure function of t, and the right-hand side is a pure function of x, and x and t are independent variables. The only way a pure function of t can equal a pure function of x is for both sides to equal the same constant, say, $(-\lambda^2)$. Then Equation 13.37 can now be expressed as two ODEs, which are

$$G' + a\lambda^2 G = 0 \tag{13.38}$$

$$F'' + \lambda^2 F = 0 \tag{13.39}$$

The boundary conditions in Equations 13.33 and 13.34 become

$$F'(0)G(t) = 0$$

or

$$F'(0) = 0 \tag{13.40}$$

and

$$F'(L)G(t) + \frac{h}{k}F(L)G(t) = 0$$

or

$$F'(L) + \frac{h}{k}F(L) = 0 \tag{13.41}$$

The function that will satisfy Equation 13.39 is of the form

$$F = ce^{\beta x}$$

Substituting this form into Equation 13.39 gives

$$\beta^2 ce^{\beta x} + \lambda^2 ce^{\beta x} = 0$$

Then,

$$\beta = \pm\lambda i, \quad \text{where } i = \sqrt{-1}$$

Knowing that $e^{ix} = \cos(x) + i\sin(x)$, F becomes

$$F(x) = A\cos(\lambda x) + B\sin(\lambda x) \tag{13.42}$$

where A and B are arbitrary constants to be determined by boundary conditions.
Taking the derivative with respect to x gives

$$F'(x) = -\lambda A\sin(\lambda x) + \lambda B\cos(\lambda x) \tag{13.43}$$

Applying Equation 13.40 to Equation 13.43 gives $B = 0$. Thus,

$$F(x) = A \cos(\lambda x) \tag{13.44}$$

and

$$F'(x) = -\lambda A \sin(\lambda x) \tag{13.45}$$

Applying Equations 13.44 and 13.45 to Equation 13.41 gives

$$-\lambda A \sin(\lambda L) + \frac{h}{k} A \cos(\lambda L) = 0$$

or

$$\tan(\lambda L) - \frac{hL}{k(\lambda L)} = 0 \tag{13.46}$$

A plot of $\tan(\delta)$ and $hL/k\delta$ vs. δ, where $\delta = \lambda L$, is shown in Figure 13.4.

It can be seen that there are an infinite number of roots that satisfy Equation 13.46, say, $\delta_1, \delta_2, \delta_3, \ldots, \delta_n = \lambda_1 L, \lambda_2 L, \lambda_3 L, \ldots, \lambda_n L$, giving an infinite number of solutions, each satisfying the PDE and the boundary conditions. The sum of the solutions also satisfies the differential equation; thus,

$$F(x) = \sum_{n=1}^{\infty} A_n \cos(\lambda_n x)$$

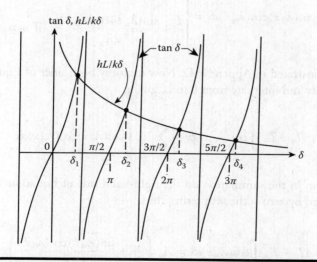

Figure 13.4 Plot of tan δ and hL/kδ vs. δ.

The solution to Equation 13.38 can readily be obtained by separating the variables giving

$$\frac{dG}{G} = -a\lambda_n^2 dt$$

Integrating and taking the antilog gives

$$G = C\exp(-a\lambda_n^2 t) \tag{13.47}$$

where C is a constant of integration. The constant C can be absorbed into the constants A_n. Thus, the general solution is

$$\vartheta(x,t) = \sum_{n=1}^{\infty} A_n \cos(\lambda_n x)\exp(-a\lambda_n^2 t) \tag{13.48}$$

The initial condition now needs to be applied. Applying Equation 13.32 to Equation 13.48 gives

$$T_0 - T_\infty = \sum_{n=1}^{\infty} A_n \cos(\lambda_n x) \tag{13.49}$$

It can be shown that the functions $\cos(\lambda_n x)$ are orthogonal, that is,

$$\int_0^L \cos(\lambda_n x)\cos(\lambda_m x)dx = \begin{cases} 0, & \text{if } m \neq n \\ \dfrac{L}{2} + \dfrac{\sin(\lambda_m L)\cos(\lambda_m L)}{2\lambda_m}, & \text{if } m = n \end{cases}$$

This is demonstrated in Appendix D. Now multiply both sides of Equation 13.49 by $\cos(\lambda_m x)dx$ and integrate from 0 to L, giving

$$(T_0 - T_\infty)\int_0^L \cos(\lambda_m x)dx = \sum_{n=1}^{\infty} A_n \int_0^L \cos(\lambda_n x)\cos(\lambda_m x)dx \tag{13.50}$$

The only term in the summation on the right-hand side of Equation 13.50 that is not multiplied by zero is the mth term. Thus,

$$(T_0 - T_\infty)\int_0^L \cos(\lambda_m x)dx = A_m\left(\frac{L}{2} + \frac{\sin(\lambda_m L)\cos(\lambda_m L)}{2\lambda_m}\right) \tag{13.51}$$

or

$$A_m = \frac{2(T_0 - T_\infty)\sin(\lambda_m L)}{\lambda_m L + \sin(\lambda_m L)\cos(\lambda_m L)} \tag{13.52}$$

Since m is an index from 0, 1, 2, ..., we can replace m with n. Substituting Equation 13.52 into Equation 13.48 gives

$$\vartheta(x,t) = 2(T_0 - T_\infty)\sum_{n=1}^{\infty}\left(\frac{\sin(\lambda_n L)\cos(\lambda_n x)}{\lambda_n L + \sin(\lambda_n L)\cos(\lambda_n L)}e^{-a\lambda_n^2 t}\right) \tag{13.53}$$

See Appendix D for the proof of the orthogonality of the $\cos(\lambda_n x)$ functions.

13.2.3 Unsteady Heat Transfer in 2-D

Consider a bar having a rectangular cross section, initially at temperature, T_0, that is suddenly immersed in a huge bath at a temperature T_∞ (see Figure 13.5). The governing PDE is

$$\frac{1}{\alpha}\frac{\partial T}{\partial t} = \frac{\partial^2 T}{\partial x^2} + \frac{\partial^2 T}{\partial y^2} \tag{13.54}$$

To obtain homogeneous boundary conditions, let $\vartheta(x,y,t) = T(x,y,t) - T_\infty$, then the PDE and the initial condition and boundary conditions in variable ϑ are

$$\frac{1}{\alpha}\frac{\partial \vartheta}{\partial t} = \frac{\partial^2 \vartheta}{\partial x^2} + \frac{\partial^2 \vartheta}{\partial y^2} \tag{13.55}$$

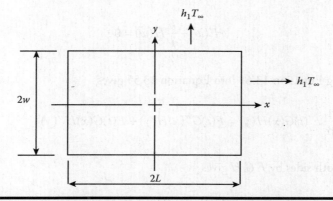

Figure 13.5   Bar, having a rectangular cross section, suddenly immersed in a huge bath.

$$\vartheta(x, y, 0) = T_0 - T_\infty \tag{13.56}$$

$$\frac{\partial \vartheta}{\partial x}(0, y, t) = 0 \quad \text{(due to symmetry)} \tag{13.57}$$

$$\frac{\partial \vartheta}{\partial y}(x, 0, t) = 0 \quad \text{(due to symmetry)} \tag{13.58}$$

$$\frac{\partial \vartheta}{\partial x}(L, y, t) + \frac{h}{k}\vartheta(L, y, t) = 0 \tag{13.59}$$

$$\frac{\partial \vartheta}{\partial y}(x, w, t) + \frac{h}{k}\vartheta(x, w, t) = 0 \tag{13.60}$$

Assume that ϑ can be expressed as the product of three single-variable functions:

$$\vartheta(x, y, t) = F(t)G(x)H(y) \tag{13.61}$$

The boundary conditions and initial condition expressed by Equations 13.56 through 13.60 become

$$G'(0) = 0 \tag{13.62}$$

$$H'(0) = 0 \tag{13.63}$$

$$G'(L) + \frac{h}{k}G(L) = 0 \tag{13.64}$$

$$H'(w) + \frac{h}{k}H(w) = 0 \tag{13.65}$$

Substituting Equation 13.61 into Equation 13.55 gives

$$\frac{1}{\alpha}F'(t)G(x)H(y) = F(t)G''(x)H(y) + F(t)G(x)H''(y) \tag{13.66}$$

Dividing both sides by $F\,G\,H$ gives

$$\frac{1}{\alpha}\frac{F'(t)}{F(t)} = \frac{G''(x)}{G(x)} + \frac{H''(y)}{H(y)} \tag{13.67}$$

The left-hand side is only a function of t and the right-hand side is only a function of x and y. Since x, y, and t are independent variables, the only way that the left-hand side can equal the right-hand side is for both to be equal to the same constant, say, $(-\lambda^2)$. That is,

$$\frac{1}{\alpha}\frac{F'(t)}{F(t)} = \frac{G''(x)}{G(x)} + \frac{H''(y)}{H(y)} = -\lambda^2 \tag{13.68}$$

This gives the following equation for $F(t)$

$$F' + \alpha\lambda^2 F = 0 \tag{13.69}$$

The solution of Equation 13.69 is

$$F = Ce^{-\alpha\lambda^2 t} \tag{13.70}$$

The right-hand side of Equation 13.68 can be written as

$$\frac{G''(x)}{G(x)} + \lambda^2 = -\frac{H''(y)}{H(y)} \tag{13.71}$$

Again, the only way for the left-hand side to equal the right-hand side is for both to equal the same constant, say, β^2, then

$$\frac{G''(x)}{G(x)} + \lambda^2 = -\frac{H''(y)}{H(y)} = \beta^2$$

This gives

$$G'' + (\lambda^2 - \beta^2)G = 0 \tag{13.72}$$

and

$$H'' + \beta^2 H = 0 \tag{13.73}$$

The solution for G is

$$G = A_1 \cos\left(\sqrt{\lambda^2 - \beta^2}\, x\right) + A_2 \sin\left(\sqrt{\lambda^2 - \beta^2}\, x\right) \tag{13.74}$$

and

$$G' = \sqrt{\lambda^2 - \beta^2}\left\{-A_1 \sin\left(\sqrt{\lambda^2 - \beta^2}\, x\right) + A_2 \cos\left(\sqrt{\lambda^2 - \beta^2}\, x\right)\right\} \tag{13.75}$$

Applying the boundary condition Equation 13.62, $(G'(0) = 0)$, gives $A_2 = 0$.

Applying the boundary condition Equation 13.64, $\left(G'(L) + \dfrac{h}{k} G(L) = 0 \right)$, gives

$$-\sqrt{\lambda^2 - \beta^2}\, A_1 \sin\left(\sqrt{\lambda^2 - \beta^2}\, L\right) + \frac{h}{k} A_1 \cos\left(\sqrt{\lambda^2 - \beta^2}\, L\right) = 0$$

or

$$\tan\left(\sqrt{\lambda^2 - \beta^2}\, L\right) = \frac{hL}{k} \times \frac{1}{\sqrt{\lambda^2 - \beta^2}\, L} \tag{13.76}$$

Let $\gamma L = \sqrt{\lambda^2 - \beta^2}\, L$, then Equation 13.76 can be written as

$$\tan(\gamma L) = \frac{hL}{k} \times \frac{1}{\gamma L} \tag{13.77}$$

As seen in Figure 13.4, there are in infinite number of γL's that satisfy Equation 13.77:

$$(\gamma L)_1, (\gamma L)_2, (\gamma L)_3, \ldots, \quad \text{then} \quad \gamma_i = \frac{(\gamma L)_i}{L}, \quad \text{and}$$

$$G(x) = \sum_{i=1}^{\infty} A_i \cos(\gamma_i x) \tag{13.78}$$

Returning to Equation 13.73, and noting the similarity between the G function and the H function, we can determine that

$$H(y) = \sum_{j=1}^{\infty} B_j \cos(\beta_j y) \tag{13.79}$$

By the definition of γ, we see that

$$\lambda_{ij}^2 = \gamma_i^2 + \beta_j^2 \tag{13.80}$$

Combining Equations 13.70, 13.78 through 13.80 and replacing A_i, B_j and C by a single coefficient a_{ij}, we obtain

$$\vartheta(x, y, t) = \sum_i \sum_j a_{ij} \exp(\alpha \lambda_{ij}^2 t) \cos(\gamma_i x) \cos(\beta_j y) \tag{13.81}$$

The initial condition provides the means for determining the coefficients a_{ij}, that is,

$$\vartheta(x, y, 0) = T_0 - T_\infty \tag{13.82}$$

or

$$T_0 - T_\infty = \sum_i \sum_j a_{ij} \cos(\gamma_i x) \cos(\beta_j y) \tag{13.83}$$

Now multiply both sides of Equation 13.83 by $\cos(\gamma_n x)\cos(\beta_m y)dxdy$ and integrate for x from 0 to L and y from 0 to w.

As in Section 13.2.2 and Appendix D, the functions $\cos(\gamma_i x)$ and $\cos(\beta_j y)$ are orthogonal. Thus,

$$\int_0^L \cos(\gamma_i x)\cos(\gamma_n x)dx = \begin{cases} 0, & \text{if } i \neq n \\ \dfrac{L}{2} + \dfrac{\sin(\gamma_n L)\cos(\gamma_n L)}{2\gamma_n}, & \text{if } i = n \end{cases} \tag{13.84}$$

and

$$\int_0^w \cos(\beta_j y)\cos(\beta_m y)dx = \begin{cases} 0, & \text{if } j \neq m \\ \dfrac{L}{2} + \dfrac{\sin(\beta_m w)\cos(\beta_m w)}{2\beta_m}, & \text{if } j = m \end{cases} \tag{13.85}$$

Also,

$$(T_0 - T_\infty)\int_0^L \int_0^w \cos(\gamma_n x)\cos(\beta_m y)dxdy = \frac{\sin(\gamma_n L)}{\gamma_n} \times \frac{\sin(\beta_m w)}{\beta_m} \tag{13.86}$$

The orthogonality of the $\cos(\gamma_i x)$ and $\cos(\beta_j y)$ functions eliminates the summation signs giving

$$a_{nm} = \frac{(T_0 - T_\infty)\left(\dfrac{\sin(\gamma_n L)}{\gamma_n} \times \dfrac{\sin(\beta_m w)}{\beta_m}\right)}{\left(\dfrac{L}{2} + \dfrac{\sin(\gamma_n L)\cos(\gamma_n L)}{2\gamma_n}\right) \times \left(\dfrac{w}{2} + \dfrac{\sin(\beta_m w)\cos(\beta_m w)}{2\beta_m}\right)} \tag{13.87}$$

Finally,

$$\vartheta(x,y,t)=\sum_{n=1}^{\infty}\sum_{m=1}^{\infty}\frac{(T_0-T_\infty)\left(\dfrac{\sin(\gamma_nL)}{\gamma_n}\times\dfrac{\sin(\beta_mw)}{\beta_m}\right)\exp(-\alpha\lambda_{nm}^2t)\cos(\gamma_nx)\cos(\beta_my)}{\left(\dfrac{L}{2}+\dfrac{\sin(\gamma_nL)\cos(\gamma_nL)}{2\gamma_n}\right)\times\left(\dfrac{w}{2}+\dfrac{\sin(\beta_mw)\cos(\beta_mw)}{2\beta_m}\right)}$$

(13.88)

13.3 Review of Finite-Difference Formulas

We will need the following finite-difference formulas for Section 13.4.

These formulas were derived in Section 8.2. Given $y = y(x)$ and for a uniform subdivision on the x-axis

$$y_i'=\frac{y_{i+1}-y_i}{\Delta x}\qquad\text{forward-difference formula for }y'(x_i)$$

$$y_i'=\frac{y_i-y_{i-1}}{\Delta x}\qquad\text{backward-difference formula for }y'(x_i)$$

$$y_i''=\frac{y_{i+1}+y_{i-1}-2y_i}{\Delta x^2}\qquad\text{central-difference formula for }y''(x_i)$$

$$y_i'=\frac{3y_i-4y_{i-1}+y_{i-2}}{2\Delta x}\qquad\text{backward-difference formula for }y'(x_i)\text{ of order }(\Delta x^2)$$

$$y_i'=\frac{-y_{i+2}+4y_{i+1}-3y_i}{2\Delta x}\qquad\text{forward-difference formula for }y'(x_i)\text{ of order }(\Delta x^2)$$

13.4 Finite-Difference Methods Applied to Partial Differential Equations

We wish to solve the problem described in Section 13.2.2 by finite-difference method. In that problem, we considered a thick plate, initially at temperature, T_0, which is suddenly immersed in a huge bath at temperature T_∞ (see Figure 13.3).

The governing PDE for the temperature field, $T(x, t)$, with the following initial and boundary conditions are

$$\frac{1}{\alpha}\frac{\partial T}{\partial t}=\frac{\partial^2 T}{\partial x^2}$$

(13.89)

$$T(x,0) = T_0$$

(13.90)

$$\frac{\partial T}{\partial x}(0,t) = 0 \qquad\qquad (13.91)$$

$$\frac{\partial T}{\partial x}(L,t) + \frac{h}{k}[T(L,t) - T_\infty] = 0 \qquad\qquad (13.92)$$

where
  α is the thermal diffusivity of the plate and $2L$ is the plate thickness
  h is the convective heat transfer coefficient
  k is the thermal conductivity of the plate material

To solve this heat transfer problem numerically, subdivide the x and t domains into I and J subdivisions, respectively, giving

$$x_1, x_2, x_3, \ldots, x_{I+1} \quad \text{and} \quad t_1, t_2, t_3, \ldots, t_{J+1}$$

There are two finite-difference numerical methods for solving this problem, the explicit method and the implicit method. The explicit method has a stability problem if the following condition is not satisfied:

$$\beta = \frac{\alpha \Delta t}{\Delta x^2} < 0.5$$

The implicit method does not have a stability problem.

13.4.1 *Explicit Method*

Writing the governing PDE at (x_i, t_j) using the forward finite-difference formula for $\frac{\partial T}{\partial t}(x_i, t_j)$ and the central-difference formula for $\frac{\partial^2 T}{\partial x^2}(x_i, t_j)$ gives

$$\frac{\partial T}{\partial t}(x_i, t_j) \approx \frac{1}{\Delta t}[T(x_i, t_{j+1}) - T(x_i, t_j)] \qquad\qquad (13.93)$$

and

$$\frac{\partial^2 T}{\partial x^2}(x_i, t_j) = \frac{1}{\Delta x^2}[T(x_{i+1}, t_j) + T(x_{i-1}, t_j) - 2T(x_i, t_j)] \qquad\qquad (13.94)$$

To simplify the notation, use

$$T(x_i, t_j) = T_i^j$$

then

$$\frac{\partial T}{\partial t}(x_i, t_j) \approx \frac{1}{\Delta t}\left[T_i^{j+1} - T_i^j\right] \tag{13.95}$$

$$\frac{\partial^2 T}{\partial x^2}(x_i, t_j) = \frac{1}{\Delta x^2}\left[T_{i+1}^j + T_{i-1}^j - 2T_i^j\right] \tag{13.96}$$

The governing PDE becomes

$$\frac{1}{\alpha\Delta t}\left[T_i^{j+1} - T_i^j\right] = \frac{1}{\Delta x^2}\left[T_{i+1}^j + T_{i-1}^j - 2T_i^j\right] \tag{13.97}$$

Solving for T_i^{j+1} gives

$$T_i^{j+1} = T_i^j + \frac{\alpha\Delta t}{\Delta x^2}\left[T_{i+1}^j + T_{i-1}^j - 2T_i^j\right] \tag{13.98}$$

Equation 13.98 is valid for $i = 2, 3, \ldots, I$.

The initial condition reduces to

$$T_i^1 = T_o \quad \text{for} \quad i = 1, 2, 3, \ldots, I+1$$

The boundary condition $\dfrac{\partial T}{\partial x}(0, t) = 0$ is also valid at $t + \Delta t$.

Using the forward-difference formula of order Δx^2 gives

$$\frac{-T_3^{j+1} + 4T_2^{j+1} - 3T_1^{j+1}}{2\Delta x} = 0$$

Solving for T_1^{j+1} gives

$$T_1^{j+1} = \frac{1}{3}\left[4T_2^{j+1} - T_3^{j+1}\right] \tag{13.99}$$

The boundary condition $\dfrac{\partial T}{\partial x}(L,t) + \dfrac{h}{k}(T(L,t) - T_\infty) = 0$ is also valid at $t + \Delta t$.

Using the backward-difference formula for $\dfrac{\partial T}{\partial x}(L,t)$ of order Δx^2 gives

$$\frac{3T_{I+1}^{j+1} - 4T_I^{j+1} + T_{I-1}^{j+1}}{2\Delta x} + \frac{h}{k}\left(T_{I+1}^{j+1} - T_\infty\right) = 0$$

Solving for T_{I+1}^{j+1} gives

$$T_{I+1}^{j+1} = \frac{k}{3k + 2h\Delta x}\left(4T_I^{j+1} - T_{I-1}^{j+1} + \frac{2h\Delta x}{k}T_\infty\right) \tag{13.100}$$

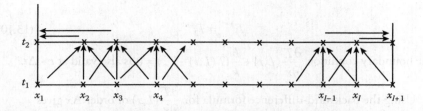

Figure 13.6 **Order of calculations and marching in time.**

The solution is obtained by "marching" in time. A sketch of the order of calculations is shown in Figure 13.6.

Finally, the amount of heat transfer per unit surface area, Q, that occurs in time t_f is given by

$$Q = -2k \int_0^{t_f} \frac{\partial T}{\partial x}(L,t)dt = -\frac{k}{\Delta x} \int_0^{t_f} [3T_{I+1}(t) - 4T_I(t) + T_{I-1}(t)]dt \qquad (13.101)$$

13.4.2 *Implicit Method*

We now write the governing PDE using the forward finite-difference formula for $\partial T/\partial t$ and the central-difference formula for $\partial^2 T/\partial x^2$, but take the time position at $j+1$ giving

$$\frac{1}{\alpha} \frac{T_i^{j+1} - T_i^j}{\Delta t} = \frac{T_{i+1}^{j+1} + T_{i-1}^{j+1} - 2T_i^{j+1}}{\Delta x^2}$$

Solving for T_i^{j+1} gives

$$T_i^{j+1} = T_i^j + \frac{\alpha \Delta t}{\Delta x^2 + 2\alpha \Delta t}(T_{i+1}^{j+1} + T_{i-1}^{j+1}) \qquad (13.102)$$

Equation 13.102 is valid for $i = 2, 3, …, I$. There are three unknowns in Equation 13.102: T_i^{j+1}, T_{i+1}^{j+1}, and T_{i-1}^{j+1}. Term T_i^j is assumed to be known. So far the set fits into a tri-diagonal system. The boundary conditions need to be checked to see if they also fit into a tri-diagonal system. The initial condition is

$$T_i^1 = T_0, \quad \text{valid for } i = 1, 2, 3, …, I+1 \qquad (13.103)$$

The boundary condition $\dfrac{\partial T}{\partial x}(0,t) = 0$ is also valid at $t + \Delta t$.

Using the forward-difference formula of order Δx gives

$$\frac{T_2^{j+1} - T_1^{j+1}}{\Delta x} = 0$$

or

$$T_1^{j+1} = T_2^{j+1} \tag{13.104}$$

The boundary condition $\dfrac{\partial T}{\partial x}(L,t) + \dfrac{h}{k}(T(L,t) - T_\infty) = 0$ is also valid at $t + \Delta t$.

Using the backward-difference formula for $\dfrac{\partial T}{\partial x}(L,t)$ of order Δx gives

$$\frac{T_{I+1}^{j+1} - T_I^{j+1}}{\Delta x} + \frac{h}{k}(T_{I+1}^{j+1} - T_\infty) = 0$$

Solving for T_{I+1}^{j+1} gives

$$T_{I+1}^{j+1} = \frac{k}{k + h\Delta x}T_I^{j+1} + \frac{h\Delta x}{k + h\Delta x}T_\infty \tag{13.105}$$

Equations 13.102 through 13.105 fall into a tri-diagonal system, allowing for a solution of all temperatures at t^{j+1} by the method described in Section 8.3. A complete solution can be obtained by marching in time.

13.5 The Gauss–Seidel Method

The Gauss–Seidel iteration method may be used to solve Laplace's Equation. Consider the steady-state heat conduction problem of the slab shown in Figure 13.7. For the derivation of the heat conduction equation, see Appendix B.

The governing partial differential equation for the temperature distribution is:

$$\frac{\partial^2 T}{\partial x^2} + \frac{\partial^2 T}{\partial y^2} = 0 \tag{13.106}$$

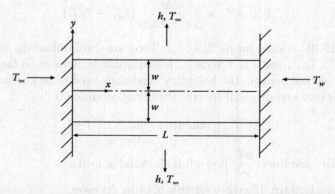

Figure 13.7 Slab subjected to heat transfer.

The boundary conditions are:

$$T(0, y) = T_\infty \tag{13.107}$$

$$T(L, y) = T_w \tag{13.108}$$

$$\frac{\partial T}{\partial y}(x, 0) = 0 \text{ (by symmetry)} \tag{13.109}$$

$$\frac{\partial T}{\partial y}(x, w) + \frac{h}{k}(T(x, w) - T_\infty) = 0 \tag{13.110}$$

The finite-difference form of the partial differential equation is:

$$\frac{1}{(\Delta x)^2}\left\{ T(x + \Delta x, y) - 2T(x, y) + T(x - \Delta x, y) \right\}$$

$$+ \frac{1}{(\Delta y)^2}\left\{ T(x, y + \Delta y) - 2T(x, y) + (T(x, y - \Delta y) \right\} = 0 \tag{13.111}$$

Now subdivide the x domain into N subdivisions and the y domain into M subdivisions giving positions (x_n, y_m), where $n = 1, 2, ..., N+1$ and $m = 1, 2, ..., M+1$ and $\Delta x = \dfrac{L}{N}$ and $\Delta y = \dfrac{w}{M}$.

Let $T(x_n, y_m) = T_{n,m}$, then the finite difference form of the partial differential equation is:

$$T_{n,m} = \frac{1}{2(1+\beta^2)}\left(T_{n,m+1} + T_{n,m-1} + \beta^2 T_{n+1,m} + \beta^2 T_{n-1,m} \right) \tag{13.112}$$

where $\beta = \Delta y / \Delta x$.

The above equation is valid at all interior points. Thus, it is valid for $n = 2, 3, ..., N$ and $m = 2, 3, ..., M$. There are $(N-1)(M-1)$ such equations.

The finite-difference form for the boundary conditions are:

$$T_{1,m} = T_\infty \quad \text{for } m = 1, 2, 3, \ldots, M+1 \tag{13.113}$$

$$T_{N+1,m} = T_w \quad \text{for } m = 1, 2, 3, \ldots, M+1 \tag{13.114}$$

Using the forward difference formula for $\frac{\partial T}{\partial y}(x, 0)$ of order $(\Delta y)^2$, the boundary condition $\frac{\partial T}{\partial y}(x, 0) = 0$ becomes:

$$T_{n,1} = \frac{1}{3}(4T_{n,2} - T_{n,3}) \tag{13.115}$$

Using the backward-difference formula for $\frac{\partial T}{\partial y}(x, w)$ of order $(\Delta y)^2$, the boundary condition $\frac{\partial T}{\partial y}(x, w) + \frac{h}{k}\left(T(x, w) - T_\infty\right) = 0$ becomes

$$T_{n,M+1} = \frac{1}{3 + \dfrac{2h\Delta y}{k}}\left\{ 4T_{n,M} - T_{n,M-1} + \frac{2h\Delta y T_\infty}{k} \right\} \tag{13.116}$$

Equations 13.111 through 13.116 represent the finite-difference equations describing the temperature distribution in the slab.

Method of solution:

1. Assume a set of values for $T_{n,m}$, say $T_{n,m}^1$ for $n = 2, 3, \ldots, N$ and $m = 1, 2, 3, \ldots, M+1$.
2. Successively substitute into Equations 13.109 through 13.116, obtaining a new set of values for $T_{n,m}$, say $T_{n,m}^2$, using the updated values in the equations when available. Note: $T_{1,m}$ and $T_{N+1,m}$ are fixed for all m (Equations 13.113 and 13.114).
3. Repeat process until $\left| T_{n,m}^2 - T_{n,m}^1 \right| < \varepsilon$ for all n, m.

Faster convergence may be obtained by over-relaxing the set of equations. This is done by adding and subtracting $T_{n,m}^1$ from Equation (13.111) and introducing a relaxation parameter, ω, giving

$$T_{n,m}^2 = T_{n,m}^1 + \frac{\omega}{2(1+\beta^2)}\left(T_{n,m+1} + T_{n,m-1} + \beta^2 T_{n+1,m} + \beta^2 T_{n-1,m} - \frac{2(1+\beta^2)}{\omega}T_{n,m}^1 \right) \tag{13.117}$$

where $1 < \omega < 2$. A similar procedure is carried out for Equations 13.115 and 13.116 giving

$$T_{n,1}^2 = T_{n,1}^1 + \frac{\omega}{3}\left(4T_{n,2} - T_{n,3} - \frac{3}{\omega}T_{n,1}^1\right) \qquad (13.118)$$

and

$$T_{n,M+1}^2 = T_{n,M+1}^1 + \frac{\omega k}{3k + 2h\,\Delta y}\left\{4T_{n,M} - T_{n,M-1} + \frac{2h\,\Delta y\,T_\infty}{k} - \frac{3k + 2h\,\Delta y}{\omega k}T_{n,M+1}^1\right\} \qquad (13.119)$$

The method of solution described earlier is still valid, except Equations 13.117, 13.118 and 13.119 are substituted for Equations 13.112, 13.115 and 13.116 respectively. Sometimes in order to get convergence, one might have to under-relax; that is, take $0 < \omega < 1$.

Projects

P13.1 Use MATLAB to solve the vibrating string problem discussed in Section 13.2 for the initial condition shown in Figure P13.1. Plot Y vs. x at the following times: $t = 0.0001$ s, $t = 1.0$ s, $t = 10$ s, and $t = 100$ s. Use the following parameters in your solution:

   $R = 8240$ kg/m^3, $T_o = 90$ N, d = diameter of string = 0.16 cm, $L = 1$ m, and $h = 6$ cm.

   Note: ρ (kg/m) $= R$ (kg/m^3)A, where A is the cross-sectional area of the string.

P13.2 Rearrange Equation 13.53 to read

$$\text{TRATIO} = \frac{T[(x/L),t] - T_\infty}{T_0 - T_\infty} = 2\sum_{n=1}^{\infty}\frac{\sin(\delta_n)\cos[\delta_n(x/L)]e^{-a\delta_n^2 t/L^2}}{\cos(\delta_n)\sin(\delta_n) + \delta_n} \qquad (\text{P13.2a})$$

   where $\delta_n = \lambda_n L$

   Write a MATLAB program using 50 δ_n values and solve for TRATIO for the following parameters: $h = 890.0$ W/m^2-°C, $k = 386.0$ W/m-°C, $L = 0.5$ m, $a = 11.234 \times 10^{-5}$ m^2/s, $T_0 = 300$°C, $T_\infty = 30$°C, $x/L = 0.0, 0.2, 0.4, 0.6, 0.8, 1.0$, and $t = 0, 20, 40, \ldots 400$ s. Print out results in table form as shown in Table P13.2. Also create a plot of *TRATIO* vs. t for $x/L = 1.0, 0.8, 0.6, 0.4, 0.2, 0.0$.

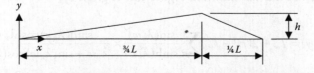

Figure P13.1   Initial string displacement.

Table P13.2 Temperature Ratio (*TRATIO*) vs. Time (*t*)

	X/L					
Time (s)	0.0	0.2	0.4	0.6	0.8	1.0
0	1.0	1.0	1.0	1.0	1.0	1.0
20	—	—	—	—	—	—
40	—	—	—	—	—	—
⋮	⋮	⋮	⋮	⋮	⋮	⋮
400	—	—	—	—	—	—

P13.3 Write a MATLAB program to solve numerically, by the explicit method, the problem described in Section 13.4.1. Use the parameters described in P13.2, that is, $h = 890.0$ W/m²-°C, $k = 386.0$ W/m-°C, $L = 0.5$ m, $a = 11.234 \times 10^{-5}$ m²/s, $T_0 = 300°C$, and $T_\infty = 30°C$. Take $\Delta x = 0.005$ m and $\Delta t = 0.1$ s. Carry the calculations to 400 s. To compare the results obtained by this numerical method with the results obtained by the closed-form solution (P13.2), write your answer in the form

$$\text{TRATIO} = \frac{T[(x/L),t] - T_\infty}{T_0 - T_\infty} \qquad (P13.3)$$

for $x/L = 1.0, 0.8, 0.6, 0.4, 0.2, 0.0$ and for $t = 0, 10$ s, 20 s, …, 400 s. Print out a table as shown in Table P13.2. Also create a plot of TRATIO vs. t for $x/L = 1.0, 0.8, 0.6, 0.4, 0.2, 0.0$.

P13.4 If you did Projects P13.2 and P13.3, superimpose the solution obtained in P13.2 on the plot created in Project P13.3. The resulting plot should be similar to the plot shown in Figure P13.4.

P13.5 Repeat Project P13.3, but this time use the implicit method. If you also did Project P13.2, superimpose the solution obtained in P13.2 on your plot. The resulting plot should be similar to the plot shown in Figure P13.4.

P13.6 Using separation of variables, show that the steady-state temperature distribution in the slab shown in Figure 13.7 is given by:

$$T(x,y) = T_\infty + 2(T_w - T_\infty) \sum_1^\infty \frac{\sin(\lambda_n w)}{\sinh(\lambda_n L)} \times \frac{\sinh(\lambda_n x)\cos(\lambda_n y)}{\lambda_n w + \sin(\lambda_n w)\cos(\lambda_n w)}$$

$$(P13.6a)$$

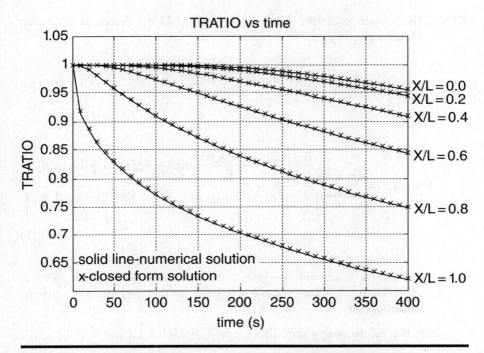

Figure P13.4 Plot of TRATIO vs. time. Numerical solution is by the explicit method.

where λ_n is determined by:

$$\tan(\lambda w) - \frac{hw}{k} \times \frac{1}{\lambda w} = 0 \qquad \text{(P13.6b)}$$

Use only the first 10 λ_n values in Equation P13.6a.

a. Print out a table for the first 10 eigenvalues (λ_n) applicable to this problem.
b. Print out a table of $T(x, y)$ at every second x position and every second y position.
c. Create a plot of $T(x, 0)$ and $T(x, w)$ vs. x, both on the same graph.

Use the following values:

$$L = 1.0 \text{ m}, \quad w = 0.2 \text{ m}, \quad T_w = 300°\text{C}, \quad T = 50°\text{C}, \quad k = 386 \text{ W/m-}°\text{C},$$

$$h = 800 \text{ W/m}^2\text{-}°\text{C}$$

P13.7 The temperature ratio, TR (x, y, t), of the 2-D bar described in Section 13.2.3 is given by

$$TR(x, y, t) = \frac{\vartheta(x, y, t)}{T_0 - T_\infty}$$ (P13.7a)

Then,

$$TR(x, y, t) = \sum_{n=1}^{\infty} \sum_{m=1}^{\infty} \frac{\left(\dfrac{\sin(\gamma_n L)}{\gamma_n} \times \dfrac{\sin(\beta_m w)}{\beta_m} \right) \exp(-\alpha \lambda_{nm}^2 t) \cos(\gamma_n x) \cos(\beta_m y)}{\left(\dfrac{L}{2} + \dfrac{\sin(\gamma_n L)\cos(\gamma_n L)}{2\gamma_n} \right) \times \left(\dfrac{w}{2} + \dfrac{\sin(\beta_m w)\cos(\beta_m w)}{2\beta_m} \right)}$$

(P13.7b)

Develop a computer program in MATLAB to evaluate TR at $x/L = 0.0, 0.5, 1.0$ and $y/w = 0.0, 0.5, 1.0$ for $0 \le t \le 400$ s. Create the following plots:

a. Plot on the same graph, TR vs. t for $\dfrac{x}{L} = 0.0, 0.5, 1.0$ and $\dfrac{y}{w} = 0.0$.

b. Plot on the same graph, TR vs. t for $\dfrac{x}{L} = 0.0, 0.5, 1.0$ and $\dfrac{y}{w} = 0.5$.

c. Plot on the same graph, TR vs. t for $\dfrac{x}{L} = 0.0, 0.5, 1.0$ and $\dfrac{y}{w} = 1.0$.

d. Plot on the same graph, TR vs. t for $\dfrac{y}{w} = 0.0, 0.5, 1.0$ and $\dfrac{x}{L} = 0.0$.

e. Plot on the same graph, TR vs. t for $\dfrac{y}{w} = 0.0, 0.5, 1.0$ and $\dfrac{x}{L} = 0.5$.

f. Plot on the same graph, TR vs. t for $\dfrac{y}{w} = 0.0, 0.5, 1.0$ and $\dfrac{x}{L} = 1.0$.

Use the following parameters:

$$L = w = 0.5 \text{ m}, \quad h = 890 \frac{\text{W}}{\text{m}^2 \cdot {}^\circ\text{C}}, \quad k = 386 \frac{\text{W}}{\text{m} \cdot {}^\circ\text{C}}, \quad \alpha = 11.234 \times 10^{-5} \frac{\text{m}^2}{\text{s}}.$$

P13.8 A freezer compartment in a refrigerator suddenly looses power due to a power failure resulting from a hurricane. The initial temperature in the freezer compartment is -18°C and the outside temperature is 26°C. The freezer wall is constructed of two stainless steel plates separated by fiberglass insulation (see Figure P13.8a). The interior dimension of the freezer is 0.75 m × 0.75 m × 0.5 m. A finite-difference numerical analysis may be used to

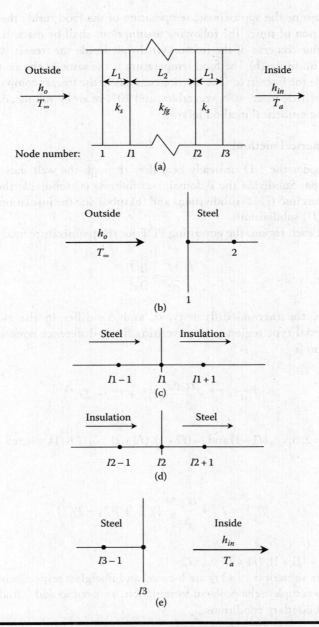

Figure P13.8 **(a) Freezer wall geometry. (b) Outside air–steel interface. (c) Left steel–insulation interface. (d) Right insulation–steel interface. (e) Steel–inside air interface.**

determine the approximate temperature of the food inside the freezer as a function of time. The following assumptions shall be made regarding the interior contents of the freezer: (a) the air inside the freezer is well mixed and uniform; (b) The food temperature is the same as the air temperature inside the freezer; (c) The interior contents of the freezer compartment consist of 40% meat, 40% vegetables, and 20% of air by volume. A description of the numerical method follows:

Numerical method

Consider the 1-D unsteady heat flow through the wall shown in Figure P13.8a. Subdivide the x domains as follows: (a) subdivide the steel plate regions into $(I1-1)$ subdivisions and (b) subdivide the insulation region into $(I2-I1)$ subdivisions.

In each region, the governing PDE for the temperature field is

$$\frac{1}{\alpha}\frac{\partial T}{\partial t} = \frac{\partial^2 T}{\partial x^2} \tag{P13.8a}$$

Since the thermal diffusivity, α, and Δx differ in the two different material type regions, the governing finite-difference equation for each region is

$$T_{i+1}^n = T_i^n + \frac{\alpha_s \Delta t}{\Delta x_1^2}\left(T_{i+1}^n + T_{i-1}^n - 2T_i^n\right) \tag{P13.8b}$$

for $i = 2, 3, \ldots, (I1-1)$ and $i = (I2+1), (I2+2), \ldots, (I3-1)$, where $T(x_i, t_n) = T_i^n$ and

$$T_i^{n+1} = T_i^n + \frac{\alpha_{fg} \Delta t}{\Delta x_2^2}\left(T_{i+1}^n + T_{i-1}^n - 2T_i^n\right) \tag{P13.8c}$$

for $i = (I1+1), (I1+2), \ldots, (I2-1)$.

The subscripts s and fg are for steel and fiberglass respectively.

To complete the problem formulation, we need to add initial conditions and boundary conditions.

Initial conditions

We will assume that at the start of the power loss, the freezer compartment was in a steady-state condition; thus, the initial temperature in each region is linear. We can determine the heat rate per unit surface area, q, entering the freezer compartment from the outside by:

$$q = \frac{(T_{inf} - T_a^1)}{\dfrac{1}{h_o} + \dfrac{1}{h_{in}} + \dfrac{L_{fg}}{k_{fg}} + \dfrac{2L_s}{k_s}} \tag{P13.8d}$$

where h_o is the convective heat coefficient in the outside air; h_i is the convective heat transfer coefficient of the air inside the freezer compartment; L_{fg} is the thickness of the fiberglass; k_{fg} is the thermal conductivity of the fiberglass; L_s is the thickness of each steel plate; k_s is the thermal conductivity of steel; T_{inf} is the temperature of outside air far from the freezer and T_a is the air temperature inside the freezer compartment.

Then,

$$T_1^1 = T_{inf} - \frac{q}{h_o} \tag{P13.8e}$$

$$T_{I1}^1 = T_1^1 - \frac{qL_1}{k_s} \tag{P13.8f}$$

$$T_{I2}^1 = T_{I1}^1 - \frac{qL_2}{k_{fg}} \tag{P13.8g}$$

$$T_{I3}^1 = T_a^1 + \frac{q}{h_{in}} \tag{P13.8h}$$

The initial temperature for each region can be expressed as a linear relationship. We can use similar triangles to determine $T(x_i, 1) = T_i^1$.

In region 1 (left steel plate)

$$\frac{T_1^1 - T_i^1}{T_1^1 - T_{n1}^1} = \frac{x_1 - x_i}{x_1 - x_{n1}}$$

or

$$T_i^1 = T_1^1 - (T_1^1 - T_{n1}^1) \times \frac{x_i - x_1}{x_{n1} - x_1} \tag{P13.8i}$$

Similarly for region 2 (fiber glass insulation),

$$T_i^1 = T_{n1}^1 - (T_{n1}^1 - T_{n2}^1) \times \frac{x_i - x_{n1}}{x_{n2} - x_{n1}}$$

For region 3 (right steel plate),

$$T_i^1 = T_{n2}^1 - (T_{n2}^1 - T_{n3}^1) \times \frac{x_i - x_{n2}}{x_{n3} - x_{n2}}$$

Boundary conditions

a. At the left air–steel interface (see Figure P13.8b):

The rate that heat is carried to the wall by convection per unit surface area = the rate that heat enters the steel plate by conduction per unit surface area; that is,

$$h_o[T_\infty - T(x_1, t)] = -\vec{\mathbf{q}}(x_1, t) \cdot (-\hat{\mathbf{e}}_x) = -k_s \frac{\partial T}{\partial x}(x_1, t)$$

The boundary condition is valid at t_n and t_{n+1}. The simplest finite-difference form of the preceding equation is

$$h_o\left(T_\infty - T_1^{n+1}\right) = -k_s \frac{T_2^{n+1} - T_1^{n+1}}{\Delta x_1}$$

Solving for T_1^{n+1} gives

$$T_1^{n+1} = \frac{h_o T_\infty + \dfrac{k_s}{\Delta x_1} T_2^{n+1}}{h_o + \dfrac{k_s}{\Delta x_1}} \tag{P13.8j}$$

b. At the left steel–insulation interface (see Figure P13.8c):

The rate that heat flows out of the steel plate per unit surface area = the rate that heat flows into the insulation per unit surface area. Expressed mathematically,

$$\vec{\mathbf{q}}(I1^-, t) \cdot \hat{\mathbf{e}}_x = -\vec{\mathbf{q}}(I1^+, t) \cdot (-\hat{\mathbf{e}}_x)$$

or

$$-k_1 \frac{\partial T}{\partial x}(I1^-, t) = -k_{fg} \frac{\partial T}{\partial x}(I1^+, t)$$

The boundary condition is valid at t_n and t_{n+1}. The simplest finite-difference form of the preceding equation is

$$-k_s \frac{T_{I1}^{n+1} - T_{I1-1}^{n+1}}{\Delta x_1} = -k_{fg} \frac{T_{I1+1}^{n+1} - T_{I1}^{n+1}}{\Delta x_2}$$

Solving for T_{I1}^{n+1} gives

$$T_{I1}^{n+1} = \frac{\dfrac{k_s}{\Delta x_1} T_{I1-1}^{n+1} + \dfrac{k_{fg}}{\Delta x_2} T_{I1+1}^{n+1}}{\dfrac{k_s}{\Delta x_1} + \dfrac{k_{fg}}{\Delta x_2}} \qquad \text{(P13.8k)}$$

c. At the right insulation–steel interface (see Figure P13.8d):

The rate that heat flows out of insulation per unit surface area
    = the rate that heat flows into the steel plate per unit surface area.

Similarly to the equations developed at node $I1$, the equation at node $I2$ is

$$-k_{fg} \frac{T_{I2}^{n+1} - T_{I2-1}^{n+1}}{\Delta x_2} = -k_s \frac{T_{I2+1}^{n+1} - T_{I2}^{n+1}}{\Delta x_1}$$

Solving for T_{I2}^{n+1} gives

$$T_{I2}^{n+1} = \frac{\dfrac{k_{fg}}{\Delta x_2} T_{I2-1}^{n+1} + \dfrac{k_s}{\Delta x_1} T_{I2+1}^{n+1}}{\dfrac{k_{fg}}{\Delta x_2} + \dfrac{k_s}{\Delta x_1}} \qquad \text{(P13.8l)}$$

d. At the right steel–air interface (see Figure P13.8e):

The rate that heat leaves the steel plate by conduction per unit surface = the rate that heat enters the freezer by convection

$$\vec{q}(x_{I3}, t) \cdot \hat{e}_x = -k_s \frac{\partial T}{\partial x}(x_{I3}, t)\hat{e}_x \cdot \hat{e}_x = h_{in}[T(x_{I3}, t) - T_a(t)]$$

The boundary condition is valid at t_n and t_{n+1}. The simplest difference form of the preceding equation is

$$-k_s \frac{T_{I3}^{n+1} - T_{I3-1}^{n+1}}{\Delta x_3} = h_{in}\left[T_{I3}^{n+1} - T_a^{n+1}\right]$$

Solving for T_{I3}^{n+1} gives

$$T_{I3}^{n+1} = \frac{\dfrac{k_s}{\Delta x_3} T_{I3-1}^{n+1} + h_{in} T_a^{n+1}}{h_{in} + \dfrac{k_s}{\Delta x_3}} \qquad \text{(P13.8m)}$$

Inside the freezer

Heat transfer into the freezer is slow. It has been assumed that all material inside the freezer is at the same uniform temperature. The freezer interior consists of meat, vegetables, and air. The finite-difference formula for the temperature inside the safe is

$$T_a^{n+1} = T_a^n + \frac{A_s h_{in} \Delta t \left(T_{I3}^n - T_a^n\right)}{\rho_a V_a c_{v,a} + \rho_m V_m c_m + \rho_{veg} V_{veg} c_{veg}} \tag{P13.8n}$$

where
  ρ is the density
  V is the volume
  c is the specific heat
  Subscripts: a is air, m is meats, veg is vegetables

This numerical method has a stability criteria; that is, in each region

$$r = \frac{\alpha \Delta t}{\Delta x^2} < \frac{1}{2}$$

Method of solution

Since T_i^1, for $i = 1, 2, ..., I3$ and T_a^1 are known, T_i^2 can be obtained for all i by marching as indicated in the following procedure:

1. Solve for T_i^2 by Equation P13.8b for $i = 2, 3, ..., I1 - 1$.
2. Solve for T_i^2 by Equation P13.8c for $i = I1 + 1, I1 + 2, I1 + 3, ..., I2 - 1$.
3. Solve for T_i^2 by Equation P13.8b for $i = I2 + 1, I2 + 2, ..., I3 - 1$.
4. Solve for T_1^2 by Equation P13.8j.
5. Solve for T_{I1}^2 by Equation P13.8k.
6. Solve for T_{I2}^2 by Equation P13.8L.
7. Solve for T_a^2 by Equation P13.8n.
8. Solve for T_{I3}^2 by Equation P13.8m.
9. Use a counter and an if statement to determine when to print out temperature values.
10. Use a loop to reset $T_i^1 = T_i^2$ for all i.
11. Repeat process until $t =$ the specified time for the study.

Develop a computer program in MATLAB to solve the wall temperature distribution and the temperature inside the freezer. Use the following constants for the problem:

$$k_s = 61.0 \text{ W/(m-°C)}, \quad k_{fg} = 0.038 \text{ W/(m-°C)},$$

$$\alpha_{fg} = 1.6 \times 10^{-7} \text{ m}^2/\text{s}, \quad \alpha_s = 1.665 \times 10^{-5} \text{ m}^2/\text{s}$$

$$\rho_a = 1.394 \text{ kg/m}^3, \quad \rho_m = 961.0 \text{ kg/m}^3, \quad \rho_{veg} = 770 \text{ kg/m}^3$$

$$c_{v,a} = 0.716 \times 10^3 \text{ W/(kg-°C)}, \quad c_m = 1.93 \times 10^3 \text{ W/(kg-°C)},$$

$$c_{veg} = 1.97 \times 10^3 \text{ W/(kg-°C)}$$

$$h_o = 20 \text{ W/(m}^2\text{-°C)}, \quad h_{in} = 10 \text{ W/(m}^2\text{-°C)}, \quad T_\infty = 26.0°\text{C},$$

$$T_a^1 = -18°\text{C}$$

$$L_1 = 0.005 \text{ m}, \quad L_2 = 0.02 \text{ m}, \quad I1 = 6, \quad I2 = 86, \quad I3 = 91,$$

$$dx_1 = 0.001 \text{ m}, \quad dx_2 = 0.00025 \text{ m}, \quad dt = 0.02 \text{ s},$$

$$V_{int} = 0.75 \times 0.75 \times 0.5 \text{ m}^3$$

$$V_a = 0.2 V_{int}, \quad V_m = 0.4 V_{int}, \quad V_{veg} = 0.4 V_{int}, \quad \text{time of study} = 7200 \text{ s}.$$

Neglecting heat transfer from the bottom surface, which is in contact with the refrigeration compartment, $A_s = 2.0625 \text{ m}^2$.

Print out a table in steps of 100 seconds as shown in Table P13.8.

Note: We developed the program as described above and ran the study for 7200 seconds (three hours). We found that in the time span of the study, the temperature change inside the freezer compartment was approximately 3.5°C. As a test, run the program for 7200 seconds with only air inside the freezer compartment and see if that significantly increases the temperature change inside the freezer compartment. To make the change, you will need to replace Equation P13.8n with the following equation:

$$T_a^{n+1} = T_a^n + \frac{A_s h_{in} \Delta t \left(T_{I3}^n - T_a^n\right)}{\rho_a V_a c_{v,a}}$$

Table P13.8 Temperature Distribution Table

Time (s)	Exterior Surface x_1	x_{I1}	x Position (m) $\frac{1}{2}(x_{I1} + x_{I2})$	x_{I2}	Interior Surface x_{I3}	Air Temp. (°C)
0	—	—	—	—	—	−18
100	—	—	—	—	—	—
200	—	—	—	—	—	—
⋮	⋮	⋮	⋮	⋮	⋮	⋮
7200	—	—	—	—	—	—

Note: Print out temperatures every 100 s.

P13.9 Determine the temperature distribution in the slab shown in Figure 13.7 by the Gauss–Seidel method described in Section 13.5. Use the following values:

$$L = 1.0 \text{ m}, \ w = 0.2 \text{ m}, \ T_w = 300°C, \ T_\infty = 50°C, \ h = 800 \text{ W/m}^2\text{-}°C,$$
$$k = 386 \text{ W/m-}°C, \ N = 200, \ M = 50, \ \varepsilon = 0.001$$

As an initial guess, take the centerline temperature to be the solution of heat flow through a wall; that is,

$$T_{n,1}^1 = T_\infty + (T_w - T_\infty) \times \frac{x_n}{L}, \quad \text{for } n = 2,3,\ldots, N$$

and for positions other than the centerline, assume a temperature of a flow through a wall with a convective boundary condition; that is,

$$T_{n,M+1}^1 = \frac{w}{k - hw}\left(\frac{k}{w} T_{n,1}^1 - h T_\infty \right)$$

and

$$T_{n,m}^1 = T_{n,1}^1 + \left(T_{n,M+1}^1 - T_{n,1}^1 \right) \times \frac{y_m}{w}, \quad \text{for } n = 2,3,\ldots,N, \quad m = 2,3,\ldots,M$$

Once convergence has been achieved, define $T1_n = T_{n,1}^2$, $T2_n = T_{n,M/2+1}^2$, and $T3_n = T_{n,M+1}^2$ for $n = 1, 2, 3, \ldots, N + 1$. Construct a table for temperatures $T1$, $T2$ and $T3$ for $n = 1, 6, 11, \ldots, N + 1$. Also create plots of $T1$, $T2$, and $T3$ all on the same page.

Chapter 14

Laplace Transforms

14.1 Introduction

Laplace transforms [1,2] can be used to solve ordinary differential equations (ODEs) and partial differential equations (PDEs).

The method transforms an ODE to an algebraic equation in the Laplace domain that can be manipulated into a form such that the inverse transform can be obtained from tables. The inverse transform is the solution to the differential equation. The inverse transform can also be obtained by residue theory in complex variables (which is beyond the scope of this textbook). The method is applicable to problems where the independent variable domain is from (0 to ∞). The method is particularly useful for linear, nonhomogeneous differential equations, such as vibration problems where the forcing function is piecewise continuous. In electrical engineering, the method is often used to solve circuit problems containing capacitors or inductors that contain at least one differential relationship.

14.2 Laplace Transform and Inverse Transform

Let $f(t)$ be a function defined for all $t \geq 0$, then

$$\mathcal{L}\big(f(t)\big) = F(s) = \int_0^\infty e^{-st} f(t)dt \qquad (14.1)$$

$F(s)$ is called the *Laplace transform* of $f(t)$, and the symbol $\mathcal{L}$ is used to indicate the transform operation. By convention, we use lowercase letters for t domain functions and uppercase for their Laplace domain (or "*s* domain") counterparts.

The *inverse Laplace transform* of $F(s)$ is defined to be the function $f(t)$ such that

$$\mathcal{L}^{-1}\big(F(s)\big) = f(t) \tag{14.2}$$

Tables have been created that contain both $f(t)$ and the corresponding $F(s)$ (see Table 14.1 at the end of this chapter). Often, we will need to rearrange the transform into a form (or several forms) that are available in the table of Laplace transforms. We now derive the Laplace transforms for some common functions to illustrate how a table of Laplace transforms might be created.

■ Laplace transform of constant 1

$$\text{Let } f(t) = 1; \quad t \geq 0,$$

then

$$\mathcal{L}(1) = \int_0^\infty e^{-st} dt = -\left[\frac{e^{-st}}{s}\right]_0^\infty = \frac{1}{s} \tag{14.3}$$

■ Laplace transform of exponential

$$\text{Let } f(t) = e^{at}$$

$$\mathcal{L}(e^{at}) = \int_0^\infty e^{at} e^{-st} dt = \int_0^\infty e^{-(s-a)t} dt = \left[-\frac{e^{-(s-a)t}}{(s-a)}\right]_0^\infty = \frac{1}{s-a} \tag{14.4}$$

■ Linearity of Laplace transforms

$$\mathcal{L}\big(af(t) + bg(t)\big) = a\mathcal{L}\big(f(t)\big) + b\mathcal{L}\big(g(t)\big) \tag{14.5}$$

■ Laplace transform of complex exponential, sine, and cosine functions

$$\text{Let } f(t) = e^{i\omega t}$$

$$\mathcal{L}(e^{i\omega t}) = \frac{1}{s - i\omega} = \frac{1}{s - i\omega} \times \frac{s + i\omega}{s + i\omega} = \frac{s + i\omega}{s^2 + \omega^2}$$

$$\mathcal{L}(e^{i\omega t}) = \frac{s}{s^2 + \omega^2} + i\frac{\omega}{s^2 + \omega^2} \tag{14.6}$$

$$\mathcal{L}(e^{i\omega t}) = \mathcal{L}(\cos \omega t) + i\mathcal{L}(\sin \omega t) \tag{14.7}$$

Equating the real and imaginary components of Equations 14.6 and 14.7 gives

$$\mathcal{L}(\cos \omega t) = \frac{s}{s^2 + \omega^2} \tag{14.8}$$

$$\mathcal{L}(\sin \omega t) = \frac{\omega}{s^2 + \omega^2} \tag{14.9}$$

■ Laplace transform of a function, $f(t)$, times the exponential e^{at}
  If $\mathcal{L}(f(t)) = F(s)$, then

$$F(s-a) = \int_0^\infty f(t)e^{-(s-a)t}\,dt = \int_0^\infty f(t)e^{at}e^{-st}\,dt \tag{14.10}$$

or

$$F(s-a) = \mathcal{L}(f(t)e^{at}) \tag{14.11}$$

Combining Equations 14.8 and 14.11, we obtain

$$\mathcal{L}(e^{at}\cos \omega t) = \frac{s-a}{(s-a)^2 + \omega^2} \tag{14.12}$$

Similarly,

$$\mathcal{L}(e^{at}\sin \omega t) = \frac{\omega}{(s-a)^2 + \omega^2} \tag{14.13}$$

■ Laplace transform of t^{n+1}

$$\mathcal{L}(t^{n+1}) = \int_0^\infty t^{n+1}e^{-st}\,dt$$

To solve, integrate by parts. Let $u = t^{n+1}$ and $dv = e^{-st}\,dt$. Then, $du = (n+1)t^n\,dt$ and $v = -\dfrac{e^{-st}}{s}$. The formula for integration by parts is

$$\int u\,dv = uv - \int v\,du \tag{14.14}$$

Thus,

$$\int_0^\infty t^{n+1} e^{-st} dt = -t^{n+1} \frac{e^{-st}}{s} \Bigg|_{t=0}^\infty + \frac{n+1}{s} \int_0^\infty t^n e^{-st} dt = \frac{n+1}{s} \mathcal{L}(t^n) \qquad (14.15)$$

From Equation 14.15, we can see that

$$\mathcal{L}(t^n) = \frac{n}{s} \mathcal{L}(t^{n+1})$$

$$\mathcal{L}(t^{n-1}) = \frac{n-1}{s} \mathcal{L}(t^{n-2})$$

$$\vdots$$

$$\mathcal{L}(t) = \frac{1}{s} \mathcal{L}(t^0) = \frac{1}{s} \mathcal{L}(1) = \frac{1}{s^2}$$

Exercises

E14.1 Determine $\mathcal{L}(e^{2t} \cos t)$.

E14.2 Determine $\mathcal{L}(e^{-3t} \sin 2t)$.

E14.3 Determine $\mathcal{L}^{-1}\left(\dfrac{1}{(s+3)(s+2)}\right)$.

14.3 Transforms of Derivatives

(a) $\mathcal{L}(f') = \int_0^\infty f'(t) e^{-st} dt$

Let $dv = \dfrac{df}{dt} dt$, then $v = f$.

Let $u = e^{-st}$, then $du = -se^{-st} dt$:

$$\int_0^\infty f'(t) e^{-st} dt = [fe^{-st}]_0^\infty + s \int_0^\infty fe^{-st} dt = -f(0) + s\mathcal{L}(f) \qquad (14.16)$$

(b) $\mathcal{L}(f'') = \int\limits_0^\infty f''(t)e^{-st}\,dt$

Let $dv = \dfrac{df'}{dt}\,dt$, then $v = f'$.

Let $u = e^{-st}$, then $du = -s\,e^{-st}\,dt$:

$$\int\limits_0^\infty f''(t)e^{-st}\,dt = [f'e^{-st}]_0^\infty + s\int\limits_0^\infty f'e^{-st}\,dt = -f'(0) - sf(0) + s^2\mathcal{L}(f) \quad (14.17)$$

By Equations 14.16 and 14.17, we can see the pattern for the Laplace transform of the nth derivative, that is,

$$\mathcal{L}(f^{(n)}) = s^n\mathcal{L}(f) - s^{n-1}f(0) - s^{n-2}f'(0) - \cdots - f^{(n-1)}(0) \quad (14.18)$$

14.4 Ordinary Differential Equations, Initial Value Problem

Consider the differential equation arising from a spring–dashpot–mass system with a driving force (see Projects P2.10 and P2.11). The governing differential equation for the motion of the mass, $y(t)$, is

$$y'' + \frac{c}{m}y' + \frac{k}{m}y = \frac{F_0(t)}{m} \quad (14.19)$$

where
m is the mass (kg)
k is the spring constant (N/m)
c is the damping coefficient (N-s/m)
F_0 is the driving force (N)

Take the initial conditions to be

$$y(0) = \alpha, \quad y'(0) = \beta$$

Let $p = \dfrac{c}{m}$, $q = \dfrac{k}{m}$, and $r = \dfrac{F_0(t)}{m}$, then Equation 14.19 becomes

$$y'' + py' + qy = r(t) \tag{14.20}$$

The Laplace transform of each of the terms in Equation 14.20 follows:

$$\mathcal{L}(y'') = s^2 \mathcal{L}(y) - sy(0) - y'(0)$$

$$\mathcal{L}(y') = s\mathcal{L}(y) - y(0)$$

Let $\mathcal{L}(y) = Y$ and $\mathcal{L}(r) = R$, then the Laplace transform becomes

$$(s^2 Y - s\alpha - \beta) + p(sY - \alpha) + qY = R$$

or

$$(s^2 + ps + q)Y = (s + p)\alpha + \beta + R$$

Then,

$$Y(s) = \frac{(s + p)\alpha + \beta + R(s)}{s^2 + ps + q} \tag{14.21}$$

By the use of partial fractions and the Laplace transform tables, we can obtain the inverse transform, $\mathcal{L}^{-1}(Y(s)) = y(t)$.

Another example involving differential equations that can be solved by Laplace transforms arises in parallel RLC circuits. The governing differential equation for the inductor current i is

$$i'' + \frac{1}{RC}i' + \frac{1}{LC}i = \frac{I_o(t)}{LC} \tag{14.22}$$

where
  L is the inductance (H)
  R is the resistance (Ω)
  C is the capacitance (F)
  I_o is the driving current (A)

We shall assume that the initial conditions are

$$i(0) = \alpha \quad \text{and} \quad i'(0) = \beta$$

First, we convert this equation into the following general form:

Let $p = \dfrac{1}{RC}$, $q = \dfrac{1}{LC}$, and $r(t) = \dfrac{I_o(t)}{LC}$. Then, Equation 14.22 becomes

$$i'' + pi' + qi = r(t) \tag{14.23}$$

Equation 14.23 is the same as Equation 14.20 and can be solved in the same manner.

Example 14.1

Use Laplace transforms to solve the following differential equation (no damping and an exponentially decaying driving force):

$$y'' + y = 5e^{-t}$$
$$\tag{14.24}$$
$$y(0) = 2, \quad y'(0) = 0$$

This problem fits the general form of Equation 14.21, with

$$p = 0, \quad q = 1, \quad \alpha = 2, \quad \beta = 0 \quad \text{and} \quad r = 5e^{-t} \quad \text{and} \quad \mathcal{L}(5e^{-t}) = \frac{5}{s+1}$$

Then,

$$Y(s) = \frac{2s}{s^2 + 1} + \frac{5}{(s+1)(s^2 + 1)} \tag{14.25}$$

For the second term on the right side of Equation 14.25, we need to decompose the term by the method of partial fractions, that is,

$$\frac{5}{(s+1)(s^2 + 1)} = \frac{A}{s+1} + \frac{Bs + C}{s^2 + 1} = \frac{A(s^2 + 1) + (Bs + C)(s+1)}{(s+1)(s^2 + 1)}$$

then

$$A + B = 0, \quad C + B = 0, \quad A + C = 5 \rightarrow A = \frac{5}{2}; \quad B = -\frac{5}{2}, \quad C = \frac{5}{2}$$

Therefore,

$$\frac{5}{(s+1)(s^2 - 1)} = \frac{5}{2(s+1)} - \frac{5}{2}\frac{(s-1)}{(s^2 + 1)}$$

Thus,

$$Y(s) = \frac{2s}{(s^2+1)} + \frac{5}{2(s+1)} - \frac{5s}{2(s^2+1)} + \frac{5}{2(s^2+1)} \qquad (14.26)$$

From Table 14.1,

$$\mathcal{L}^{-1}\left(\frac{s}{s^2+1}\right) = \cos t$$

$$\mathcal{L}^{-1}\left(\frac{1}{s^2+1}\right) = \sin t$$

$$\mathcal{L}^{-1}\left(\frac{1}{s+1}\right) = e^{-t}$$

Therefore, the solution to Equation 14.22 is

$$y(t) = -\frac{1}{2}\cos t + \frac{5}{2}\sin t + \frac{5}{2}e^{-t} \qquad (14.27)$$

A plot of y vs. t is shown in Figure 14.1.

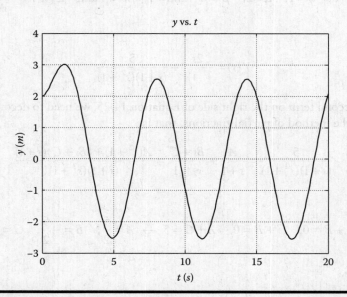

Figure 14.1   Plot of *y* vs. *t* for Example 14.1.

14.5 MATLAB®'s `residue` Function

MATLAB®'s `residue` function reduces two polynomials of the form $B(s)/A(s)$ into partial fractions. If there are no multiple roots, the partial fraction expansion is of the form

$$\frac{B(s)}{A(s)} = \frac{R(1)}{s - P(1)} + \frac{R(2)}{s - P(2)} + \cdots + \frac{R(n)}{s - P(n)} + K(s) \qquad (14.28)$$

The residue function takes two input arguments, B and A, where vectors B and A specify the coefficients of the numerator and denominator polynomials in descending powers of s. The output of the function contains three arguments, R(residues), P(poles), and K. To call the `residue` function, use [R, P, K] = `residue(B, A)`.

The residues are returned in the column vector R, the pole locations in column vector P, and the direct terms in row vector K. The number of poles, n, is

$$n = \text{length}(A) - 1 = \text{length}(R) = \text{length}(P).$$

The direct term coefficient vector, K, is empty if `length(B)` < `length(A)`; otherwise,

$$\text{length}(K) = \text{length}(B) - \text{length}(A) + 1.$$

Example 14.2

Let us apply MATLAB's residue function to the second term on the right side of Equation 14.25, which is

$$\frac{5}{(s+1)(s^2 + 1)} = \frac{5}{s^3 + s^2 + s + 1}$$

Then $B = 5$ and $A = [1\ 1\ 1\ 1]$.

The program using MATLAB's residue function follows:

```
% Example_14_2.m
% This example is a test of MATLAB's residue function.
% A = s^3+s^2+s+1 = [1 1 1 1], B = 5.
clear; clc;
[R,P,K] = residue(5,[1 1 1 1])
```

Program results

```
R =
    2.5000 + 0.0000i
   -1.2500 - 1.2500i
   -1.2500 + 1.2500i
P =
   -1.0000 + 0.0000i
   -0.0000 + 1.0000i
   -0.0000 - 1.0000i
K =
    []
>>
```

Thus,

$$\frac{5}{s^3 + s^2 + s + 1} = \frac{2.5}{s+1} + \frac{-1.25 - 1.25i}{s-i} + \frac{-1.25 + 1.25i}{s+i}$$

From Table 14.1,

$$\mathcal{L}^{-1}\left(\frac{1}{s-a}\right) = e^{at}$$

Therefore,

$$\mathcal{L}^{-1}\left(\frac{5}{s^3 + s^2 + s + 1}\right) = 2.5e^{-t} - (1.25 + 1.25i)e^{it} - (1.25 - 1.25i)e^{-it}$$

$$\mathcal{L}^{-1}\left(\frac{5}{s^3 + s^2 + s + 1}\right) = 2.5e^{-t} - (1.25 + 1.25i)(\cos t + i\sin t)$$

$$- (1.25 - 1.25i)(\cos t - i\sin t)$$

$$\mathcal{L}^{-1}\left(\frac{5}{s^3 + s^2 + s + 1}\right) = 2.5e^{-t} - 2.5\cos t + 2.5\sin t$$

Adding $\mathcal{L}^{-1}\left(\dfrac{2s}{s^2 + 1}\right) = 2\cos t$ gives the solution to Equation 14.25, which is

$$y(t) = -0.5\cos t + 2.5\sin t + 2.5e^{-t}$$

We see that we got the same answer as before using MATLAB's residue function.

Exercise

E14.4 Using Laplace transforms and MATLAB's `residue` function, determine the solution of the following differential equation:

$$y'' + 3y' + 2y = 5\sin 2t$$

$$y(0) = 1, \quad y'(0) = -4$$

E14.5 Solve the following differential equation by Laplace transforms:

$$y'' + 3y' - 2y = 3e^{-t}, \quad y(0) = 1, \quad y'(0) = 2$$

14.6 Unit Step Function

The unit step function, $u(t - a)$, is useful in analyzing beams and electrical circuits and is defined as follows (see Figure 14.2):

$$u(t - a) = \begin{cases} 0, & \text{if } t < a \\ 1, & \text{if } t \geq a \end{cases} \tag{14.29}$$

The Laplace transform of the unit step function is

$$\mathcal{L}(u(t - a)) = \int_0^\infty u(t - a)e^{-st}\,dt = \int_0^a 0 \cdot e^{-st}\,dt + \int_a^\infty 1 \cdot e^{-st}\,dt$$

$$\mathcal{L}(u(t - a)) = -\frac{1}{s}\left[e^{-st}\right]_a^\infty = -\frac{1}{s}[0 - e^{-as}] = \frac{1}{s}e^{-as} \tag{14.30}$$

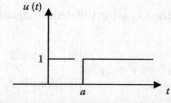

Figure 14.2 Unit step function.

The function

$$f(t-a)u(t-a) = \begin{cases} 0, & \text{if } t < a \\ f(t-a), & \text{if } t \geq a \end{cases}$$

has the Laplace transform $e^{-as}F(s)$, where $\mathcal{L}(f(t)) = F(s)$.

Proof

$$e^{-as}F(s) = e^{-as}\int_0^\infty f(\tau)e^{-s\tau}\,d\tau = \int_0^\infty f(\tau)e^{-s(\tau+a)}\,d\tau$$

Let $t = \tau + a$, then $dt = d\tau$.
　When $\tau = 0$, $t = a$, and when $\tau = \infty$, $t = \infty$.
Therefore,

$$e^{-as}F(s) = e^{-as}\int_a^\infty f(t-a)e^{-s(t-a)}\,dt = \int_a^\infty f(t-a)e^{-st}\,dt$$

$$= \int_0^a 0e^{-st}\,dt + \int_a^\infty f(t-a)e^{-st}\,dt$$

$$e^{-as}F(s) = \int_0^\infty u(t-a)f(t-a)e^{-st}\,dt$$

$$e^{-as}F(s) = \mathcal{L}(u(t-a)f(t-a)) \tag{14.31}$$

Example 14.3

Determine the solution of the following differential equation:

$$y'' + 3y' + 2y = \begin{cases} 5t, & \text{for } t < 2 \\ 0, & \text{for } t \geq 2 \end{cases} \tag{14.32}$$

$$y(0) = 1, \quad y'(0) = 0$$

Applying the standard form of Equation 14.20 for this problem, $p = 3$, $q = 2$, $\alpha = 1$, $\beta = 0$, and

$$r(t) = \begin{cases} 5t, & \text{for } t < 2 \\ 0, & \text{for } t \geq 2 \end{cases}$$

$$r(t) = 5tu(t) - 5tu(t-2) = 5tu(t) - 5(t-2)u(t-2) - 10u(t-2)$$

$$R(s) = \mathcal{L}(r(t)) = \mathcal{L}(5tu(t) - 5(t-2)u(t-2) - 10u(t-2))$$

$$R(s) = \frac{5}{s^2} - \frac{5e^{-2s}}{s^2} - \frac{10e^{-2s}}{s}$$

$$(s^2 + ps + q)Y(s) = (s + p)\alpha + \beta + R$$

$$Y(s) = \frac{(s+p)\alpha + \beta}{s^2 + ps + q} + \frac{R(s)}{s^2 + ps + q}$$

$$Y(s) = \frac{s}{(s^2 + 3s + 2)} + \frac{3}{(s^2 + 3s + 2)} + \frac{5}{s^2(s^2 + 3s + 2)}$$

$$- \frac{5e^{-2s}}{s^2(s^2 + 3s + 2)} - \frac{10e^{-2s}}{s(s^2 + 3s + 2)}$$

$$\text{Let } Y(s) = Y_1 + Y_2 + Y_3 + Y_4 + Y_5$$

where

$$Y_1 = \frac{s}{(s+2)(s+1)}$$

$$Y_2 = \frac{3}{(s+2)(s+1)}$$

$$Y_3 = \frac{5}{s^2(s+2)(s+1)}$$

$$Y_4 = -\frac{5e^{-2s}}{s^2(s+2)(s+1)}$$

$$Y_5 = -\frac{10e^{-2s}}{s(s+2)(s+1)}$$

By the use of partial fractions, we can determine that

$$\mathcal{L}^{-1}(Y_1) = 2e^{-2t} - e^{-t}$$

$$\mathcal{L}^{-1}(Y_2) = -3e^{-2t} + 3e^{-t}$$

$$\mathcal{L}^{-1}(Y_3) = \frac{5}{2}t - \frac{15}{4} - \frac{5}{4}e^{-2t} + 5e^{-t}$$

$$\mathcal{L}^{-1}(Y_4) = -\left\{ \frac{5}{2}t - 5 - \frac{15}{4} - \frac{5}{4}e^4 e^{-2t} + 5e^2 e^{-t} \right\} u(t-2)$$

$$\mathcal{L}^{-1}(Y_5) = -10\left\{ \frac{1}{2} + \frac{1}{2}e^4 e^{-2t} - e^2 e^{-t} \right\} u(t-2)$$

Summing these five terms gives

$$y(t) = \begin{cases} \dfrac{5}{2}t - \dfrac{15}{4} - \dfrac{9}{4}e^{-2t} + 7e^{-t}, & \text{for } t < 2 \\[2ex] -\left(\dfrac{9}{4} + \dfrac{15}{4}e^4 \right)e^{-2t} + (7 + 5e^2)e^{-t}, & \text{for } t \geq 2 \end{cases}$$

This problem can also be solved numerically by the use of `ode45` function in MATLAB. The MATLAB program follows:

```
% Example_14_3.m
% This program solves a system of 2 differential equations
% by using ode45 function. The problem is to determine the
% y(t)positions of a mass in a mass-spring-dashpot system
% Y1 = y, Y2 = v, Y1' = Y2, Y2' = 5*t-3Y2-2Y1, for t < 2,
% Y2' = -3Y2-2Y1, for t > = 2,
% y(0) = 1.0, y'(0) = 0
clear; clc;
initial = [1.0 0.0];
tspan = 0.0:0.1:4;
[t,Y] = ode45('dYdt_laplace',tspan,initial);
for i = 1:length(tspan)
    if tspan(i) < 2
        ylp(i) = 2.5*t(i)-15/4-9/4*exp(-2*t(i))+7*exp(-t(i));
    end
```

```
    if t(i) >= 2
        ylp(i) = -(9/4+15/4*exp(4))*exp(-2*t(i))+...
            (7+5*exp(2))*exp(-t(i));
    end
end
size(Y)
plot(t,Y(:,1),'x',t,ylp), xlabel('t'),ylabel('y,ylp'),grid,
title('y(ode45) & y(Laplace) vs. t'), legend('ode45','Laplace');
```

```
% dYdt_laplace.m
% This function works with Example_14_3.m
% Y(1) = y, Y(2) = v
% Y1' = Y(2), Y2' = 5*t-3Y2-2Y1, if t<2
% Y2' = -3Y2-2Y1, if t >= 2
function Yprime = dYdt_laplace(t,Y)
Yprime = zeros(2,1);
Yprime(1) = Y(2);
if t < 2
    Yprime(2) = 5*t-3.0*Y(2)-2.0*Y(1);
else
    Yprime(2) = -3.0*Y(2)-2.0*Y(1);
end
```

A comparison of the `ode45` and the Laplace transform solutions is shown in Figure 14.3.

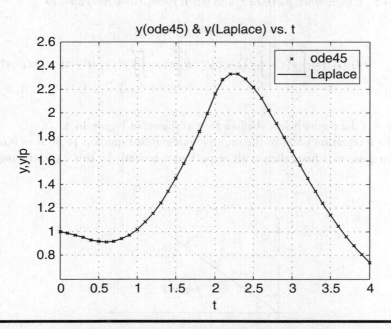

Figure 14.3 Comparison between solution obtained by Laplace transforms and `ode45`.

14.7 Convolution

Given two transforms $F(s)$ and $G(s)$ whose inverse transforms are $f(t)$ and $g(t)$, respectively, then

$$\mathcal{L}^{-1}\left(F(s)G(s)\right) = \int_0^t f(\tau)g(t-\tau)d\tau = \int_0^t g(\tau)f(t-\tau)d\tau \qquad (14.33)$$

Proof:

$$F(s)G(s) = \left(\int_0^\infty f(p)e^{-ps}\,dp\right)\left(\int_0^\infty g(\tau)e^{-\tau s}\,d\tau\right) \qquad (14.34)$$

$$F(s)G(s) = \int_0^\infty g(\tau)\left\{\int_0^\infty f(p)e^{-(p+\tau)s}\,dp\right\}d\tau \qquad (14.35)$$

Let $p + \tau = t$, then when $p = 0$, $t = \tau$, and when $p = \infty$, $t = \infty$. Also $dp = dt$.

$$F(s)G(s) = \int_0^\infty g(\tau)\left\{\int_\tau^\infty f(t-\tau)e^{-ts}\,dt\right\}d\tau = \iint_R g(\tau)f(t-\tau)e^{-ts}\,dt\,d\tau \qquad (14.36)$$

where R is the region below the line $t = \tau$ as shown in Figure 14.4.

The integration order of the multiple integral in Equation 14.36 is to integrate with respect to t first, then with respect to τ second. In this case, the order of

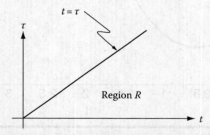

Figure 14.4   Integration region for convolution in the (τ, t) plane.

integration doesn't matter. So we can integrate with respect to τ first, then with respect to t second. This gives

$$F(s)G(s) = \iint_R g(\tau)f(t-\tau)e^{-st}\,dt\,d\tau$$

$$= \int_0^\infty e^{-st}\left\{\int_0^t g(\tau)f(t-\tau)d\tau\right\}dt = \mathcal{L}\left(\int_0^t g(\tau)f(t-\tau)d\tau\right)$$

Thus,

$$\mathcal{L}^{-1}\left(F(s)G(s)\right) = \int_0^t g(\tau)f(t-\tau)d\tau \tag{14.37}$$

We can reverse the roles of f and g, giving

$$\mathcal{L}^{-1}\left(F(s)G(s)\right) = \int_0^t g(t-\tau)f(\tau)d\tau$$

Convolution is frequently denoted by the "*" operator, and thus,

$$\mathcal{L}^{-1}\left(F(s)G(s)\right) = f(t) * g(t) = g(t) * f(t)$$

Example 14.4

Let us obtain the inverse of Y_3 in Example 14.3 using convolution. Y_3 is

$$Y_3 = \frac{5}{s^2(s+2)(s+1)} \tag{14.38}$$

Let $F(s) = \dfrac{5}{s^2}$ and $G(s) = \dfrac{1}{(s+2)(s+1)}$

From Table 14.1, items 2 and 16 are

$$f(t) = 5t \quad \text{and} \quad g(t) = e^{-t} - e^{-2t}$$

By the convolution formula,

$$\mathcal{L}^{-1}\left(F(s)G(s)\right) = \int_0^t g(\tau)f(t-\tau)d\tau = 5\int_0^t \left(e^{-\tau} - e^{-2\tau}\right)(t-\tau)d\tau \tag{14.39}$$

From integral tables,

$$\int e^{ax}\,dx = \frac{e^{ax}}{a} \quad \text{and} \quad \int xe^{ax}\,dx = \frac{e^{ax}}{a^2}(ax-1) \tag{14.40}$$

Thus,

$$\mathcal{L}^{-1}\left(F(s)G(s)\right) = 5\left(t\int_0^t e^{-\tau}\,d\tau + \int_0^t \tau e^{-2\tau}\,d\tau - t\int_0^t e^{-2\tau}\,d\tau - \int_0^t \tau e^{-\tau}\,d\tau \right)$$

$$\mathcal{L}^{-1}\left(F(s)G(s)\right) = 5\left\{ t\left(\frac{e^{-\tau}}{-1}\right)_0^t + \left(\frac{e^{-2\tau}}{4}(-2\tau-1)\right)_0^t - t\left(\frac{e^{-2\tau}}{-2}\right)_0^t - \left(\frac{e^{-\tau}}{1}(-\tau-1)\right)_0^t \right\}$$

$$= 5\left\{ -t(e^{-t}-1) - \frac{t}{2}e^{-2t} - \frac{1}{4}e^{-2t} + \frac{1}{4} + \frac{t}{2}e^{-2t} - \frac{t}{2} + te^{-t} + e^{-t} - 1 \right\}$$

$$\mathcal{L}^{-1}\left(F(s)G(s)\right) = \frac{5}{2}t - \frac{15}{4} - \frac{5}{4}e^{-2t} + 5e^{-t} \tag{14.41}$$

Comparing the solution of Y_3 obtained in Example 14.3 with the solution obtained by the convolution formula, we see that we got the same answer.

Exercise

E14.6 Determine $\mathcal{L}^{-1}\left(\dfrac{1}{(s+1)(s^2+3s+2)}\right)$ by convolution in the time domain.

14.8 Laplace Transforms Applied to Circuits

In circuit theory, capacitors and inductors have constituent relations that particularly lend themselves to analysis in the Laplace domain. Starting with the capacitor,

$$i_C = C\frac{dv_C}{dt}$$

where
  i_C is the capacitor current
  v_C is the capacitor voltage
  C is the capacitor value (F)

Applying Equation 14.18, we can rewrite this equation in the Laplace domain as

$$I_C(s) = C\big(sV_C(s) - v_C(0)\big)$$

If we neglect for the moment the initial condition, this becomes

$$V_C(s) = I_C(s)\frac{1}{Cs} \tag{14.42}$$

Note that Equation 14.42 resembles Ohm's law $(v = iR)$ and we thus define the *complex impedance* Z_C of a capacitor to be

$$Z_C(s) = \frac{1}{Cs} \tag{14.43}$$

By similar analysis with the constituent relation for inductors $v_L = L(di_L/dt)$, the complex impedance Z_L of an inductor L (neglecting initial conditions) is shown to be

$$Z_L(s) = Ls \tag{14.44}$$

Finally, for resistors, the Laplace transform of Ohm's law implies

$$Z_R(s) = R \tag{14.45}$$

Because we have previously proven that Laplace transforms are linear, we can now solve complicated linear circuits directly in the Laplace domain by using the same algebraic techniques used to solve the input/output relationships for resistor-only circuits. For example, the parallel RLC circuit as redrawn in Figure 14.5 with complex impedances can be solved with one equation:

$$V_C(s) = I_o(s)(Z_R(s) \| Z_C(s) \| Z_L(s))$$

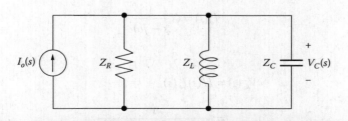

Figure 14.5   RLC circuit using impedances.

where the parallel operator $(\|)$ indicates the calculation for resistors (or impedances) in parallel. Then,

$$\frac{V_C(s)}{I_o(s)} = \frac{1}{\dfrac{1}{Z_R} + \dfrac{1}{Z_C} + \dfrac{1}{Z_L}}$$

$$= \frac{1}{\dfrac{1}{R} + Cs + \dfrac{1}{Ls}}$$

$$= \frac{1}{C} \times \frac{s}{s^2 + \dfrac{1}{RC}s + \dfrac{1}{LC}} \tag{14.46}$$

We also define

$$H(s) = \frac{V_C(s)}{I_o(s)} = \frac{1}{C} \times \frac{s}{s^2 + \dfrac{1}{RC}s + \dfrac{1}{LC}} \tag{14.47}$$

as the *transfer function* of the circuit, that is, the ratio of the output variable (in this case V_C) to the input driving function (in this case I_o).

Let us now assume a sinusoidal form for $i_o(t)$ such that

$$i_o(t) = Ce^{j\omega t}$$

where $e^{j\omega t} (= \cos \omega t + j \sin \omega t)$ is a sum of sines and cosines of frequency ω and $j = \sqrt{-1}$. For convenience, we have also added a constant factor of C in order to simplify the math later. Then,

$$I_o(s) = C\frac{1}{s - j\omega}$$

and

$$V_C(s) = H(s)I_o(s)$$

$$= \frac{s}{(s + s_1)(s + s_2)(s - j\omega)} \tag{14.48}$$

where s_1 and s_2 are the zeros of the denominator of Equation 14.48 and are calculated from R, L, and C. We also assume for this analysis that s_1 and s_2 are real. Rearranging Equation 14.48 by partial fractions, we get

$$V_C(s) = \frac{\dfrac{s_1}{(s_1 + s_2)(s_1 - j\omega)}}{s + s_1} + \frac{\dfrac{s_2}{(s_2 + s_1)(s_2 - j\omega)}}{s + s_2} + \frac{\dfrac{j\omega}{(j\omega + s_1)(j\omega + s_2)}}{s - j\omega}$$

By inverse transform, we get

$$v_C(t) = \frac{s_1}{(s_1 + s_2)(s_1 - j\omega)} e^{-s_1 t} + \frac{s_2}{(s_2 + s_1)(s_2 - j\omega)} e^{-s_2 t} + H(j\omega)e^{j\omega t} \quad (14.49)$$

Although Equation 14.49 looks complicated, note that the first two terms contain negative exponentials, and thus, in the *steady state* ($t \to \infty$), this equation reduces to

$$\lim_{t \to \infty} v_c(t) = H(j\omega)e^{j\omega t} \quad (14.50)$$

The meaning of Equation 14.50 is that for any sinusoidal input (of frequency ω) to the circuit, we can calculate the value of the steady-state output (also of frequency ω) by calculating $H(j\omega)$. This derivation can be generalized for determining the *frequency response* of any linear circuit by finding the transfer function $H(s)$ analytically and then letting $s = j\omega$ to find its frequency-dependent behavior. Frequency response is traditionally plotted as two graphs showing the magnitude $|H(j\omega)|$ and phase $\angle H(j\omega)$, both vs. frequency. The magnitude plot is typically on log-log axes, and the phase plot is typically on semilog axes (linear phase vs. log frequency). Together, the plots are referred to as a *Bode plot*.

Example 14.5

Let

$$H(s) = \frac{1000}{s + 1000}$$

Plot the magnitude and phase of $H(j\omega)$ over the interval $\omega = [1 \times 10^{-1}, 1 \times 10^{6}]$ rad/s.

```
% Example_14_5.m
% Plot magnitude and phase of H(jw)=1000/(s+1000)
% for w = [1e-1,1e6].
% First, create a logarithmic point set for the frequency
% range from 1 to 1e6 rad/sec with ten points in each decade.
for k = 1:60
    w(k) = 10^(k/10);
end
% Compute the magnitude for H(jw)
H_mag = abs(1000./(j*w + 1000));
% Compute the phase for H(jw), and convert to degrees
H_phase = (180/pi) * angle(1000./(j*w + 1000));
% Plot magnitude on log-log axes
subplot(2,1,1);
loglog(w,H_mag);
axis([10^0 10^6 10^-3 2*10^0]);
ylabel('mag H(jw)');
title('Bode Plot');
% Plot phase on log-linear axes
subplot(2,1,2);
semilogx(w,H_phase);
axis([10^0 10^6 -100 10]);
xlabel('Frequency (rad/sec)');
ylabel('phase H(jw)');
```

The solution is shown in Figure 14.6.

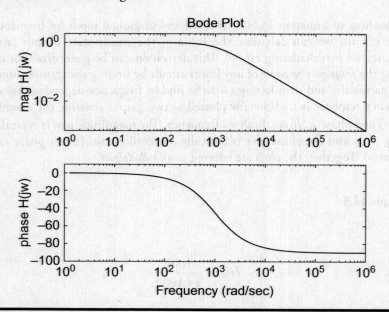

Figure 14.6   Bode plot for $H(j\omega) = 1000/(j\omega + 1000)$.

14.9 Delta Function

The *delta function* was formalized by physicist Paul Dirac to represent the action of an impulse, such as a bat striking a baseball. The function is defined as follows:

$$\delta(t-a) = \begin{cases} \infty & \text{for } t = a \\ 0 & \text{otherwise} \end{cases} \qquad (14.51)$$

such that

$$\int_0^\infty \delta(t-a)dt = 1 \qquad (14.52)$$

Although there are several different ways to develop this concept, we will use the one in Ref. [2]. Let us define a function *f* as follows:

$$f(t-a) = \begin{cases} \dfrac{1}{k} & \text{for } a \leq t \leq a + k \\ 0 & \text{otherwise} \end{cases} \qquad (14.53)$$

Then,

$$\lim_{k \to 0} \left(f(t-a) \right) = \begin{cases} \infty & \text{for } t = a \\ 0 & \text{otherwise} \end{cases} = \delta(t-a)$$

Now integrate $f(t-a)$ from 0 to ∞, giving

$$I = \int_0^\infty f(t-a)dt = \int_a^{a+k} \frac{1}{k}dt = \frac{1}{k}(a + k - a) = 1 \qquad (14.54)$$

Taking the limit of Equation 14.54 as $k \to 0$ gives Equation 14.52.

To obtain the Laplace transform, we can represent function *f* as two step functions, that is,

$$f(t-a) = \frac{1}{k} \left(u(t-a) - u(t-(a+k)) \right)$$

Then,

$$\mathcal{L}\big(f(t-a)\big) = \frac{1}{ks}\Big(e^{-as} - e^{-(a+k)s}\Big) = e^{-as}\left(\frac{1-e^{-ks}}{ks}\right) \tag{14.55}$$

In order to make $f(t-a) = \delta(t-a)$, we need to take the limit of $f(t-a)$ as $k \to 0$. To obtain the Laplace transform of $\delta(t-a)$, use L'Hôpital's rule, that is,

$$\lim_{k\to 0}\left(e^{-as}\frac{1-e^{-ks}}{ks}\right) = \lim_{k\to 0}\left(e^{-as} \times \frac{\dfrac{d}{dk}(1-e^{-ks})}{\dfrac{d}{dk}(ks)}\right) = \lim_{k\to 0}\left(e^{-as}\frac{se^{-ks}}{s}\right) = e^{-as} \tag{14.56}$$

Therefore,

$$\mathcal{L}\big(\delta(t-a)\big) = e^{-as}$$

The delta function has many useful modeling applications in mechanics and in circuit theory.

Example 14.6

Suppose we consider the mass–spring–dashpot system, initially at rest that is suddenly subjected to an impulse force. Let's use the same numbers as the left-hand side of Example 14.3, that is,

$$y'' + 3y' + 2y = 5\delta(t) \tag{14.57}$$
$$y(0) = 0, \quad y'(0) = 0$$

Comparing Equation 14.57 to the standard form of Equation 14.20, gives $p = 3$, $q = 2$, $\alpha = 0$, $\beta = 0$, and $r(t) = 5\delta(t)$:

$$R(s) = \mathcal{L}\big(r(t)\big) = 5$$

$$(s^2 + ps + q)Y(s) = (s + p)\alpha + \beta + R$$

$$Y(s) = \frac{(s+p)\alpha + \beta}{s^2 + ps + q} + \frac{R(s)}{s^2 + ps + q}$$

$$Y(s) = \frac{5}{(s^2 + 3s + 2)}$$

From Example 14.3, we see that $Y(s) = Y_2$ with 5 replacing 3 in the numerator. Thus, the solution is

$$y(t) = -5e^{-2t} + 5e^{-t}$$

We can also use the delta function to examine the behavior of a circuit subjected to a pulse input. In Equation 14.47, we defined $H(s)$ as the ratio of the circuit output to the circuit input. If we assume that the input is a delta function, then

$$V_C(s) = H(s)I_o(s)$$

$$= H(s)(1) \qquad (14.58)$$

where we have assumed $i_o(t) = \delta(t)$ with corresponding Laplace transform of unity. If we inverse transform Equation 14.58, we get

$$v_C(t) = h(t)$$

In other words, if the circuit input is a delta function, then the circuit output is simply the inverse transform of the transfer function $H(s)$, which is $h(t)$, also known as the *impulse response*. Conversely, if we can accurately calculate (or measure) the impulse response $h(t)$ of a circuit, then we can obtain the transfer function $H(s)$ and thus predict the circuit behavior for *any* arbitrary input waveform. This concept is illustrated in the following example.

Example 14.7

A circuit has an impulse response of

$$h(t) = 10e^{-2t}\cos 10t$$

What is the unit step response $r(t)$ of the circuit?

Transforming $h(t)$ into $H(s)$, we obtain

$$H(s) = 10\left(\frac{s+2}{(s+2)^2 + 100}\right) \qquad (14.59)$$

where we have applied line 26 in the Laplace transform table to obtain Equation 14.59. In order to obtain the step response, we first multiply $H(s)$ by the unit step transform $1/s$ to obtain $R(s)$:

$$R(s) = 10\left(\frac{s+2}{(s+2)^2 + 100}\right)\left(\frac{1}{s}\right)$$

We now calculate the inverse transform as follows:

$$r(t) = \mathcal{L}^{-1}\{R(s)\}$$

$$= 10\mathcal{L}^{-1}\left\{\frac{s+2}{s^3 + 4s^2 + 104s}\right\}$$

$$= 10\mathcal{L}^{-1}\left\{\frac{-0.0096 - 0.0481j}{s+(2-10j)} + \frac{-0.0096 + 0.0481j}{s+(2+10j)} + \frac{0.0192}{s}\right\}$$

(calculated using `residue`)

$$= 10(-0.0096e^{-2t+10jt} - 0.0481je^{-2t+10jt} - 0.0096e^{-2t-10jt}$$

$$+ 0.0481je^{-2t-10jt} + 0.0192u(t))$$

$$= 0.192(-e^{-2t}\cos 10t + 5e^{-2t}\sin 10t + u(t))$$

14.10 Laplace Transforms Applied to Partial Differential Equations

We will now show how Laplace transforms may be used for solving PDEs.

Let us obtain the Laplace transform of the following PDE:

$$A\frac{\partial^2 \theta}{\partial x^2} + B\frac{\partial^2 \theta}{\partial t^2} + C\frac{\partial \theta}{\partial x} + D\frac{\partial \theta}{\partial t} + E\theta = f(x,t) \qquad (14.60)$$

where $\theta = \theta(x, t)$ and $0 \leq t \leq \infty$.

We begin by multiplying each term of Equation 14.60 by e^{-st} and integrating from 0 to ∞:

The first term becomes

$$\mathcal{L}_{t \to s}\left(A\frac{\partial^2 \theta}{\partial x^2}\right) = \int_0^\infty A\frac{\partial^2 \theta}{\partial x^2}e^{-st}\,dt = A\frac{d^2}{dx^2}\int_0^\infty \theta e^{-st}\,dt = A\frac{d^2}{dx^2}\Theta(x,s)$$

where $\Theta(x,s)$ is the Laplace transform of $\theta(x,t)$. The remaining terms become

$$\mathcal{L}_{t \to s}\left(B\frac{\partial^2 \theta}{\partial t^2}\right) = \int_0^\infty B\frac{\partial^2 \theta}{\partial t^2}e^{-st}\,dt = B\left(s^2\Theta - s\theta(x,0) - \frac{d\theta}{dt}(x,0)\right)$$

$$\mathcal{L}_{t \to s}\left(C\frac{\partial \theta}{\partial x}\right) = \int_0^\infty C\frac{\partial \theta}{\partial x}e^{-st}\,dt = C\frac{d}{dx}\int_0^\infty \theta e^{-st}\,dt = C\frac{d}{dx}\Theta(x,s)$$

$$\underset{t \to s}{\mathscr{L}}\left(D\frac{\partial \theta}{\partial t} \right) = \int_0^\infty D\frac{\partial \theta}{\partial t}e^{-st}\,dt = D(s\Theta(x,s) - \theta(x,0))$$

$$\underset{t \to s}{\mathscr{L}}\left(E\theta \right) = \int_0^\infty E\theta e^{-st}\,dt = E\big(\Theta(x,s)\big)$$

$$\underset{t \to s}{\mathscr{L}}\left(f \right) = \int_0^\infty fe^{-st}\,dt = F(x,s)$$

From the above relations, it can be seen that Equation 14.60 becomes an ODE with respect to x. In this equation, s is considered a constant. That is,

$$A\frac{d^2\Theta}{dx^2} + C\frac{d\Theta}{dx} + (Bs^2 + Ds + E)\Theta = F(x,s) + \big(Bs + D\big)\theta(x,0) + B\frac{\partial\theta}{\partial t}(x,0)$$

(14.61)

We also needs to take the Laplace transforms of the boundary conditions. Suppose that

$$\theta(0,t) = g_1(t) \quad \text{and} \quad \frac{\partial\theta}{\partial x}(L,t) = g_2(t)$$

Then,

$$\Theta(0,s) = \mathscr{L}(g_1)$$

(14.62)

$$\underset{t \to s}{\mathscr{L}}\left(\frac{\partial\theta}{\partial x}(L,t) \right) = \lim_{x \to L}\frac{d}{dx}\int_0^\infty \theta(x,t)e^{-st}\,dt = \lim_{x \to L}\frac{d}{dx}\Theta(x,s)$$

Thus,

$$\frac{d\Theta}{dx}(L,s) = \mathscr{L}(g_2)$$

(14.63)

The initial conditions $\theta(x,0)$ and $\dfrac{\partial\theta}{\partial t}(x,0)$ are directly entered into Equation 14.61.

Example 14.8

Consider a semi-infinite slab, initially at a uniform temperature, T_i, which is suddenly subjected to temperature T_0 at its free surface (see Figure 14.7).

The governing PDE is

$$\frac{1}{a}\frac{\partial T}{\partial t} = \frac{\partial^2 T}{\partial x^2} \tag{14.64}$$

where a is the thermal diffusivity of the slab material.

Initial condition

$$T(x,0) = T_i \tag{14.65}$$

Boundary conditions

$$T(0,t) = T_0 \tag{14.66}$$

$$T(\infty,t) = T_i \tag{14.67}$$

To simplify the method for obtaining the solution, let $\theta(x,t) = T(x,t) - T_i$, then the equations reduce to

$$\frac{1}{a}\frac{\partial \theta}{\partial t} = \frac{\partial^2 \theta}{\partial x^2} \tag{14.68}$$

$$\theta(x,0) = 0 \tag{14.69}$$

$$\theta(0,t) = T_0 - T_i \tag{14.70}$$

$$\theta(\infty,t) = 0 \tag{14.71}$$

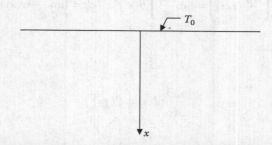

Figure 14.7 Semi-infinite slab suddenly subjected to a temperature change at its surface.

Taking the Laplace transform of both sides of Equations 14.68, 14.70, and 14.71 gives

$$\frac{1}{a}(s\Theta - \theta(x,0)) = \frac{d^2\Theta}{dx^2} \tag{14.72}$$

$$\Theta(\infty, s) = 0 \tag{14.73}$$

$$\Theta(0,s) = \frac{T_0 - T_i}{s} \tag{14.74}$$

where $\Theta = \underset{t \to s}{\mathcal{L}}(\theta(x,t))$.

Equation 14.72 becomes

$$\frac{d^2\Theta}{dx^2} - \frac{s}{a}\Theta = 0 \tag{14.75}$$

The solution is

$$\Theta(x,s) = c_1 e^{\sqrt{(s/a)}\,x} + c_2 e^{-\sqrt{(s/a)}\,x} \tag{14.76}$$

Appling boundary condition, Equation 14.73 gives

$$\Theta(\infty, s) = 0 \rightarrow c_1 = 0$$

Then,

$$\Theta = c_2 e^{-\sqrt{(s/a)}\,x}$$

Applying boundary condition, Equation 14.74 gives

$$\Theta(0,s) = \frac{T_0 - T_i}{s} = c_2(1)$$

Thus,

$$\Theta = \frac{T_0 - T_i}{s}e^{-\sqrt{s/a}\,x} \tag{14.77}$$

From Table 14.1, line 49

$$\mathcal{L}^{-1}\left(\frac{1}{s}e^{-k\sqrt{s}}\right) = \operatorname{erfc}\left(\frac{k}{2\sqrt{t}}\right) \tag{14.78}$$

Therefore,

$$\theta(x,t) = (T_0 - T_i)\operatorname{erfc}\left(\frac{x}{2\sqrt{at}}\right) \tag{14.79}$$

or

$$T(x,t) - T_i = T_0 - T_i - (T_0 - T_i)\operatorname{erf}\left(\frac{x}{2\sqrt{at}}\right)$$

$$\frac{T(x,t) - T_0}{T_i - T_0} = \operatorname{erf}\left(\frac{x}{2\sqrt{at}}\right) \tag{14.80}$$

Exercises

E14.7   Determine $\mathcal{L}(\sin^2 \omega t)$.

E14.8   Determine the Laplace transform of the function shown graphically in Figure E14.8.

E14.9   Determine $\mathcal{L}^{-1}\left(\dfrac{2s+1}{(s^2+9)(s-2)}\right)$.

E14.10 Determine $\mathcal{L}^{-1}\left(\dfrac{1}{(s^2+3s+2)}\right)$.

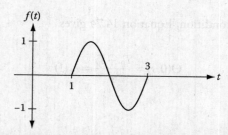

Figure E14.8   *f(t) consists of one cycle of a sine wave over the interval $1 \le t \le 3$ s.*

Projects

P14.1 Figure P14.1 shows a series RLC circuit driven by an input voltage $v_i(t)$ and three defined output voltages $v_R(t)$, $v_L(t)$, and $v_C(t)$.

1. Solve the circuit using complex impedances to find the three transfer functions $H_R(s) = V_R(s)/V_i(s)$, $H_L(s) = V_L(s)/V_i(s)$, and $H_C(s) = V_C(s)/V_i(s)$. Put the transfer functions in the following standard form:

$$H(s) = \frac{K_1 s^2 + K_2 s + K_3}{s^2 + K_4 s + K_5}$$

where the constants K_1 through K_5 are real. Note that the five constants are scaled such that the highest-order term in the denominator should have no constant.

2. Create the Bode plots for $H_R(j\omega)$, $H_L(j\omega)$, and $H_C(j\omega)$. Use these component values: $R = 1$ kΩ, $L = 0.1$ mH, and $C = 0.1$ μF. Plot over the range $\omega = [1, 1 \times 10^{10}]$ rad/s.

3. The three transfer functions represent three types of filters: *high pass*, *low pass*, and *bandpass*. Which transfer function corresponds to which filter type?

P14.2 A *common emitter amplifier* is shown in Figure P14.2a with input voltage v_{in} and output voltage v_{out}. In these types of single-transistor circuits, the voltages of interest ride atop DC components (V_{BIAS} and V_{OFFSET}) that are used to *bias* the transistor into its linear (i.e., amplification) mode. V_{BIAS} is the non-varying DC component of the input, which is adjusted to obtain the desired operating point, I_{BIAS}, of the transistor. V_{OFFSET} is

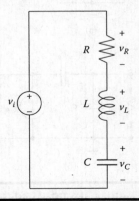

Figure P14.1   Series *RLC* circuit.

the DC component of the output. The voltages of interest are the varying components of the input v_{in} and v_{out}.

Figure P14.2b shows the small-signal *hybrid-pi* model of the bipolar transistor [3]. In the small-signal model, the DC components (V_{CC}, V_{BIAS}, and V_{OFFSET}) are ignored and we can use Kirchhoff's current law (KCL) to solve for the transfer function in terms of complex impedances. Writing KCL equations at nodes v_π and v_{out} gives

$$\frac{(v_{in} - v_\pi)}{r_b} - \frac{v_\pi}{r_\pi \| \dfrac{1}{c_\pi s}} + \frac{(v_{out} - v_\pi)}{\dfrac{1}{c_\mu s}} = 0 \tag{P14.2a}$$

$$\frac{(v_{out} - v_\pi)}{\dfrac{1}{c_\mu s}} + g_m v_\pi + \frac{v_{out}}{r_o \| R_L} = 0 \tag{P14.2b}$$

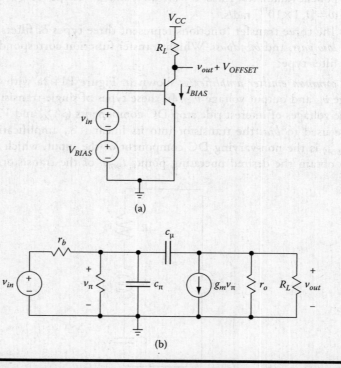

(a)

(b)

Figure P14.2 **(a) Common emitter amplifier with input voltage v_{in} and output voltage v_{out}. (b) Small-signal model of the common emitter amplifier.**

where

v_π is the small-signal base–emitter voltage

r_π is the small-signal base–emitter resistance

r_b is the base resistance

c_π and c_μ are the base–emitter and collector–emitter capacitances

r_o is the small-signal collector–emitter resistance

g_m is the small-signal transconductance

R_L is the load resistance

Solving these equations gives the transfer function of the system:

$$H(s) = \frac{v_{out}}{v_{in}} = \frac{\dfrac{1}{r_b}(c_\mu s - g_m)}{c_\pi c_\mu s^2 + \left(\dfrac{c_\pi}{r_o \parallel R_L} + \dfrac{c_\mu}{r_o \parallel R_L \parallel r_b \parallel r_\pi \parallel \dfrac{1}{g_m}}\right)s + \dfrac{1}{(r_b \parallel r_\pi)(r_o \parallel R_L)}}$$

(P14.2c)

If we assume that the transistor is model 2N3904, we can obtain numerical values for the circuit parameters from the manufacturer data sheet for this component [4]. These are summarized in Table P14.2. We will also assume that V_{BIAS} is adjusted to obtain a transistor collector current (I_{BIAS}) of 1 milliamp and that $R_L = 5$ kΩ. Because the 2N3904 is a discrete device (and not part of an integrated circuit), we will also

Table P14.2 Parameters for 2N3904 NPN Transistor

Transistor Parameter	Typical Value
$h_{FE}(=\beta)$	125
$C_{obo}(=c_\mu)$	3 pF
$C_{ibo}(=c_\pi)$	4 pF
$h_{ie}(=r_\pi)$	4 kΩ
$h_{oe}(=1/r_o)$	10 μmho

assume a small base resistance $r_b = 25 \ \Omega$. Also, assume that $g_m = I_o/V_T$, where $V_T = kT/q = 0.026$ mV.

1. Create the Bode plot for the transfer function of Equation P14.2a using the parameters in Table P14.2. Plot over the frequency interval $\omega = [1 \times 10^5, 1 \times 10^{11}]$ rad/s.

2. A common trick for hand analysis of this circuit is to use the *Miller approximation* where we assume alternative values for c_μ and c_π as follows:

$$c'_\mu = 0$$

$$c'_\pi = c_\pi + (g_m(r_o \ \| \ R_L) + 1)c_\mu$$

Substitute these approximated values for c_μ and c_π into Equation P14.2a and find the approximate transfer function $H_{\text{Miller}}(s)$.

3. Plot $H_{\text{Miller}}(s)$ on the same set of axes that you already used to plot $H(s)$. Under what circumstances is the Miller approximation reasonable to use?

P14.3 Consider the same problem as Example 14.8 except replace the boundary condition at $x = 0$ with the convection boundary condition, that is,

$$\frac{\partial T}{\partial x}(0,t) - \frac{h}{k}(T(0,t) - T_\infty) = 0 \tag{P14.3a}$$

Again, let $\theta(x,t) = T(x,t) - T_i$, then the PDE, the initial condition, and the boundary condition at $x = \infty$ are same as in Example 14.8, that is,

$$\frac{1}{a}\frac{\partial\theta}{\partial t} = \frac{\partial^2\theta}{\partial x^2} \tag{P14.3b}$$

$$\theta(x,0) = 0 \tag{P14.3c}$$

$$\theta(\infty,t) = 0 \tag{P14.3d}$$

By adding and subtracting T_i to the terms inside the parenthesis of the second term in Equation P14.2a, the boundary condition becomes

$$\frac{\partial\theta}{\partial x}(0,t) - \frac{h}{k}[\theta(0,t) + T_i - T_\infty] = 0 \tag{P14.3e}$$

Determine $\theta(x, t)$ by Laplace transforms.

Table 14.1   Table of Laplace Transforms

	$F(s)$	$f(t)$
1	$\dfrac{1}{s}$	1
2	$\dfrac{1}{s^2}$	t
3	$\dfrac{1}{s^n}$ $(n = 1, 2, \ldots)$	$\dfrac{t^{n-1}}{(n-1)!}$
4	$\dfrac{1}{\sqrt{s}}$	$\dfrac{1}{\sqrt{\pi t}}$
5	$s^{-(3/2)}$	$2\sqrt{\dfrac{t}{\pi}}$
6	$s^{-(n+1/2)}$ $(n = 1, 2, \ldots)$	$\dfrac{2^n\, t^{n-1/2}}{1 \times 3 \times 5 \ldots (2n-1)\sqrt{\pi}}$
7	$\dfrac{1}{s-a}$	e^{at}
8	$\dfrac{1}{(s-a)^2}$	te^{at}
9	$\dfrac{1}{(s-a)^n}$ $(n = 1, 2, \ldots)$	$\dfrac{1}{(n-1)!}t^{n-1}e^{at}$
10	$\dfrac{\Gamma(k)}{(s-a)^k}$ $(k > 0)$	$t^{k-1}e^{at}$
11	$\dfrac{1}{(s-a)(s-b)}$ $(a \neq b)$	$\dfrac{1}{(a-b)}(e^{at} - e^{bt})$
12	$\dfrac{s}{(s-a)(s-b)}$ $(a \neq b)$	$\dfrac{1}{(a-b)}(ae^{at} - be^{bt})$
13	$\dfrac{1}{(s-a)(s-b)(s-c)}$	$-\dfrac{(b-a)e^{at} + (c-a)e^{bt} + (a-b)e^{ct}}{(a-b)(b-c)(c-a)}$

(continued)

Table 14.1 (continued)   Table of Laplace Transforms

	$F(s)$	$f(t)$
14	$\dfrac{1}{s^2 + a^2}$	$\dfrac{1}{a}\sin at$
15	$\dfrac{s}{s^2 + a^2}$	$\cos at$
16	$\dfrac{1}{s^2 - a^2}$	$\dfrac{1}{a}\sinh at$
17	$\dfrac{s}{s^2 - a^2}$	$\cosh at$
18	$\dfrac{1}{s(s^2 + a^2)}$	$\dfrac{1}{a^2}(1 - \cos at)$
19	$\dfrac{1}{s^2(s^2 + a^2)}$	$\dfrac{1}{a^3}(at - \sin at)$
20	$\dfrac{1}{(s^2 + a^2)^2}$	$\dfrac{1}{2a^3}(\sin at - at\cos at)$
21	$\dfrac{s}{(s^2 + a^2)^2}$	$\dfrac{t}{2a}\sin at$
22	$\dfrac{s^2}{(s^2 + a^2)^2}$	$\dfrac{1}{2a}(\sin at + at\cos at)$
23	$\dfrac{s^2 - a^2}{(s^2 + a^2)^2}$	$t\cos at$
24	$\dfrac{s}{(s^2 + a^2)(s^2 + b^2)}$ $(a^2 \neq b^2)$	$\dfrac{\cos at - \cos bt}{b^2 - a^2}$
25	$\dfrac{1}{(s - a)^2 + b^2}$	$\dfrac{1}{b}e^{at}\sin bt$
26	$\dfrac{s - a}{(s - a)^2 + b^2}$	$e^{at}\cos bt$
27	$\dfrac{3a^2}{s^3 + a^3}$	$e^{-at} - e^{at/2}\left(\cos\dfrac{at\sqrt{3}}{2} - \sqrt{3}\sin\dfrac{at\sqrt{3}}{2}\right)$
28	$\dfrac{4a^3}{s^4 + 4a^4}$	$\sin at\cosh at - \cos at\sinh at$

Table 14.1 (continued)   Table of Laplace Transforms

	$F(s)$	$f(t)$
29	$\dfrac{s}{s^4 + 4a^4}$	$\dfrac{1}{2a^2}\sin at \sinh at$
30	$\dfrac{1}{s^4 - a^4}$	$\dfrac{1}{2a^3}(\sinh at - \sin at)$
31	$\dfrac{s}{s^4 - a^4}$	$\dfrac{1}{2a^2}(\cosh at - \cos at)$
32	$\dfrac{8a^3 s^2}{(s^2 + a^2)^3}$	$(1 + a^2 t^2)\sin at - at\cos at$
33	$\dfrac{1}{s}\left(\dfrac{s-1}{s}\right)^n$	$L_n(t) = \dfrac{e^t}{n!}\dfrac{d^n}{dt^n}(t^n e^{-t})$
34	$\dfrac{s}{(s-a)^{3/2}}$	$\dfrac{1}{\sqrt{\pi t}}e^{at}(1 + 2at)$
35	$\sqrt{s-a} - \sqrt{s-b}$	$\dfrac{1}{2\sqrt{\pi t^3}}(e^{bt} - e^{at})$
36	$\dfrac{1}{\sqrt{s}+a}$	$\dfrac{1}{\sqrt{\pi t}} - ae^{a^2 t}\operatorname{erfc}\left(a\sqrt{t}\right)$
37	$\dfrac{\sqrt{s}}{s - a^2}$	$\dfrac{1}{\sqrt{\pi t}} + ae^{a^2 t}\operatorname{erfc}\left(a\sqrt{t}\right)$
38	$\dfrac{1}{\sqrt{s}(s - a^2)}$	$\dfrac{1}{a}e^{a^2 t}\operatorname{erf}(a\sqrt{t})$
39	$\dfrac{1}{\sqrt{s}\left(\sqrt{s}+a\right)}$	$e^{a^2 t}\operatorname{erfc}\left(a\sqrt{t}\right)$
40	$\dfrac{1}{(s+a)\sqrt{s+b}}$	$\dfrac{1}{\sqrt{b-a}}e^{-at}\operatorname{erf}\left(\sqrt{b-a}\sqrt{t}\right)$
41	$\dfrac{b^2 - a^2}{\sqrt{s}\left(s - a^2\right)\left(\sqrt{s}+b\right)}$	$e^{a^2 t}\left[\dfrac{b}{a}\operatorname{erf}\left(a\sqrt{t}\right) - 1\right]$

(continued)

Table 14.1 (continued)   Table of Laplace Transforms

	$F(s)$	$f(t)$
42	$\dfrac{1}{\sqrt{s^2 + a^2}}$	$J_0(at)$
43	$\dfrac{1}{s}e^{-k/s}$	$J_0\left(2\sqrt{kt}\right)$
44	$\dfrac{1}{\sqrt{s}}e^{-k/s}$	$\dfrac{1}{\sqrt{\pi t}}\cos 2\sqrt{kt}$
45	$\dfrac{1}{\sqrt{s}}e^{k/s}$	$\dfrac{1}{\sqrt{\pi t}}\cosh 2\sqrt{kt}$
46	$\dfrac{1}{s^{3/2}}e^{-k/s}$	$\dfrac{1}{\sqrt{\pi k}}\sin 2\sqrt{kt}$
47	$\dfrac{1}{s^{3/2}}e^{k/s}$	$\dfrac{1}{\sqrt{\pi k}}\sinh 2\sqrt{kt}$
48	$e^{-k\sqrt{s}}\quad (k>0)$	$\dfrac{k}{2\sqrt{\pi t^3}}\exp\left(-\dfrac{k^2}{4t}\right)$
49	$\dfrac{1}{s}e^{-k\sqrt{s}}\quad (k\geq 0)$	$\mathrm{erfc}\left(\dfrac{k}{2\sqrt{t}}\right)$
50	$\dfrac{1}{\sqrt{s}}e^{-k\sqrt{s}}\quad (k\geq 0)$	$\dfrac{1}{\sqrt{\pi t}}\exp\left(-\dfrac{k^2}{4t}\right)$
51	$s^{-3/2}e^{-k\sqrt{s}}\quad (k\geq 0)$	$2\sqrt{\dfrac{t}{\pi}}\exp\left(-\dfrac{k^2}{4t}\right)-k\,\mathrm{erfc}\left(\dfrac{k}{2\sqrt{t}}\right)$
52	$\dfrac{ae^{-k\sqrt{s}}}{s\left(a+\sqrt{s}\right)}\quad (k\geq 0)$	$-e^{ak}e^{a^2t}\mathrm{erfc}\left(a\sqrt{t}+\dfrac{k}{2\sqrt{t}}\right)+\mathrm{erfc}\left(\dfrac{k}{2\sqrt{t}}\right)$
53	$\dfrac{e^{-k\sqrt{s}}}{\sqrt{s}\left(a+\sqrt{s}\right)}\quad (k\geq 0)$	$e^{ak}e^{a^2t}\mathrm{erfc}\left(a\sqrt{t}+\dfrac{k}{2\sqrt{t}}\right)$

Table 14.1 (continued) Table of Laplace Transforms

	$F(s)$	$f(t)$
54	$\log\left(\dfrac{s-a}{s-b}\right)$	$\dfrac{1}{t}\left(e^{bt}-e^{at}\right)$
55	$\log\left(\dfrac{s^2+a^2}{s^2}\right)$	$\dfrac{2}{t}\left(1-\cos at\right)$
56	$\log\left(\dfrac{s^2-a^2}{s^2}\right)$	$\dfrac{2}{t}\left(1-\cosh at\right)$
57	$\arctan\dfrac{k}{s}$	$\dfrac{1}{t}\sin kt$

References

1. Churchill, R.V., *Operational Mathematics*, 2nd edn., McGraw-Hill, New York, 1958.
2. Kreyszig, E., *Advanced Engineering Mathematics*, 8th edn., John Wiley & Sons, New York, 1999.
3. Sedra, A.S. and Smith, K.C., *Microelectronic Circuits*, 3rd edn., Saunders College Publishing, Philadelphia, PA, 1991.
4. Fairchild Semiconductor data sheet, 2N3904/MMBT3904/PZT3904 NPN general purpose amplifier, from www.fairchildsemi.com, 2011.

Review Answers

Review 1.1

1. List several ways engineers use the computer.
   a. To solve mathematical models of physical phenomena
   b. Storing and reducing experimental data
   c. Controlling machines
   d. Communicating with other engineers on a particular project
2. List several areas of interest for engineers.
   a. Designing new products
   b. Improving performance of existing products
   c. Improving manufacturing efficiency
   d. Minimizing costs of production
   e. Minimizing power consumption
   f. Maximizing product yield
   g. Minimizing time to market
   h. Research and development of new products
3. List several methods that can be used in the design of new products.
   a. Full-scale experiments
   b. Small-scale model experiments
   c. A mathematical model describing the phenomenon of interest
4. Which method mentioned in item 3 is the least expensive?
   The mathematical model is the least expensive.
5. For engineers, what is the principal advantage of MATLAB® over some of the other computer programming platforms?
   MATLAB has built-in functions that solve many different types of mathematical problems that other computer platforms do not have.
6. List several items that are recommended in developing a computer program.
   a. List the algebraic equations involved in the project.
   b. Create a flow chart or a program outline.

    c. Write the program using the list of algebraic equations and the program outline or flow chart.

    d. Run the program and correct any syntax errors.

7. List several items that can be considered building blocks available in developing a computer program in MATLAB.

    a. Arithmetic statements

    b. Input/output statements

    c. Loop statements (`for` loop and `while` loop)

    d. Alternative path statements (`if`, `elseif`)

    e. Functions (built-in and self-written)

Review 2.1

1. What are the two alternative ways to start the MATLAB program?

If available, start the MATLAB program by double-clicking on the MATLAB icon on the Window's desktop. If not available, go to the Window's *Start* menu, click on *All Programs*, find the *MATLAB* program among the list of available programs, and double-click on it. This will open up the MATLAB desktop.

2. What are the windows in the MATLAB's default desktop?

The main windows are the Command window, Command History, Current Folder, and Workspace.

3. It is best to write a MATLAB script (program) in the Script Window. From MATLAB's default desktop how does one open the script window?

To open the script window, click on the *New Script* icon in the Toolstrip in MATLAB's desktop.

4. After you have completed writing a script in the appropriate window, what is the next step?

Click on the Save icon in the Toolstrip. In the window that opens, select the folder in which the script is to be saved and in the *File Name* box type in the name of the script with the *.m* extension.

5. Name two ways to execute a script.

    1. In the script window, click on the *Run* icon (green arrow) in the Toolstrip.

    2. In the command window after the MATLAB prompt (>>) sign, type in the script name (without the *.m* extension).

6. What happens if you attempt to execute a script and the script is not in the folder listed in the Current Folder Toolbar?

A dialog box will open giving you the choice of changing the folder listed in the Current Folder Toolbar or adding the folder containing your script to the MATLAB path.

7. In MATLAB, what is the file name extension for saved scripts?

The file name extension is *.m*.

Review 2.2

1. Are command statements and variable names case sensitive?
   Yes.
2. What is the purpose of placing a semicolon at the end of a command statement or a variable assignment?
   Placing a semicolon at the end of a command statement or a variable assignment suppresses the command statement or the variable assignment from
   being echoed to the screen.
3. How does one establish a comment line in a script?
   Placing a % sign in front of a statement makes it a comment line.
4. What is the command that will clear the Command window?
   The command `clc` clears the Command window.
5. What is the basic data structure in MATLAB?
   The basic data structure in MATLAB is a matrix.
6. Name two functions of the colon operator.
   1. The colon operator may be used to create a new matrix from an existing
      matrix.
   2. The colon operator can also be used to generate a series of numbers.
7. List the arithmetic operators in MATLAB.
   The arithmetic operators in MATLAB are
   +   addition
   −   subtraction
   *   multiplication
   /   division
   ^   exponentiation
8. What is MATLAB's command for
   a. π.   `pi`
   b. e.   `exp`
   c. ln.   `log`
   d. Sine function in radians.   `sin()`
   e. Sine function in degrees.   `sind()`
   f. $\sin^{-1}$ function.   `asin()`
   g. The number of elements in a vector.   `length()`
   h. The size of a matrix (the number of rows and columns).   `size()`
   i. The sum of the elements in a vector.   `sum()`
   j. The maximum of the elements in a vector.   `max()`
   k. Preallocating the size of a 3×3 matrix.   `zeros(3)`

Review 2.3

1. If you enter a variable assignment without placing a semicolon after the
   assignment, what happens?
   The variable assignment will be echoed to the screen.

2. Name two commands that will result in printing to the screen.
   1. `fprintf`
   2. `display`
3. What format string will move the cursor to the next line?
   `\n`.
4. What is the format that will print a variable to 10 spaces and to three decimal places?
   `%10.3f`
5. What are the commands necessary to print to a file?
   1. `fo = fopen('file_name','w');`
   2. `fprintf(fo,'format \n',variables);`
6. Name three commands that can be used in a script to input data into the workspace.
   1. Using the `input` statement
   2. Using the `load filename` statement
   3. Using the `fscanf` statement

Review 2.4

1. What is the objective in using a `for` loop?
   The objective of a `for` loop is to repeat a series of statements with just a few lines of code.
2. What is the syntax of a `for` loop?
   `for` index variable = starting value: step size: final value
3. Should table headings that are not to be repeated be inside a `for` loop?
   No.
4. If the index of a `for` loop is used to select an element of a vector or a matrix, what variable type should the `for` loop index be?
   It should be an integer.
5. What other statement type can be used to create a loop?
   The `while` loop.
6. What is the major difference between a `for` loop and a `while` loop?
   The syntax of the `for` loop generates its own index. If a program requires an index, the `while` loop requires program statements that generate an index.

Review 2.5

1. What is the command that will produce a linear graph?
   For function y vs. x, use `plot(x,y)`.
2. What are the commands that will label the x- and y-axis and provide a title to a plot?
   `xlabel('x'), ylabel('y'), title('y vs. x')`

3. When there is more than one function plotted on a graph, what are the ways to identify which curve goes with which function?

Each curve can be given a different color or a different line type. In each case, you can use the `legend` command to identify which curve goes with which function. You can also use the `text` command to label each of the curves.

Review 2.6

1. What statement is frequently used to establish two alternate paths?
The `if-else` statements.
2. What series of statements is used to establish several alternate paths?
The `if-elseif-else` statements.
3. List the various types of logic statements that can be used with the `if-else` and the `if-elseif-else` ladder.
`a < b`, `a > b`, `a <= b`, `a >= b`, `a == b`, and `a ~= b`.
4. Is the `else` statement required with the `if-else` and with the `if-elseif-else` ladder?
No.
5. What statement group is an alternative to the `if-elseif-else` ladder?
The `switch` statement.

Review 2.7

1. What is the name of the function that will allow you to plot several graphs on one page?
The name of the function is `subplot`.
2. How does one enter Greek symbols into a plot?
Use the `\Greek` symbol name (see Appendix A).
3. What are the commands that will allow you to enter text onto a plot once the plot has been created?
In the plot window, click on the *Insert* option in the menu bar.
   A dropdown menu will appear that contains the following options:
   X Label, Y Label, Title, TextBox, and others. Click on the item that you wish to enter on the plot. If you select the *TextBox* option, a crosshair will appear and you can drag it to the location where you wish to start the text, and then type in the text that you want to enter into your plot.

Review 3.1

1. What is the series expansion for e^x?

$$e^x = 1 + x + \frac{x^2}{2!} + \frac{x^3}{3!} + \frac{x^4}{4!} + \cdots + \frac{x^n}{n!}$$

2. What are the two different approaches for evaluating the terms in the series for e^x?
   a. Determining $term(n + 1) = term(n) \times x/n$.
   b. Taking $term(n) = x^n/n!$
3. When is it appropriate to write a self-written function?
   a. It is appropriate when the program requires a series of statements that need to be repeated several times.
   b. The program is quite complicated, and you wish to break it up into smaller segments.
   c. The program contains a MATLAB function that requires a self-written function to define the problem of interest.
4. A self-written function usually has both an input and an output. Where does the input come from? Where does the output go to?
   The input is defined in the calling program.
   The output goes to the calling program or to another function or to the screen.
5. If a self-written function has more than one output, how must the output be presented?
   In the function statement, the output variables must be enclosed by brackets and separated by commas.
6. How does a self-written function communicate with the calling program?
   The self-written function communicates with the calling program through the function's input arguments and output.
7. What can be said about variables in the self-written function that are not in the input or output arguments of the function?
   All variables not in the input arguments or output are local in both the function and the calling program.
8. Do the variable names in the input arguments and output between the calling program and the function have to be the same?
   The variable names in the input arguments and output in the function need not be the same as those in the calling program but have to be in the same position in both the calling program and in the function for the one-to-one correspondence.

Review 3.2

1. If a programmer wishes to write a self-written function but does not wish to create an additional *.m* file, what can the programmer do and what is the constraint?
   The programmer can write an *anonymous* function, which is included in the main program and not as a separate *.m* file. The constraint is that it needs to be a single statement.
2. What is the name of MATLAB's function that does interpolation?
   The function name is `interp1`.

3. What are the inputs to MATLAB's interpolation function?

MATLAB's `interp1` function has three arguments, say, (X,Y,Xi), where (X,Y) are a set of known (*x*, *y*) data points and Xi is the set of *x* values at which the set of *y* values, Yi, are to be determined by interpolation. Arrays X and Y must be of the same length. If Xi is a vector, then Yi will be a vector.

4. What are the outputs from MATLAB's interpolation function?

The output from MATLAB's `interp1` function is the interpolated values of y.

Review 3.3

1. Suppose you wish to assign a matrix consisting of string elements, what are the conditions that need to be followed in setting up this matrix?

The conditions are as follows: (a) Each string row needs to be enclosed by single quotation marks, (b) each string row must have the same number of columns, and (c) the entire matrix must be enclosed by brackets.

2. Suppose our independent variable is *x* and the *x* domain is subdivided into small intervals and we wish to determine which interval contains an item of interest. What is the most efficient way to determine the correct interval?

Use a `for` loop testing each interval from the first one to the last one and an `if` statement to determine whether or not the item of interest lies within the selected interval.

Example

```
x = 0:10;
y = [0 1 4 9 16 25 36 49 64 81 100];
x1=23;
% y1 is the average of the y values at the beginning and end
% of the interval.
for i=1:length(x)-1
        if (x1 >= x(i) && x1 < x(i+1))
                y1 = 0.5*(y(i)+y(i+1));
                break;
        end
end
```

Review 4.1

1. If matrix **C** = **A** + **B**, what must be true about matrices **A** and **B**?

Matrices **A** and **B** must have the same number of rows and the same number of columns.

2. If **C** = **AB**, what must be true about matrices **A** and **B**?

The number of columns in **A** must be the same as the number of rows in **B**.

3. What command in MATLAB is used to obtain the inverse of a matrix?

In MATLAB, the inverse of matrix **A** is `inv(A)`.

4. What symbol is used in MATLAB to transpose a matrix?
   In MATLAB, the transpose of matrix **A** is written as A'.
5. Is the dot product of two vectors a scalar or a vector?
   It is a scalar $\left(\mathbf{A} \cdot \mathbf{B} = \sum a_i b_i\right)$.
6. Is the element-by-element multiplication of two vectors a scalar or a vector?
   It is a vector. If the length of vectors **A** and **B** is 3, then $\mathbf{C} = \mathbf{A} \cdot * \mathbf{B} = [a_1 b_1 \quad a_2 b_2 \quad a_3 b_3]$.
7. Does the use of MATLAB's sum command on a vector produce a scalar or a vector?
   The sum command on a vector produces a scalar. If $\mathbf{A} = [a_1 \quad a_2 \quad a_3]$, then sum (A) $= a_1 + a_2 + a_3$
8. Does the use of MATLAB's sum command on a matrix of two or more columns produce a scalar or a vector?

   It produces a vector. If $\mathbf{B} = \begin{bmatrix} b_{11} & b_{12} & b_{13} \\ b_{21} & b_{22} & b_{23} \\ b_{31} & b_{32} & b_{33} \end{bmatrix}$,

   then

$$\text{sum (B)} = [(b_{11} + b_{21} + b_{31}) \quad (b_{12} + b_{22} + b_{32}) \quad (b_{13} + b_{23} + b_{33})]$$

Review 4.2

Given a set of linear equations in the form **AX** = **C**, where **A** is the coefficient matrix and **X** and **C** are column vectors, what are the two ways for solving for **X** in MATLAB?
a. X = inv(A)*C.
b. X = A\C.

Review 5.1

1. What is the name of the MATLAB function for determining the roots of a transcendental equation of the form $f(x) = 0$?
   The name of MATLAB's function to obtain the roots of a transcendental equation is fzero.
2. In MATLAB's function for determining the roots of a transcendental equation, how does one define the function whose roots are to be determined?
   A self-written function should describe the function whose roots are to be obtained. The name of this self-written function should be entered as the first argument in MATLAB's fzero function.

3. If you suspect that there is more than one real root, what method should be used in combination with the MATLAB's function to obtain the roots.

If you suspect that there is more than one root, you should use the search method in combination with MATLAB's `fzero` function. The search method is used to obtain a small interval in which a root lies and MATLAB's `fzero` function determines the root that lies in that interval. The function values at the ends of this interval should differ in sign.

4. For the case described in item 3, what can you say about the second argument in MATLAB's function to obtain the roots?

The second argument to be entered in the `fzero` function should be a vector of length 2 specifying the endpoints of the intervals that contain the roots.

5. What is the purpose of the `global` statement?

Variables listed in the `global` statement will be common to both the calling program and the called function. Therefore, variables defined in the calling program will be available in the called function, despite the fact that these variables are neither an input nor an output argument in the called function. Of course, the reverse is also true. A value determined in the called function would also become available in the calling program.

6. If the function $f(x)$ is a polynomial, what MATLAB function should you use to obtain its roots?

You should use MATLAB's `roots` function. If the polynomial has complex roots, MATLAB's `roots` function will give the complex roots, while MATLAB's `fzero` function will only give the real roots.

Review 6.1

1. What is the formula for evaluating the integral, $I = \int_A^B f(x)dx$, by the trapezoidal rule?

$$I = \int_{x_1}^{x_{N+1}} f(x)dx \approx \left(\frac{1}{2}f_1 + f_2 + f_3 + \cdots + f_N + \frac{1}{2}f_{N+1} \right)\Delta x$$

where
$x_1 = x_2$
$x_{N+1} = B$
N = the number of sub-divisions on the x axis

2. What is the formula for evaluating the integral, $I = \int_A^B f(x)dx$, by Simpson's rule?

$$I = \int_{x_1}^{x_{N+1}} f(x)dx \approx \frac{\Delta x}{3}[f_1 + 4f_2 + 2f_3 + 4f_4 + 2f_5 + \cdots + 4f_N + f_{N+1}]$$

Review 6.2

1. What is the name of MATLAB's function for integration of a single-variable function?
   MATLAB's function for integrating a single-variable function is quad.
2. In MATLAB's function for integration, how do you define the function to be integrated?
   You need to write a self-written function that describes the integrand.
   The name of this function should be entered as the first argument in the quad function.
3. If the integrand contains nonlinear terms, how must they be treated?
   Nonlinear terms need to be entered as element-by-element multiplication or division. Terms involving exponents also need to be treated as an element-by-element operation.
4. Will MATLAB's quad function treat improper integrals?
   Yes.

Appendix A: Special Characters in MATLAB® Plots

MATLAB® allows the use of Greek and special characters in its plot headings and labels. The method for doing this is based on the TeX formatting language [1] and is summarized in this appendix.

MATLAB provides the functions `title`, `xlabel`, `ylabel`, and `text` for adding labels to plots. These labels can include Greek and special characters by applying the character sequences as shown in Table A.1. These sequences all begin with the backslash character (\) and can be embedded in any text string argument to `title`, `xlabel`, `ylabel`, and `text`. Subscripts and superscripts may also be applied by using the _ and ^ operators. For example, the sequence V_o is written as `V _ o`, and 10^6 is written as `10^6`. If the subscripts or superscripts are multiple characters, then use curly braces to delimit the string to be subscripted, for example, V_{out} is generated with `V _ {out}`.

Example A.1

The following MATLAB script shows how to use special characters in a plot.

```
% Example_A_1.m
% This script shows example usage of special characters in
% MATLAB plots.
% Plot a 1MHz sine wave over the interval 0<t<2 microsec
t = 0:2e-8:2e-6;
fo = 1e6;
xout = sin(2*pi*fo*t);
plot(t*1e6,xout);
title('Plot of sin(2\pif_{o}t) for f_{o} = 10^6 Hz');
xlabel('Time (\mus)');
ylabel('x_{out}(t)');
text(1.5,0.3,'\omega = 2\pi \times f_{o}');
```

Program results

See Figure A.1

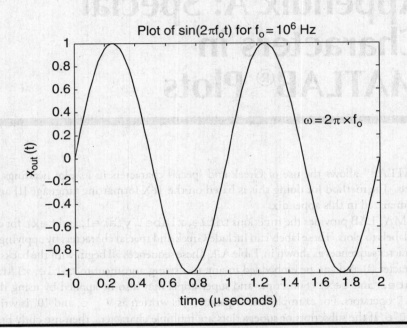

Figure A.1 **Example usage of Greek letters, special characters, subscripts, and superscripts in a MATLAB® plot.**

Table A.1 Special Symbols for Use in MATLAB® Plots

Character Sequence	Symbol	Character Sequence	Symbol	Character Sequence	Symbol
\alpha	α	\upsilon	υ	\sim	~
\beta	β	\phi	ϕ	\leq	≠
\gamma	γ	\chi	χ	\infty	∞
\delta	δ	\psi	ψ	\clubsuit	♣
\epsilon	ε	\omega	ω	\diamondsuit	◆
\zeta	ζ	\Gamma	Γ	\heartsuit	♥
\eta	η	\Delta	Δ	\spadesuit	♠
\theta	θ	\Theta	Θ	\leftrightarrow	↔
\vartheta	ϑ	\Lambda	Λ	\leftarrow	←
\iota	ι	\Xi	Ξ	\uparrow	↑
\kappa	κ	\Pi	Π	\rightarrow	→
\lambda	λ	\Sigma	Σ	\downarrow	↓
\mu	μ	\Upsilon	Υ	\circ	°
\nu	ν	\Phi	Φ	\pm	±
\xi	ξ	\Psi	Ψ	\geq	≥
\pi	π	\Omega	Ω	\propto	∝
\rho	ρ	\forall	$\forall$	\partial	∂
\sigma	σ	\exists	$\exists$	\bullet	•
\varsigma	ς	\ni	$\ni$	\div	÷
\tau	τ	\cong	$\cong$	\neq	≠
\equiv	≡	\approx	≈	\aleph	ℵ
\Im	$\Im$	\Re	$\Re$	\wp	℘
\otimes	⊗	\oplus	⊕	\oslash	ø
\cap	∩	\cup	∪	\supseteq	⊇
\supset	⊃	\subseteq	⊆	\subset	⊂

(continued)

Table A.1 (continued) Special Symbols for Use in MATLAB® Plots

Character Sequence	Symbol	Character Sequence	Symbol	Character Sequence	Symbol
\int	∫	\in	∈	\o	°
\rfloor	⌋	\lceil	⌈	\nabla	∇
\lfloor	⌊	\cdot	·	\ldots	⋯
\perp	⊥	\neg	¬	\prime	′
\wedge	∧	\times	×	\0	ø
\rceil	⌉	\surd	√	\mid		
\vee	∨	\varpi	ϖ	\copyright	©
\langle	⟨	\rangle	⟩		

Reference

1. Knuth, D.E., *The TeXbook*, Addison Wesley, New York, 1984.

Appendix B: Derivation of the Heat Transfer Equation in Solids

B.1 Heat Flux Vector and Fourier's Heat Conduction Law

1. Heat transfer is thermal energy in transit.
2. The heat conduction equation provides the means for determining the following:
   A. The temperature distribution in a solid
   B. The time it takes to transfer a specified amount of heat
   C. The amount of heat transferred in a specified period of time

Heat flow can be represented by the heat flux vector, $\vec{\mathbf{q}}$, which is defined as follows:

Select $\Delta s_\perp$ to be perpendicular to the heat flow direction, then $|\vec{\mathbf{q}}|$ = the rate that heat flows through $\Delta s_\perp$ per unit surface area, and $\vec{\mathbf{q}}$ points in the direction of heat flow (see Figure B.1a).

If ΔQ is the rate that heat passes through $\Delta s_\perp$, then

$$|\vec{\mathbf{q}}| = \lim_{\Delta s \to 0} \frac{\Delta Q}{\Delta s_\perp} \tag{B.1}$$

If the surface area, Δs, is not perpendicular to the direction of heat flow, then the rate that heat flows through Δs, say, ΔQ, is given by

$$\Delta Q = \vec{\mathbf{q}} \cdot \hat{\mathbf{e}}_n \Delta s = \vec{\mathbf{q}} \cdot \Delta \vec{\mathbf{s}} \tag{B.2}$$

where $\Delta \vec{\mathbf{s}} = \Delta s \hat{\mathbf{e}}_n$ and $\hat{\mathbf{e}}_n$ is a unit vector perpendicular to surface Δs (see Figure B.1b).

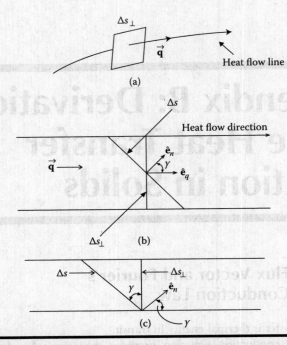

Figure B.1 **(a) Heat flux vector, $\vec{\mathbf{q}}$. (b) Heat flow through an arbitrarily-oriented surface Δs. (c) Relation between Δs and $\Delta s_\perp$.**

From Figure B.1c, it can be seen that

$$\Delta s_\perp = \Delta s \cos\gamma = \Delta s\,\hat{\mathbf{e}}_n \cdot \hat{\mathbf{e}}_q = \Delta\vec{\mathbf{s}} \cdot \hat{\mathbf{e}}_q \tag{B.3}$$

where $\hat{\mathbf{e}}_q$ is a unit vector pointing in the direction of heat flow.

The heat flow, ΔQ, through Δs is the same as the heat flow through $\Delta s_\perp$ (see Figure B.1b). Thus,

$$\Delta Q \approx \left|\vec{\mathbf{q}}\right|\Delta s_\perp = \left|\vec{\mathbf{q}}\right|\hat{\mathbf{e}}_q \cdot \Delta\vec{\mathbf{s}} = \vec{\mathbf{q}} \cdot \Delta\vec{\mathbf{s}} \tag{B.4}$$

For a finite surface area,

$$Q = \iint_s \vec{\mathbf{q}} \cdot d\vec{\mathbf{s}} \tag{B.5}$$

There are a large class of materials that obey Fourier's heat conduction law, which is

$$\vec{\mathbf{q}} = -k\nabla T \tag{B.6}$$

where
  k is the thermal conductivity of the material
  ∇T is the gradient of the temperature

In Cartesian coordinates

$$\nabla T = \frac{\partial T}{\partial x}\hat{\mathbf{e}}_x + \frac{\partial T}{\partial y}\hat{\mathbf{e}}_y + \frac{\partial T}{\partial z}\hat{\mathbf{e}}_z \qquad (B.7)$$

where $\hat{\mathbf{e}}_x$, $\hat{\mathbf{e}}_y$, $\hat{\mathbf{e}}_z$ are unit vectors in the x, y, z directions, respectively. The significance of the gradient at point P, is that its magnitude is the maximum rate of change of the variable with respect to distance at point P and it points in that direction. So Fourier's heat conduction law states that heat will flow down the steepest temperature hill available at a particular point.

B.2 Heat Conduction Equation for Stationary Solids

The heat conduction equation is based on the first law of thermodynamics and Fourier's heat conduction law. The first law of thermodynamics will be discussed first.

First Law of Thermodynamics:

  For an arbitrary system (region) within the continuum,
  The rate of increase in the total energy in the system = the rate that heat
  is added to the system plus the rate that heat is generated within the system +
  the rate of work done on the system.

We only need to consider energy forms within the system that change during the process. If any of the following phenomena occur within the material: electric current, chemical reactions, and nuclear reactions; then changes in these energy forms need to be accounted for. When this occurs, the work term is not zero and additional constitutive laws are needed. Since all these processes result in a conversion of some energy form to internal (thermal) energy, the process is accounted for by including a heat generation term in the first law of thermodynamics. Under these conditions, the work term for stationary solids in the first law of thermodynamics is zero.

We now apply our statement of the first law to the infinitesimal volume shown in Figure B.2.

Note: Evaluate surface terms at the centroid of the surface and evaluate volume terms at the centroid of the volume. The heat flux vector, $\vec{\mathbf{q}}$, can be decomposed into its components; that is, $\vec{\mathbf{q}} = q_x\hat{\mathbf{e}}_x + q_y\hat{\mathbf{e}}_y + q_z\hat{\mathbf{e}}_z$.

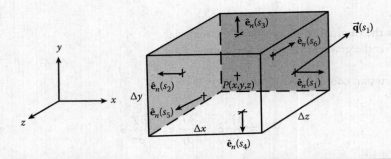

Figure B.2   Infinitesimal volume in Cartesian coordinates.

The surface areas Δs_1 through Δs_6, the unit normal vectors, and the coordinates of the area centroids of the infinitesimal volume shown in Figure B.2 are given in the following table:

Surf.	Area	$\hat{\mathbf{e}}_n$	Coord. of Centroid
s_1	$\Delta y \Delta z$	$\hat{\mathbf{e}}_x$	$\left(x+\dfrac{\Delta x}{2},y,z,t\right)$
s_2	$\Delta y \Delta z$	$-\hat{\mathbf{e}}_x$	$\left(x-\dfrac{\Delta x}{2},y,z,t\right)$
s_3	$\Delta x \Delta z$	$\hat{\mathbf{e}}_y$	$\left(x,y+\dfrac{\Delta y}{2},z,t\right)$
s_4	$\Delta x \Delta z$	$-\hat{\mathbf{e}}_y$	$\left(x,y-\dfrac{\Delta y}{2},z,t\right)$
s_5	$\Delta x \Delta y$	$\hat{\mathbf{e}}_z$	$\left(x,y,z,+\dfrac{\Delta z}{2},t\right)$
s_6	$\Delta x \Delta y$	$-\hat{\mathbf{e}}_z$	$\left(x,y,z,-\dfrac{\Delta z}{2},t\right)$

Applying first law to infinitesimal system shown gives

$$\left[\frac{\partial}{\partial t}(\rho u)\right]_{(x,y,z,t)} \Delta x \Delta y \Delta z = -\sum_{m=1}^{6}\vec{\mathbf{q}}\,(s_m,t)\cdot\vec{\mathbf{e}}_n\,(s_m)\Delta s_m + g(x,y,z,t)\Delta x \Delta y \Delta z \qquad (B.8)$$

where
   u is the internal energy per unit mass
   g is the rate of heat generation per unit volume by internal heat sources

The summation term is evaluated as follows:

$$\vec{\mathbf{q}}(s_1,t)\cdot\vec{\mathbf{e}}_n(s_1)\Delta s_1 = \vec{\mathbf{q}}\left(x+\frac{\Delta x}{2},y,z,t\right)\cdot\hat{\mathbf{e}}_x\,\Delta y\Delta z = q_x\left(x+\frac{\Delta x}{2},y,z,t\right)\Delta y\Delta z \quad\text{(B.9)}$$

$$\vec{\mathbf{q}}(s_2,t)\cdot\vec{\mathbf{e}}_n(s_2)\,\Delta s_2 = \vec{\mathbf{q}}\left(x-\frac{\Delta x}{2},y,z,t\right)\cdot(-\hat{\mathbf{e}}_x)\,\Delta y\Delta z = -q_x\left(x-\frac{\Delta x}{2},y,z,t\right)\Delta y\Delta z$$

$$\text{(B.10)}$$

Similarly,

$$\vec{\mathbf{q}}(s_3,t)\cdot\vec{\mathbf{e}}_n(s_3)\Delta s_3 = q_y\left(x,y+\frac{\Delta y}{2},z,t\right)\Delta x\Delta z \quad\text{(B.11)}$$

$$\vec{\mathbf{q}}(s_4,t)\cdot\vec{\mathbf{e}}_n(s_4)\Delta s_4 = -q_y\left(x,y-\frac{\Delta y}{2},z,t\right)\Delta x\Delta z \quad\text{(B.12)}$$

$$\vec{\mathbf{q}}(s_5,t)\cdot\vec{\mathbf{e}}_n(s_5)\Delta s_5 = q_z\left(x,y,z+\frac{\Delta z}{2},t\right)\Delta x\Delta y \quad\text{(B.13)}$$

$$\vec{\mathbf{q}}(s_6,t)\cdot\vec{\mathbf{e}}_n(s_6)\Delta s_6 = -q_z\left(x,y,z-\frac{\Delta z}{2},t\right)\Delta x\Delta y \quad\text{(B.14)}$$

Applying Equations B.9 through B.14 to Equation B.8 and dividing both sides by $\Delta x\Delta y\Delta z$ gives

$$\left[\frac{\partial}{\partial t}(\rho u)\right]_{(x,y,z,t)} = -\left\{\frac{q_x(x+(\Delta x/2),y,z,t)-q_x(x-(\Delta x/2),y,z,t)}{\Delta x}\right.$$

$$+\frac{q_y(x,y+(\Delta y/2),z,t)-q_y(x,y-(\Delta y/2),z,t)}{\Delta y}$$

$$\left.+\frac{q_z(x,y,z+(\Delta z/2),t)-q_z(x,y,z-(\Delta z/2),t)}{\Delta z}+g(x,y,z,t)\right\}$$

$$\text{(B.15)}$$

Taking limits of Equation B.15 as $\Delta x \to 0$, $\Delta y \to 0$, and $\Delta z \to 0$ on both sides of the equation gives

$$\frac{\partial}{\partial t}(\rho u) = -\left[\frac{\partial q_x}{\partial x} + \frac{\partial q_y}{\partial y} + \frac{\partial q_z}{\partial z}\right] + g = -\nabla \cdot \vec{q} + g \qquad (B.16)$$

From vector calculus

$$\vec{A} = A_x \hat{e}_x + A_y \hat{e}_y + A_z \hat{e}_z$$

$$\nabla \cdot \vec{A} = \frac{\partial A_x}{\partial x} + \frac{\partial A_y}{\partial y} + \frac{\partial A_z}{\partial z}$$

$$\nabla T = \frac{\partial T}{\partial x} \hat{e}_x + \frac{\partial T}{\partial y} \hat{e}_y + \frac{\partial T}{\partial z} \hat{e}_z$$

Applying Fourier's heat conduction law,

$$\vec{q} = -k\nabla T \qquad (B.17)$$

$$\nabla \cdot \vec{q} = -\nabla \cdot (k\nabla T) \qquad (B.18)$$

$$\nabla \cdot \vec{q} = -\left\{\frac{\partial}{\partial x}\left(k\frac{\partial T}{\partial x}\right) + \frac{\partial}{\partial y}\left(k\frac{\partial T}{\partial y}\right) + \frac{\partial}{\partial z}\left(k\frac{\partial T}{\partial z}\right)\right\} \qquad (B.19)$$

Also, $u = c_v T$, where c_v is the specific heat at constant volume.

For solids and liquids, $c_v = c_p = c$.

Since k, ρ, and c are mild functions of temperature, they frequently are taken as constants (especially if analytical techniques are used to solve the problem). The preceding equation becomes

$$\frac{1}{a}\frac{\partial T}{\partial t} = \left\{\frac{\partial^2 T}{\partial x^2} + \frac{\partial^2 T}{\partial y^2} + \frac{\partial^2 T}{\partial z^2}\right\} + g \qquad (B.20)$$

where a is the thermal diffusivity of the material $= k/\rho c$.

Appendix C: Derivation of the Beam Deflection Equation

C.1 Internal Moment, Stress, and Deflection

A horizontal beam subjected to a load will deflect from the unloaded condition as shown in Figures C.1a and C.1b. The internal moment, M, about the z-axis at any section is determined by

$$M = \iint_A \sigma_x(\bar{y})\bar{y}\,dA \tag{C.1}$$

where $\sigma_x(\bar{y})$ is the normal stress at position $\bar{y}$ at the cut section and $\bar{y}$ is measured from the neutral axis of the section. A is the cross-sectional area of the beam. The neutral axis is a section on the beam that does not elongate or shorten during the bending process.

Assuming that $\sigma_x(\bar{y})$ and $\bar{y}$ are positive, M would be in the clockwise direction (see Figure C.1c). For deflection analysis, the moment, M_d, about the z-axis is considered positive if the moment at a section is counterclockwise. Therefore,

$$M_d = -\iint_A \sigma_x(\bar{y})\bar{y}\,dA \tag{C.2}$$

The beam deflection with a positive M_d will cause an element with a negative $\bar{y}$ to elongate and an element with a positive $\bar{y}$ to shorten. Consider the section between x and $x + \Delta x$ before bending. Element GH and element NS, where NS lies on the neutral axis (see Figures C.1d and C.1e), are the same length, Δx. After bending the element, GH becomes $G'H'$ and element NS becomes $N'S'$ (see Figure C.1e).

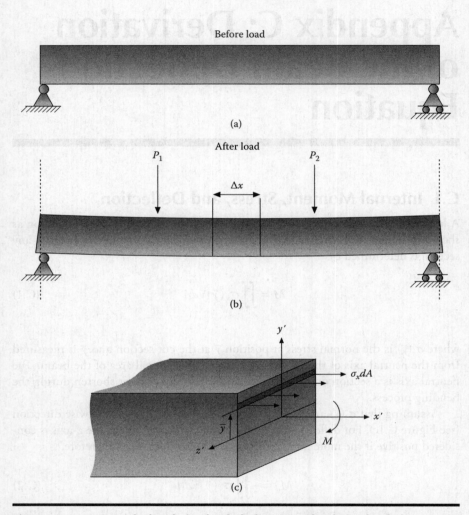

Figure C.1 **(a) Beam before being loaded. (b) Beam after being loaded. (c) Stress at beam section.**

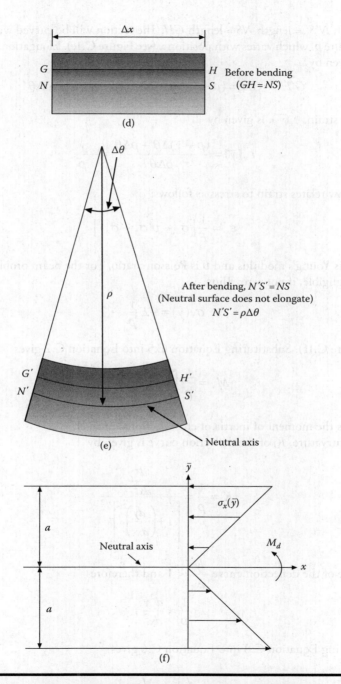

Figure C.1 (continued) (d) Beam element *GH* before beam being loaded. (e) Beam element *GH* after beam being loaded. (f) Section stress distribution.

But length $N'S'$ = length NS = length GH. The section will be curved with a radius of curvature ρ, which varies with position x (see Figure C.1e). Elongation of element GH is given by

$$G'H' - GH = (\rho - \bar{y})\Delta\theta - NS = (\rho - \bar{y})\Delta\theta - \rho\Delta\theta$$

Then the strain, $\varepsilon_x(\bar{y})$, is given by

$$\varepsilon_x(\bar{y}) = \frac{(\rho - \bar{y})\Delta\theta - \rho\Delta\theta}{\rho\Delta\theta} = -\frac{\bar{y}}{\rho} \tag{C.3}$$

Hook's law relates strain to stress as follows:

$$\varepsilon_x = \frac{1}{E}\left[\sigma_x - \upsilon(\sigma_y + \sigma_z)\right] \tag{C.4}$$

where E is Young's modulus and υ is Poisson's ratio. For the beam problem, σ_y and σ_z are negligible, if not zero. Thus,

$$\sigma_x(\bar{y}) = -E\frac{\bar{y}}{\rho} \tag{C.5}$$

(see Figure C.1f). Substituting Equation C.5 into Equation C.2 gives

$$M_d = \frac{E}{\rho}\iint_A \bar{y}^2 \, dA = \frac{EI}{\rho} \tag{C.6}$$

where I is the moment of inertia of cross-sectional area, A.

The curvature, K, of the deflection curve is given by

$$K = \frac{1}{\rho} = \frac{\dfrac{d^2y}{dx^2}}{\left[1+\left(\dfrac{dy}{dx}\right)^2\right]^{3/2}} \tag{C.7}$$

The slope of the deflection curve $\dfrac{dy}{dx} \ll 1$ and therefore

$$\frac{1}{\rho} \approx \frac{d^2y}{dx^2} \tag{C.8}$$

Substituting Equation C.8 into Equation C.6 gives

$$\frac{d^2y}{dx^2} = \frac{M_d}{EI} \tag{C.9}$$

Appendix D: Proof of Orthogonal Relationship of the $\cos(\lambda_n x)$ Functions

In Section 13.2.2, we demonstrated that the $\cos(\lambda_n x)$ functions satisfied the differential equation

$$F'' + \lambda^2 F = 0 \tag{D.1}$$

and the boundary conditions

$$F'(0) = 0 \tag{D.2}$$

$$F'(L) + \frac{h}{k} F(L) = 0 \tag{D.3}$$

Let $f_n = \cos(\lambda_n x)$ and $f_m = \cos(\lambda_m x)$

Both functions f_n and f_m satisfy the ODE, that is,

$$f_n'' + \lambda_n^2 f_n = 0 \tag{D.4}$$

and

$$f_m'' + \lambda_m^2 f_m = 0 \tag{D.5}$$

Each function satisfies the boundary conditions, that is,

$$f_n'\,(0) = 0 \quad \text{and} \quad f_m'\,(0) = 0 \tag{D.6}$$

$$f_n'(L) + \frac{h}{k} f_n(L) = 0 \quad \text{and} \quad f_m'(L) + \frac{h}{k} f_m(L) = 0 \tag{D.7}$$

Now multiply Equation D.4 by f_m and Equation D.5 by f_n and subtract the second equation from the first giving

$$(f_m f_n'' + \lambda_n^2 f_n f_m) - (f_n f_m'' + \lambda_m^2 f_n f_m) = 0 \tag{D.8}$$

or

$$f_m f_n'' - f_n f_m'' = (\lambda_m^2 - \lambda_n^2) f_n f_m \tag{D.9}$$

But

$$\frac{d}{dx}(f_m f_n') = f_m f_n'' + f_m' f_n' \tag{D.10}$$

and

$$\frac{d}{dx}(f_n f_m') = f_n f_m'' + f_m' f_n' \tag{D.11}$$

Then

$$\frac{d}{dx}(f_m f_n' - f_n f_m') = f_m f_n'' - f_n f_m'' \tag{D.12}$$

Substituting Equation D.12 into Equation D.9 gives

$$\frac{d}{dx}(f_m f_n' - f_n f_m') = (\lambda_m^2 - \lambda_n^2) f_n f_m \tag{D.13}$$

Multiplying both sides of Equation D.13 by dx and integrating from 0 to L gives

$$\int_0^L \frac{d}{dx}(f_m f_n' - f_n f_m')dx = (\lambda_m^2 - \lambda_n^2) \int_0^L f_n f_m dx \tag{D.14}$$

or

$$(\lambda_m^2 - \lambda_n^2)\int_0^L \cos(\lambda_n x)\cos(\lambda_m x)dx$$

$$= f_m(L)f_n'(L) - f_n(L)f_m'(L) - f_m(0)f_n'(0) + f_n(0)f_m'(0) \qquad \text{(D.15)}$$

But $f_n'(0) = 0$ and $f_m'(0) = 0$ and

$$f_m(L)f_n'(L) - f_n(L)f_m'(L) = -f_m(L)\frac{h}{k}f_n(L) + f_n(L)\frac{h}{k}f_m(L) = 0 \quad \text{(D.16)}$$

Thus,

$$(\lambda_m^2 - \lambda_n^2)\int_0^L \cos(\lambda_n x)\cos(\lambda_m x)dx = 0 \qquad \text{(D.17)}$$

If

$$m \neq n, \text{ then } (\lambda_m^2 - \lambda_n^2) \neq 0 \quad \text{and} \quad \int_0^L \cos(\lambda_n x)\cos(\lambda_m x)dx = 0 \qquad \text{(D.18)}$$

This demonstrates that the functions $\cos(\lambda_n x)$ for $n = 1, 2, 3, \ldots$ are orthogonal.

If $m = n$, then $(\lambda_m^2 - \lambda_n^2) = 0$ and $\int_0^L \cos^2(\lambda_m x)dx = 0$ need not be zero.

From integral tables, we can determine that

$$\int_0^L \cos^2(\lambda_m x)dx = \left(\frac{L}{2} + \frac{\sin(\lambda_m L)\cos(\lambda_m L)}{2\lambda_m}\right) \qquad \text{(D.19)}$$

Appendix E: Getting Started with MATLAB® Version R2012a

E.1 MATLAB® Updates

Mathworks normally updates their version of MATLAB® twice a year. In 2012, Mathworks introduced MATLAB version R2012b whose desktop is very different from the previous MATLAB version R2012a. Thus, getting started with MATLAB version R2012a and earlier versions is very different than getting started with MATLAB version R2012b. Sections 2.2 and 2.3 in Chapter 2, which discuss the MATLAB windows and constructing a script in MATLAB, are based on MATLAB version R2012b. Since many students may still be using an earlier version of MATLAB, this appendix discusses getting started with MATLAB version R2012a in this appendix.

E.2 MATLAB® Windows

Under Microsoft Windows, MATLAB may be started via the Start menu or by clicking on the MATLAB icon on the desktop. Upon startup, a new window will open containing the MATLAB "desktop" (not to be confused with the Windows desktop), and one or more MATLAB windows will open within the MATLAB desktop (see Figure E.1 for the default configuration).

The main windows are the Command Window, Command History, Current Folder, and Workspace. You can customize the MATLAB windows that appear upon startup by clicking on the *Desktop* option in the menu bar and in the drop-down menu that follows, checking or unchecking the windows that you wish to appear on the MATLAB desktop. Figure E.1 shows the Command Window (in the center), the Current Folder Window (on the left), the Workspace Window (on the top right), the Command History window (on the bottom right), and the Current

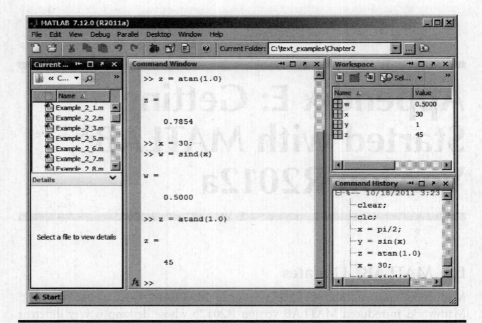

Figure E.1 MATLAB® default desktop.

Folder box (in the icon Toolbar, second from the top, just above the Command Window). These windows and the Current Folder box are summarized as follows:

- *Command Window*: in the Command Window you can enter commands and data, make calculations, and print results. You can write a script (program) in the Command Window and execute the script. However, writing a script directly into the Command Window is discouraged because it will not be saved, and if an error is made, the entire script must be retyped. By using the up arrow (↑) key on your keyboard, the previous command can be retrieved (and edited) for reexecution.
- *Command History Window*: This window lists a history of the commands that you have executed in the Command Window.
- *Current Folder Box*: This box lists the present (or active) Current Folder (also called the "Current Directory" in older versions of MATLAB). *To run a MATLAB script, the script needs to be in the folder listed in this box.* By clicking on the down arrow within the box, a dropdown menu will appear that contains the names of folders that you have previously used (see Figure E.2). If the folder containing the script of interest is not listed in the dropdown menu, you can click on the adjacent little box containing three dots, which allows you to browse for the folder containing the program of interest (see Figure E.3).

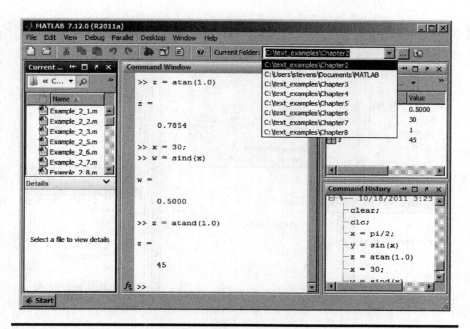

Figure E.2   Dropdown menu listing the available folders.

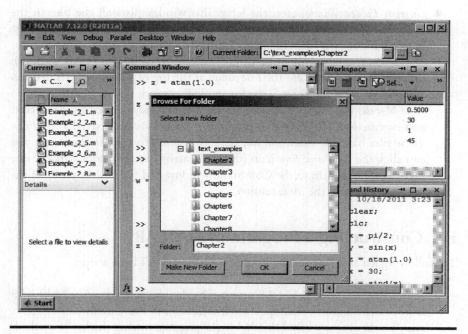

Figure E.3   Dialog box for selecting folder containing program of interest.

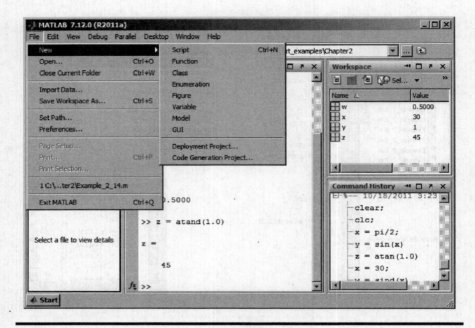

Figure E.4   Opening up the Script window.

- *Current Folder Window* (on the left): This window lists all the files in the Current Folder listed in the Current Folder box. By double-clicking on a file in this window, the file will open within MATLAB.
- *Script Window* (also called the Editor window in older MATLAB versions): To open this window, use the *File* menu at the top of the MATLAB desktop and choose *New* and then *Script* (or in older versions of MATLAB, click on *New M-File*) (see Figure E.4). The Script Window may be used to create, edit, and execute MATLAB scripts (programs). Scripts are then saved as *M-Files*. These files have the extension *.m*, such as *heat.m*. To execute the script, you can click the *Save and Run* icon (the green arrow) in the Script window (see Figure E.5) or return to the Command window and type in the name of the program (without the *.m* extension).

E.3 Constructing a Program in MATLAB®

The following list summarizes the steps for writing a MATLAB script:

1. If available, start the MATLAB program by double-clicking on the MATLAB icon on the Window's desktop. If not available, go to the Window's *Start* menu, click on *All Programs*, find the *MATLAB* program among the list of available programs and double-click on it. This will open up the MATLAB desktop.

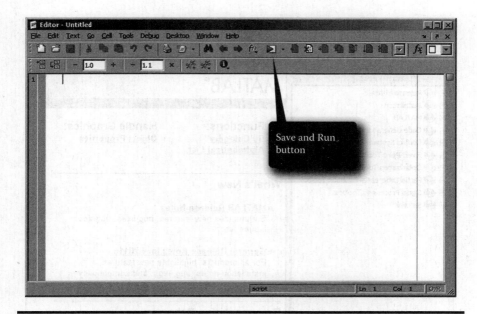

Figure E.5   Script window.

2. Click on *New Script* in the *File* menu. This brings up a new Script Window.
3. Type your program into the Script Window.
4. Save the script by clicking on the *Save* icon in the icon Toolbar or clicking on *File* in the menu bar and selecting *Save* in the dropdown menu. In the dialog box that appears, select the folder where the script is to reside and type in a file name of your own choosing. It is best to use a folder that contains *only* your own MATLAB scripts.
5. Before you can run your script, you need to go to the Current Folder box at the top of the MATLAB desktop, click on the down arrow, and in the dropdown menu, select (or browse to) the folder that contains your new script. If you try to run a script from a folder that is not listed in the Current Folder box, the MATLAB Editor will bring up a dialog box that gives you the option to change the folder listed in the Current Folder box or add your new folder to the MATLAB path.
6. You may run your script from the Script window by clicking on the *Save and Run* green arrow in the icon Toolbar (see Figure E.5) or alternatively, from the Command window by typing the script name (without the *.m* extension) after the MATLAB prompt (>>). For example, if the program has been saved as *heat.m*, then type heat after the MATLAB prompt (>>), like this:

```
>> heat
```

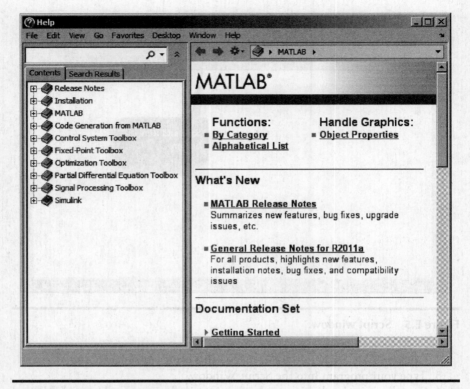

Figure E.6 Product Help window.

If you need additional help on getting started, you can click on *Help* in the menu bar in the MATLAB window and then select *Product Help* from the dropdown menu. This will bring up the help window, as shown in Figure E.6. By clicking on the little "+" box next to the MATLAB listing in the left window, you will get additional help topics, as shown in Figure E.7.

Once you select one of the help topics, the help information will be in the right-hand window. You can also type in a topic in the search box to obtain information on that topic.

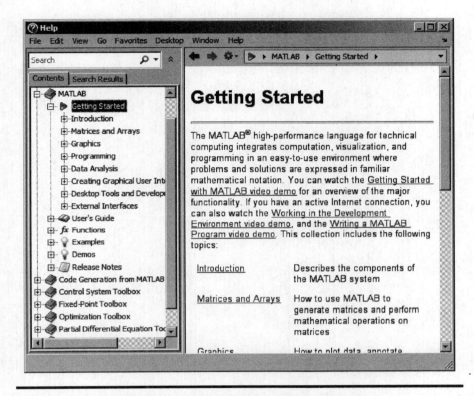

Figure E.7   Selecting MATLAB in the Product Help window.

Figure E.2 Selecting MATLAB in the Product Help window.

Function Index

This index contains a list of the MATLAB commands used in this book and the pages on which they are introduced.

Subject Index